Lecture Notes
in Control and Information Sciences   373

Editors: M. Thoma, M. Morari

Qing-Guo Wang, Zhen Ye, Wen-Jian Cai,
Chang-Chieh Hang

PID Control for Multivariable Processes

 Springer

Series Advisory Board

F. Allgöwer, P. Fleming, P. Kokotovic,
A.B. Kurzhanski, H. Kwakernaak,
A. Rantzer, J.N. Tsitsiklis

Authors

Qing-Guo Wang

Department of Electrical &
Computer Engineering
National University of Singapore
10 Kent Ridge Crescent
Singapore 119260
Republic of Singapore
Email: elewqg@nus.edu.sg

Wen-Jian Cai

School of Electrical &
Electronic Engineering
Nanyang Technological University
Singapore 639798
Republic of Singapore
Email: ewjcai@ntu.edu.sg

Zhen Ye

Department of Electrical &
Computer Engineering
National University of Singapore
10 Kent Ridge Crescent
Singapore 119260
Republic of Singapore

Chang-Chieh Hang

Department of Electrical &
Computer Engineering
National University of Singapore
10 Kent Ridge Crescent
Singapore 119260
Republic of Singapore
Email: elehcc@nus.edu.sg

ISBN 978-3-540-78481-4   e-ISBN 978-3-540-78482-1

DOI 10.1007/978-3-540-78482-1

Lecture Notes in Control and Information Sciences   ISSN 0170-8643

Library of Congress Control Number: 2008921515

© 2008 Springer-Verlag Berlin Heidelberg

This work is subject to copyright. All rights are reserved, whether the whole or part of the material is concerned, specifically the rights of translation, reprinting, reuse of illustrations, recitation, broadcasting, reproduction on microfilm or in any other way, and storage in data banks. Duplication of this publication or parts thereof is permitted only under the provisions of the German Copyright Law of September 9, 1965, in its current version, and permission for use must always be obtained from Springer. Violations are liable for prosecution under the German Copyright Law.

The use of general descriptive names, registered names, trademarks, etc. in this publication does not imply, even in the absence of a specific statement, that such names are exempt from the relevant protective laws and regulations and therefore free for general use.

Typeset & Cover Design: Scientific Publishing Services Pvt. Ltd., Chennai, India.

Printed in acid-free paper

5 4 3 2 1 0

springer.com

Preface

There are rich theories and designs for general control systems, but usually, they will not lead to PID controllers. Noting that the PID controller has been the most popular one in industry for over fifty years, we will confine our discussion here to PID control only. PID control has been an important research topic since 1950's, and causes remarkable activities for the last two decades. Most of the existing works have been on the single variable PID control and its theory and design are well established, understood and practically applied. However, most industrial processes are of multivariable nature. It is not rare that the overall multivariable PID control system could fail although each PID loop may work well. Thus, demand for addressing multivariable interactions is high for successful application of PID control in multivariable processes and it is evident from major leading control companies who all ranked the couplings of multivariable systems as the principal common problem in industry. There have been studies on PID control for multivariable processes and they provide some useful design tools for certain cases. But it is noted that the existing works are mainly for decentralized form of PID control and based on ad hoc methodologies. Obvious, multivariable PID control is much less understood and developed in comparison with the single variable case and actual need for industrial applications. Better theory and design have to be established for multivariable PID control to reach the same maturity and popularity as the single variable case.

The present monograph puts together, in a single volume, a fairly comprehensive, up-to-date and detailed treatment of PID control for multivariable processes, from paring, gain and phase margins, to various design methods and applications. The multivariable interactions are always a key issue and addressed explicitly and effectively. Both decentralized and centralized forms of PID controllers are discussed. Our design always assumes a process model. Thus, for completeness, our latest development on multivariable process identification is included as the last chapter of this text. Our table of contents can roughly give the idea of what has been contained in the book while Chapter 1 shows a more detailed chapter by chapter preview of our materials. The materials presented here are based on research results of the authors and their co-workers in the domain under the framework of transfer function, frequency response, and linear matrix inequality settings over the last five years. For presentation, we have made technical development of the results as self-contained as possible. Only knowledge of the linear control theory is assumed from readers. Illustrative examples of different degrees of complexity are given to facilitate understanding. Therefore, it is believed that the book can be accessed by graduate students, researchers and practicing engineers. To our best knowledge, this text is the first one solely focused on PID control for multivariable processes.

We would like to thank Tong Heng LEE, Maojun HE, Feng ZHENG, Chong LIN, Yong HE, Guilin WEN, Hanqin ZHOU, Min LIU, Yu ZHANG, Yong ZHANG, for their fruitful research collaborations with us, which have led to the contributions contained in the book. Our special thanks go to the Centre for Intelligent Control and Department of Electrical and Computer Engineering of the National University of Singapore for providing plenty of resources for our research work. This work was partially sponsored by the Ministry of Education's AcRF Tier 1 funding, Singapore, with the title "PID control for multivariable systems", NUS WBS No R-263-000-306-112.

<table>
<tr><td>National University of Singapore,
Nanyang Technological University,
January, 2008</td><td>Qing-Guo WANG
Zhen YE
Chang-Chieh HANG
Wen-Jian CAI</td></tr>
</table>

Contents

Acronyms

BLT	Biggest log modulus tuning
BMI	Bilinear matrix inequality
BRG	Block relative gain
DCLI	Decentralized closed-loop integrity
dDRIA	dynamic DRIA
DIC	Decentralized integral controllability
DRGA	Dynamic RGA
dRI	dynamic relative interaction
DRIA	Decomposed relative interaction array
EOP	Effective open-loop process
FFT	Fast Fourier transform
GIA	GI array
ILMI	Iterative linear matrix inequality
IMC	Internal model control
ISE	Integral square error
IV	Instrumental variable
LMI	Linear matrix inequality
LQ	Linear quadratic
LTI	Linear time invariant
MIMO	Multiple-input-multiple-output
MISO	Multiple-input-single-output
MLF	Multi-loop failure
MMF	Multiplicate model factor
NI	Niederlinski index
NMP	Non-minimum phase
NSR	Noise-to-signal ratio
PID	Proportional-integral-derivative
PRBS	Pseudo-random binary sequence
PRGA	Performance RGA
QAH_∞-γ	Quadratically admissible with H_∞ performance
QMI	Quadratic matrix inequality

XII

RGA	Relative gain array
RHP	Right half plane
RI	Relative interaction
RIA	Relative interaction array
SIMC	Simple IMC
SISO	Single-input-single-output
SLF	Single-loop failure
SOF	Static output feedback
SOFH_2	Static output feedback H_2 suboptimal control
SOFH_∞	Static output feedback H_∞ suboptimal control
SOFMOC	Static output feedback maximum output control
SOFS	Static output feedback stabilization
SOPDT	Second-order plus dead-time
SSV	Structured singular value
TITO	Two-input-two-output

1 Introduction

Classical control theory is appropriate for dealing with single-input-single-output (SISO) systems but becomes powerless for multiple-input-multiple-output (MIMO) systems because the graphical techniques were inconvenient to apply with multiple inputs and outputs. Since 1960, modern control theory, based on time-domain analysis and synthesis using state variables, has been developed to cope with the increased complexity of modern (whether SISO or MIMO) plants and the stringent requirements on accuracy, stability, and speed in industrial applications. Therefore, during the years from 1960 to 1980, optimal control of both deterministic and stochastic systems, as well as adaptive and learning control of complex systems, were well investigated. From 1980 to the present, developments in modern control theory have centered around robust control, H_2/H_∞ control, and associated topics. The result is a new control theory that blends the best features of classical and modern techniques.

Table 1.1. Publications on PID control cited by ScienceDirect©

Year	2006	2005	2004	2003	2002	2001	2000	1999	1998	1997	1996	1995
Single-loop PID	1240	1104	1093	962	720	836	955	910	825	775	800	628
Multi-loop PID	72	71	51	53	54	44	72	62	56	50	34	44
Total	1312	1175	1144	1015	774	880	1027	972	881	825	834	672

The PID controller has been the most popular one in industry for over fifty years. It also causes remarkable research activities recently, see Table 1.1 for publication statistics and Koivo [1] for a survey. Most of the existing works have been on the single-loop PID control and its theory and design is well established, understood and practically applied. They may be classified broadly into direct and indirect methods [2,3] based on past experience and heuristics and the indirect control design based on model parameters such as Ziegler-Nichols like method [4,5,6], pole-placement design [7], root-locus based methods [8], frequency response methods [9, 10, 11, 12], and optimization techniques [13, 14]. However, most industrial processes are of multivariable nature. It is not

Q.-G. Wang et al.: PID Control for Multivariable Processes, LNCIS 373, pp. 1–8, 2008.
springerlink.com
© Springer-Verlag Berlin Heidelberg 2008

rare that the overall multivariable PID control system could fail though each PID loop may work well. Thus, demand for addressing multivariable interactions is high for successful application of PID control in multivariable processes and evident from major leading control companies who all ranked the couplings of multivariable systems as the principal common problem in industry.

1.1 Multivariable Processes

A system (natural or man-made) is called a multivariable system if it has more than one variable to be controlled. Multivariable systems can be found almost everywhere. In the office, the temperature and humidity are crucial to comfort. Water level and flow rate are two key measures of a river. A robot needs six degree-of-freedoms in order to have a full range of positioning, and the same can be said to airplanes and missiles. To give the readers a more practical and concrete example of multivariable systems, let us look at the following example from chemical engineering.

Example 1.1.1. A typical process unit for refining a chemical product is shown in Fig. 1.1. First, there is a mixing of two raw materials (reactives) to feed a distillation

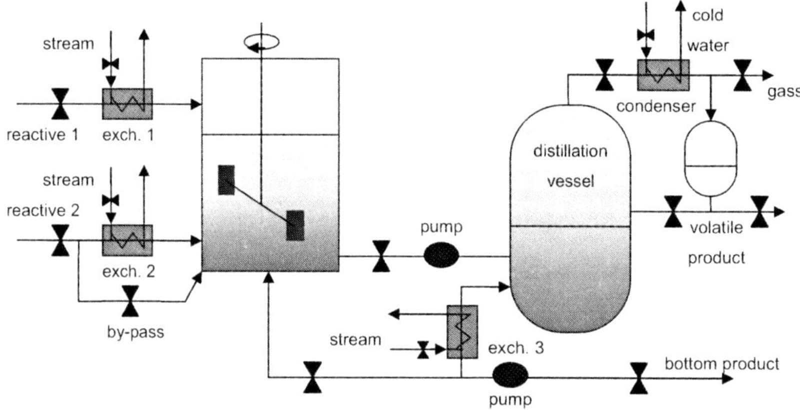

Fig. 1.1. Distillation unit

column where two final products are obtained: the head and bottom components. In order to run the unit, we must control the different flows of materials, provide adequate temperature to the inlet flows and keep the desired operating conditions in the column by adjusting its temperature, pressure and composition. Some other complementary actives are required, such as agitating the content of the mix tank or keeping the appropriate levels in all vessels, including those of auxiliary or intermediate buffers. The ultimate control goal is to obtain the best distilled products (maximum purity, less variance in concertration, ...) under the best conditions (maximum yield, minimum energy consumption, ...), also taking into account cost and pollution constraints. But before we begin to get the products, we must start up all the equipment devices, establish a regular flow of reactives, reach the nominal operating conditions and the keep the unit stable under production. Also, care should be taken about faults in any part of the unit: valves, agitator, existence of raw materials, heating systems, etc.

Some phenomena are unique to multivariable systems only, and could not occur in single variable systems. For instance, a multivariable system can have a pole and zero which coincide with each other and yet do not cancel each other. A zero of some element in a multivariable system play no role in the system properties. The most important feature with a multivariable system is possible cross-couplings or interactions between its variables, i.e., one input variable may affect other or all the output variables. They prevent the control engineer from designing each loop independently as adjusting controller parameters of one loop affects the performance of another, sometimes to the extent of destabilizing the entire system. It is the multivariable interactions that accounts for essential difference for design methodologies between single-variable and multivariable control systems. In general, multivariable control is much more difficult than single variable control, and it is the topic of interest in the present book.

1.2 Control of Multivariable Processes

It is noted that modern control theory provides rich design methodologies for MIMO systems using state feedback. Unfortunately, all state variables are usually not known or measurable in engineering practice and have no actual physical meanings, which may limit its applications in MIMO systems. Although observers can be applied to estimate the unknown/unmeasurable states, the system with state observer is completely different from the original one because their transient responses to disturbances are not the same. This often leads to the poor performance of the closed-loop system. Disturbance rejection is the primary concern in process control. With all its power and advantages, modern control was lacking in some aspects. Firstly, the guaranteed performance obtained by solving matrix design equations means that it is often possible to design a control system that works in theory without gaining any engineering intuition about the problem. Secondly, the frequency-domain techniques of classical control theory impart a great deal of intuition. Thirdly, a modern control system with any compensator dynamics can fail to be robust to disturbances, unmodelled dynamics, and measurement noise, while robustness is built in with a frequency-domain approach using notions like the gain and phase margin.

Over the last few decades, many attempts have been made to extend the classical control theory to MIMO systems. For instance, MacFarlane et al. [15] generalized the classical Nyquist and Evan's methods to MIMO systems in terms of characteristic gain and characteristic frequency methods respectively. Mayne [16] proposed the use of a sequential procedure to transfer a MIMO design into a number of SISO Nyquist designs. Bryant [17] improved this sequential technique by using a triangular decomposition via Gauss-elimination operation. The effects of closing loops in sequential order were shown to be equivalent to performing successive Gauss eliminations on the return difference matrix. This approach enables a precompensator to be designed systematically in a column-by-column way: each loop is controller by one column of the precompensator. An MIMO design is thus reduced into a number of multiple-input-single-output

(MISO) designs. The Gauss-elimination based sequential approach was later superseded by a Gauss-Jordan operation based procedure. The improved procedure offers more information than its predecessor in that the partly closed-loop responses and the inter-channel cross couplings are implicitly computed. An MIMO system can be diagonalized by inserting controllers in cascade with the system. For example, the decoupling method proposed by Falb and Wolovich [18] for state-space models belongs to such an approach, and the controllers obtained are complicated if the system is of a high order. For systems with large uncertaintics such as those found in process industry, it is difficult to see how exact decoupling can be achieved. However, total decoupling is not necessary; a system can be considered as a number of sub-systems if its inter-connection is "weak". This is the approach pursued by Rosenbrock [19], who showed that a MIMO system design can be relaxed into a number of simple decentralized classical SISO designs if it is diagonally dominant. The advantages of such an approach is that each loop can be treated as if it was an independent loop. Moreover, he showed that integrity of a MIMO system can be examined via Nyquist plots.

Despite rapid evolution of control technology over the past 50 years, PID controllers are always the most popular controllers in process industries. The proportional action (P mode) adjusts controller output according to the size of the error. The integral action (I mode) can eliminate the steady-state offset and the future trend is anticipated via the derivative action (D mode). These useful functions are sufficient for a large number of process applications and the transparency of the features leads to wide acceptance by the users. PID control has been an important research topic since 1950's, and causes remarkable activities recently. However, most of the existing works have been on the single-loop PID control while most industrial processes are of multivariable nature. It is not rare that the overall multivariable PID control system fails though each PID loop may work well. Thus, demand for addressing multivariable interactions is high for successful application of PID control in multivariable processes.

There have been several studies on multi-loop PID control. Luyben [20] employed a simple multi-loop PID tuning method in multivariable systems without considering the interaction. Palmor [21] proposed a decentralized PID controller design for two-input two-output (TITO) systems, in which the desired critical point is used to tune the PID controller by the Ziegler-Nichols rule or its modifications. Wang et al. [22] developed a decentralized PI/PID controller tuning with a lead-lag decoupler for TITO processes. Åström et al. [23] developed a decoupled PI controller for TITO processes with interacting loops and its advantage is that the interaction can be reduced substantially by using set-point weighting. Loh et al. [24] presented a PID tuning method for multi-loop systems based on relay feedback experiment. Zhang et al. [25] presented a technique for multi-loop PI controller design to achieve dominant pole placement for TITO processes. Wang et al. [26] developed a method to tune a fully cross-coupled multivariable PID controller from decentralized relay feedback with new techniques for process frequency-response matrix estimation and multivariable decoupling design. Huang et al. [27] presented a method of diagonal PID controller design based on internal model control (IMC) for multivariable temperature control system.

The developments mentioned above and others provide useful design tools for MIMO-PID controllers. However, it is obvious that MIMO-PID control is much less understood and developed, compared with the single variable case and actual need for industrial applications. Better theory and design have to be established for MIMO-PID control to reach the same maturity and popularity as the single variable case. In particular, it is noted that

(i) Unlike the SISO process, loop interactions have to be investigated carefully for MIMO processes. Therefore, loop pairing can have a heavy effect on the performance of the closed-loop system. Usually, relative gain array (RGA) method is used to give a loop pairing criterion, but sometimes, such a criterion can also lead to bad performance. Hence, new criterion for loop pairing to improve the performance of MIMO processes is needed.

(ii) For the single variable case, the gain range for P-term for closed-loop stability can be easily determined from the Nyquest test. In order to have a very first control of a given multivariable process, one always wishes to know the gain range of P control for stability. Unfortunately, there is no result in the literature for this simplest controller. In general, in order to achieve closed-loop stabilization and tracking, it is necessary to find a set of parameters of a multi-loop PID controller for any application and desirable to find the whole parameter space of the controller for advanced applications such as optimization.

(iii) There are a huge number of SISO-PID controller designs. Many of them are well known and widely utilized in practice. This is not the case for the multivariable case. There are a number of design methods for multivariable PID controllers in the literature as mentioned above. Usually, they are ad hoc in nature. Most of them are based on the following two assumptions: 1) the process can be decoupled into single variable systems; and/or 2) the process can be described by first-order plus time delay model. However, the process may be badly coupled and/or cannot be decoupled well due to simplicity of PID structure. A given process may not be well approximated by first-order plus time delay model. Some methods may not guarantee the stability of the closed-loop system. Thus, a unifying framework for analysis and design of multivariable PID control system applicable to general multivariable processes (either simple or complex) would be welcome.

(iv) For the single variable case, frequency domain stability margins such as gain and phase margins are very popular, and used for performance assessment, design specifications and robustness measure of PID control systems. There are some attempts to define multivariable system stability margins in frequency domain, but none of them is well known. One may look for better definitions which are meaningful, useful and easily checkable with clear link to the single variable case. They can then be used as performance assessment, design specifications and robustness measure of multivariable PID control systems and may lead to a large branch of tuning rules similar to the single variable case.

(v) With popularity of SISO-PID auto-tuning in commercial control systems and industrial applications, it is natural to do the same for MIMO-PID controllers. This will require simple yet effective multivariable system identification methods. Our existing works with relay and step tests using the Fast Fourier transform (FFT) are a good starting point. But they need to be further developed to eliminate the limitations such as the common frequency limit cycle with relay feedback and sequential test with step test, improve computational efficiency and robustness to measurement noises with the FFT technique. For controller tuning part, few methods are available. More and better ones are in demand.

1.3 Outline of the Book Chapters

By PID control for a multivariable process, it is meant that a multivariable process is controlled by either decentralized or centralized controller of which all elements are of PID type. For such a system, the controller is called an MIMO-PID controller while the overall system an MIMO-PID control system. The concerned case is that the process is coupled and normal single variable PID designs will fail due to such couplings, and an MIMO-PID controller has to be analyzed, designed and tuned by considering the process as an essential multivariable system.

The book assumes the pre-requisite of basic theory on linear control systems and linear algebra from readers. It can be divided into four parts. Chapters 2–4 focus on multivariable system analysis, including loop paring, loop gain and phase margins. Chapters 5–7 demonstrate multivariable PID controller design, where new methods on IMC, dominate pole placement and linear matrix inequality (LMI) are proposed. Chapter 8 shows the application of our proposed methods on synchronization. Chapter 9 presents a novel identification method for multivariable processes. More details of each chapter is highlighted as follows.

Chapter 2 presents a new control-loop configuration criterion and a novel approach for evaluating decentralized closed-loop integrity (DCLI) of multivariable control systems. For an arbitrary loop, four cases corresponding to different combinations of open and closed states with the remaining loops are investigated. A new interaction measurement, which is able to provide comprehensive description of interaction among loops, is proposed to evaluate the loop-by-loop interaction. Consequently, a new loop-paring criterion based on the new interaction measurement and the algorithm for determining loop pairings that result in minimum loop interactions in terms of interaction energy are proposed. Through applying the left-right factorization to the decomposed relative interaction array, the relative interaction to a particular loop from other loops is presented by elements summation of the decomposed relative interaction sequence. The maximum interactions from other loops under different combinations and sequences are determined by the maximum values of decomposed relative interaction sequence according to the failure index. Consequently, the necessary and sufficient conditions for DCLI of an individual loop under both single- and multiple-loop failure are provided.

Chapter 3 addresses the problem of determining the parameter ranges of proportional controllers which stabilize a given process. Accordingly, loop gain margins of multivariable systems are defined. An effective computational scheme is established by converting the considered problem to a quasi-LMI problem connected with robust stability test. The descriptor model approach is employed together with linearly parameter-dependent Lyapunov function method. Examples are given for illustration. The results are believed to facilitate real time tuning of multi-loop PID controllers for practical applications.

Chapter 4 addresses two problems. One is the definition of loop phase margins for multivariable control systems, which extends the concept of phase margin in SISO systems to MIMO systems. The other one is development of an algorithm for computing the loop phase margins. Two methods are proposed in time domain and frequency domain, respectively. For the time domain method, we first find the stabilizing ranges of loop time delay perturbations of the MIMO system using an LMI-based stability criterion derived here; Then convert these stabilizing ranges of loop time delays into the stabilizing ranges of loop phases. For the frequency domain method, the MIMO phase margin problem is converted to the problem of some simple constrained optimization with the help of unitary mapping between two complex vector space, which is then solved numerically with the Lagrange multiplier and Newton-Raphson iteration algorithm. It can provide exact margins and thus improves the LMI results by the proposed time domain method.

Chapter 5 presents a simple yet effective method to design decentralized proportional-integrated-derivative controller for multivariable processes. On the basis of structure decomposition, the dynamic relative interaction is defined and used to derive the multiplicate model factor (MMF) for an individual control loop. The MMF is approximated by a time delay function at the neighborhood of individual control loop critical frequency. An equivalent transfer function for each control loop is then obtained by combining the original loop transfer function with the approximated MMF. Consequently, appropriate controller parameters for each loop are determined by applying the single-input-single-output IMC-PID tuning rules for the equivalent transfer function. A 2×2 process is used to demonstrate a step-by-step design procedure, and the simulation results for a variety of 2×2, 3×3 and 4×4 system show that the design technique results in a better overall control system performance than those of existing design methods.

Chapter 6 demonstrates two methods of system pole placement — approximated method and guaranteed method. An analytical PID design is proposed for continuous-time delay systems to achieve approximate pole placement with dominance. Its idea is to bypass continuous infinite spectrum problem by converting a delay process to a rational discrete model and getting back continuous PID controller from its discrete form designed for the model with pole placement. This chapter also proposes two simple and easy designs which can guarantee the dominance of the assigned two poles for PID control systems. They are based on root-locus and Nyquist plot, respectively. If a solution exists, the parametrization of all the solutions is explicitly given.

Chapter 7 studies the design problem of multivariable PID controllers which guarantee the stability of the closed-loop systems, H_2 or H_∞ performance specifications, or maximum output control requirement, respectively. Algorithms based on iterative linear matrix inequality technique are developed to find the feedback gains of PID controllers corresponding to the above mentioned four cases. A numerical example on the design of PID controllers for aircraft is provided to illustrate the effectiveness of the proposed method.

Chapter 8 illustrates a strategy for fast master-slave synchronization for Lur'e systems under PD control based on the free-weighting matrix approach and the S-procedure. The purpose of the derivative action is to improve the closed-loop stability and speed synchronization response. The proposed strategy covers the existing result for the proportional control alone as a special case. Furthermore, synchronization via multivariable PID control is studied. Based on the descriptor approach, the problem od

PID controller design is transformed to that of static output feedback (SOF) controller design. The improvement of the solvability of the LMI is achieved, in comparison with the existing literature on designing PID controller based on the LMI technique. With the aid of the free-weighting matrix approach and the S-procedure, the synchronization criterion for a general Lur'e system is established based on the LMI technique. The feasibility of both methodologies is illustrated by the well-known Chua's circuit.

Chapter 9 gives an improved identification algorithm for continuous time delay systems under unknown initial conditions and disturbances for a wide range of input signals expressible as a sequence of step signals. It is based on a novel regression equation which is derived by taking into account the nature of the underlying test signal. The equation has more linearly independent functions and thus enables to identify a full process model with time delay as well as combined effects of unknown initial condition and disturbance without any iteration. Unlike the method in [28], this new algorithm requires no process data before the test starts. Based on the above identification method for SISO processes, a robust identification method is proposed for MIMO continuous-time processes with multiple time delays. Suitable multiple integrations are constructed and regression equations linear in the aggregate parameters are derived with use of the test responses and their multiple integrals. The multiple time delays are estimated by solving some algebraic equations without iteration and the other process model parameters are then recovered. Its effectiveness is demonstrated through simulation and real time test.

2 Loop Pairing Analysis

Despite the availability of sophisticated methods for designing multivariable control systems, decentralized control remains dominant in industry applications, because of its simplicity in design and ease of implementation, tuning, and maintenance with less cost [29, 30]:

(i) Hardware simplicity: The cost of implementation of a decentralized control system is significantly lower than that of a centralized controller. A centralized control system for an $n \times n$ plant consists of $n!$ individual single-input single-output transfer functions, which significantly increases the complexity of the controller hardware. Furthermore, if the controlled and/or manipulated variables are physically far apart, a full controller could require numerous expensive communication links.

(ii) Design and tuning simplicity: Decentralized controllers involve far fewer parameters, resulting in a significant reduction in the time and cost of tuning.

(iii) Flexibility in operation: A decentralized structure allows operating personnel to restructure the control system by bringing subsystems in and out of service individually, which allows the system to handle changing control objectives during different operating conditions.

However, the potential disadvantage of using the limited control structure is the deteriorated closed-loop performance caused by interactions among loops as a result of the existence of nonzero off-diagonal elements in the transfer function matrix. Therefore, the primary task in the design of decentralized control systems is to determine loop pairings that have minimum cross interactions among individual loops. Consequently, the resulting multiple control loops mostly resemble their SISO counterparts such that controller tuning can be facilitated by SISO design techniques [31]. This chapter aims to obtain a new loop paring criterion which may result in minimum loop interactions.

2.1 Introduction

Since the pioneering work of Bristol [32], the relative gain array (RGA) based techniques for control-loop configuration have found widespread industry applications, including blending, energy conservation, and distillation columns, etc [13, 33, 34, 35]. The

Q.-G. Wang et al.: PID Control for Multivariable Processes, LNCIS 373, pp. 9–38, 2008.
springerlink.com
© Springer-Verlag Berlin Heidelberg 2008

RGA-based techniques have many important advantages, such as very simple calculation because it is the only process steady-state gain matrix involved and independent scaling due to its ratio nature, etc [36]. To simultaneously consider the closed-loop properties, the RGA-based pairing rules are often used in conjunction with the Niederlinski index (NI) [37] to guarantee the system stability [31, 13, 36, 38, 39, 40]. However, it has been pointed out that this RGA- and NI-based loop-pairing criterion is a necessary and sufficient condition only for a 2×2 system; it becomes a necessary condition for 3×3 and higher dimensional systems [36, 41]. Moreover, it is very difficult to determine which pairing has less interaction between loops when the RGA values of feasible pairings have similar deviations from unity.

To overcome the limitations of a RGA-based looppairing criterion, several pairing methods have later been proposed. Witcher and McAvoy [42], as well as other authors [43, 44], defined the dynamic RGA (DRGA) to consider the effects of process dynamics and used a transfer function model instead of the steady-state gain matrix to calculate RGA, of which the denominator involved achieving perfect control at all frequencies, while the numerator was simply the open-loop transfer function. The μ-interaction measurement [45, 46, 47] is another measurement for multivariable systems under diagonal or block-diagonal feedback controllers. By employment of structured singular value (SSV) techniques, it can be used not only to predict the stability of decentralized control systems but also to determine the performance loss caused by these control structures. In particular, its steady-state value provides a sufficient condition for achieving offset-free performance with the closed-loop system. Hovd and Skogestad [41, 48] introduced performance RGA (PRGA) to solve the problem that the RGA cannot indicate the significant one-way interactions in the case in which the process transfer function matrix is triangular.

Even though some excellent techniques based on the RGA and NI principles have been proposed to measure loop interactions, there is a lack of a systematic method to treat the control structure configuration problem effectively for high-dimensional processes. To solve this problem, the following questions must be addressed: (1) What are the interaction effects to a particular loop when all other loops work together or individually? (2) What are the reverse interaction effects from a particular loop open and closed to other open and closed loops? (3) What is the feasible definition of the minimal interactions?

The flexibility to bring subsystems in and out of service is very important also for the situations when actuators or sensors in some subsystems fail. The characteristic of failure tolerance is that without readjustment to the other parts of the control system, stability can be preserved in the case of any sensor failure and/or actuator failure [49]. The RGA [32, 38], NI [37] and block relative gain (BRG) [50] are widely used for eliminating pairing that produce unstable closed-loop systems under failure conditions [36, 51, 52, 53]. Chiu and Arkun [30] introduced the concept of decentralized closed-loop integrity (DCLI) which requires that the decentralized control structure should be stabilized by a controller having integral action and should maintain its nominal stability in the face of failures in its sensors and/or actuators. A number of necessary or sufficient conditions for DCLI were also developed [30, 54]. However, the necessary and sufficient conditions for DCLI are still not available. Morari and co-workers [52, 55] defined the decentralized integral controllability (DIC) to address the operational issues, which consider the failure tolerance as a sub-problem. Physically, a decentralized integral controllable system allows the operator to reduce the controller gains independently to zero without introducing instability (as a result of positive feedback). Some necessary and/or sufficient conditions for DIC were developed [52, 53, 56, 57]. Even using only the steady state gain information, however, the calculation [58] to verify the DIC is very complicated especially for high dimension system, which is still an open problem.

2.2 Preliminaries

Throughout this chapter, it is assumed that the system we are dealing with is square ($n \times n$), open-loop stable, and nonsingular at steady state with a decentralized control structure as shown in Fig. 2.1. The transfer function matrix relating outputs and inputs of the process, its steady-state gain matrix, and individual elements are represented by $G(s)$, $G(0)$ (or simply $G \in \mathbb{R}^{n \times n}$), and g_{ij}, respectively. G^{ij} denotes the matrix G with its ith row and jth column removed. $\mathbf{r}$, $\mathbf{u}$, and $\mathbf{y}$ are vectors of references and manipulated and controlled variables, while r_i, u_i, and y_i are the ith references and manipulated and controlled variables, and $\mathbf{r}^i$, $\mathbf{u}^i$, and $\mathbf{y}^i$ denote the reference, input, and output vectors with variables r_i, u_i, and y_i removed.

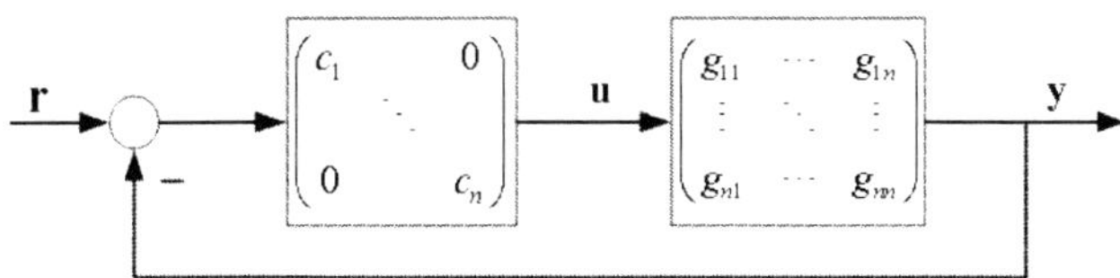

Fig. 2.1. Decentralized control of multivariable systems

Because we are investigating the interactions between an arbitrary loop y_i–u_j and all other loops of the multivariable system, the process from $\mathbf{u}$ to $\mathbf{y}$ can be explicitly expressed by

$$y_i = g_{ij}u_j + \mathbf{g}_{i*}^{ij}\mathbf{u}^j,$$
$$\mathbf{y}^i = \mathbf{g}_{*j}^{ij}u_j + G^{ij}\mathbf{u}^j, \tag{2.1}$$

where $\mathbf{g}_{i*}^{ij}$ and $\mathbf{g}_{*j}^{ij}$ denote the ith row vector and the jth column vector of matrix G with element g_{ij} removed.

2.2.1 RGA and NI

Definition 2.1. *The relative gain [32] for variable pairing y_i–u_j is defined as the ratio of two gains representing, first, the process gain in an isolated loop and, second, the apparent process gain in the same loop when all other loops are closed*

$$\lambda_{ij} = \frac{(\partial y_i / \partial u_j)_{u_{k \neq j}\text{constant}}}{(\partial y_i / \partial u_j)_{y_{l \neq i}\text{constant}}} = g_{ij}[G^{-1}(0)]_{ji}, \tag{2.2}$$

and RGA, $\Lambda(G)$, in matrix form is defined as

$$\Lambda(G) = [\lambda_{ij}] = G(0) \otimes G^{-T}(0), \tag{2.3}$$

where $\otimes$ is the hadamard product and $G^{-T}(0)$ is the transpose of the inverse of $G(0)$.

When the RGA pairing rule is followed [32], input and output variables paired with positive RGA elements that are closest to unity will result in minimum interaction from other control loops. Even though this pairing rule gives a clear indication for minimum interaction, it is often necessary (especially with 3×3 and higher dimensional systems) to use this rule in conjunction with stability considerations provided by the following theorem originally given by Niederlinski [37] and later modified by Grosdidier and Morari [36].

Theorem 2.1. *Consider an $n \times n$ multivariable system whose manipulated and controlled variables have been paired as follows: y_1–u_1, y_2–u_2, $\cdots$, y_n–u_n, resulting in a transfer function model of the form*

$$\mathbf{y} = \mathbf{Gu}.$$

Further, let each element of G, g_{ij}, be rational and open-loop stable and n individual feedback controllers (which have integral action) be designed for each loop so that each one of the resulting n feedback control loops is stable when all other $n - 1$ loops are open. Then, under closed-loop conditions in all n loops, the multiloop system will be unstable for all possible values of controller parameters (i.e., it will be "structurally monotonically unstable") if the NI defined below is negative, i.e.,

$$\mathrm{NI} = \frac{\det[G(0)]}{\prod\limits_{i=1}^{n} g_{ii}(0)} < 0. \tag{2.4}$$

One point that must be emphasized is that (2.4) is both necessary and sufficient only for 2×2 systems, but for higher dimensional systems, it provides only sufficient conditions: i.e., if (2.4) holds, then the system is definitely unstable; otherwise, the system may, or may not, be unstable because the stability will, in this case, depend on the values taken of the controller parameters.

2.2.2 RGA-Based Loop-Pairing Criterion [59]

Manipulated and controlled variables in a decentralized control system should be paired in the following way:

 (i) The paired RGA elements are closest to 1.0.
 (ii) NI is positive.
 (iii) All paired RGA elements are positive.
 (iv) Large RGA elements should be avoided.

In this criterion, both RGA and NI offer important insights into the issue of control structure selection. RGA is used to measure interactions, while NI is used as a sufficient condition to screen out the closed-loop unstable pairings. However, for higher dimensional systems, the RGA element λ_{ij} only takes the overall interaction from all other closed loops to the loop y_i–u_j into consideration; it failed to yield information on the interactions from other individual loops to the loop y_i–u_j. Consequently, the control structure configuration selected according to the RGA- and NI-based loop pairing criterion may result in an undesirable control system performance. Example 2.1 illustrates this point.

Example 2.1. Consider the process [41]

$$G(s) = \frac{1-s}{(1+5s)^2} \begin{bmatrix} 1 & -4.19 & -25.96 \\ 6.19 & 1 & -25.96 \\ 1 & 1 & 1 \end{bmatrix}.$$

The steady-state RGA is

$$\Lambda[G(0)] = \begin{bmatrix} 1 & 5 & -5 \\ -5 & 1 & 5 \\ 5 & -5 & 1 \end{bmatrix}.$$

which results in two pairing structures with positive RGA and NI values as shown in Table 2.1.

Table 2.1. Feasible pairing structure for Example 2.1

	Feasible Pairing Structure	
	1–1/2–2/3–3	1–2/2–3/3–1
RGA	1–1–1	5–5–5
NI	26.9361	0.2476

According to the RGA-based loop-pairing criterion, the pairing of 1–1/2–2/3–3 should be preferred for the zero interaction to any one loop from all other closed loops. However, the resulting closed-loop performance for the pairing 1–2/2–3/3–1 is significantly better than that of 1–1/2–2/3–3 based on the same controller tuning rule [41]. Through analysis, the main reasons can be explained as follows:

(i) For the pairing of 1–1/2–2/3–3, if loops y_2–u_2 and y_3–u_3 are closed one by one, the gain of loop y_1–u_1 will first increase by a factor of about 27 and then decrease by another factor of about 27; consequently, as reflected by the RGA, the gain of loop y_1–u_1 is not changed after all other loops are closed.

(ii) For the pairing of 1–2/2–3/3–1, if loops y_2–u_3 and y_3–u_1 are closed one by one, the gain of loop y_1–u_2 will first increase by a factor of about 1.24 and then decrease by another factor of about 6.2; consequently, as reflected by the RGA, the gain of loop y_1–u_2 is decreased by a factor of about 5 after all other loops are closed.

Hence, even the RGA demonstrates that the closed loop gain of loop y_1–u_1 in the pairing of 1–1/2–2/3–3 is unchanged; the interactions among loops are significant. On the contrary, even the gain of loop y_1–u_2 is decreased by a factor of about 5 after all other loops are closed, the interactions among loops in the pairing of 1–2/2–3/3–1 are smaller. This example reveals that both the overall interaction to the considered loop from all other closed loops and individual interactions among loops will affect the overall closed-loop performance. Therefore, a loop-by-loop analysis to establish individual interaction cause-effect relations is necessary for developing a new and effective interaction measurement.

2.2.3 DCLI

The decentralized controller $C(s)$ can be decomposed into $C(s) = N(s)K/s$, where $N(s)$ is the transfer function matrix of the dynamic compensator, which is diagonal and stable and does not contain integral action, and $K = \text{diag}\{k_i\}$, $i = 1, 2, \cdots, n$. The decentralized control configuration as shown in Fig. 2.2.

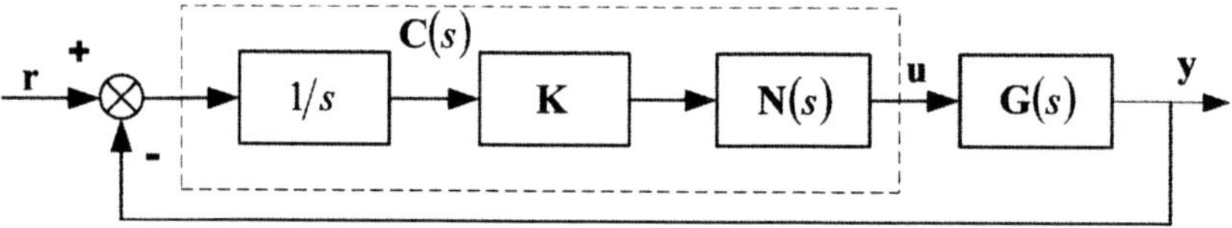

Fig. 2.2. Decentralized integral control of multivariable systems

When loop failures of an arbitrary loop in system $G(s)$ are investigated, all possible scenarios of the other $n - 1$ loops in any failure order have to be considered, which are as many as $(n - 1)!$. To effectively reflect these failed possibilities, we define a failure index M, which consist of $n - 1$ different integers $M = \{(i_1, \cdots, i_m, \cdots, i_{n-1})\}$ where $m, i_m \in [1, n - 1]$. In the design of decentralized control system, it is desirable to choose input/output-pairings such that the system possesses the property of DCLI, which is defined as follows.

Definition 2.2 ([30]). *A stable plant is said to be DCLI, if it can be stabilized by a stable decentralized controller, which contains integral action shown as Fig. 2.2, and if it remains stable after failure occurs in one or more of the feedback loops.*

The necessary conditions for a system to be DCLI is given as:

Theorem 2.2 (Necessary conditions for DCLI, [30,54]). *Given an $n \times n$ stable process $G(s)$, the closed-loop system of decentralized feedback structure possesses DCLI only if*

$$[\Lambda(G_m)]_{ii} > 0, \quad \forall m = 1, \cdots, n; i = 1, \cdots, m, \tag{2.5}$$

or

$$\text{NI}[G_m] > 0, \quad \forall m = 1, \cdots, n. \tag{2.6}$$

where G_m is an arbitrary $m \times m$ principal submatrix of G.

In theorem 2.2, either RGA or NI can be used as a necessary condition to examine the DCLI of decentralize control systems. However, the necessary and sufficient condition for DCLI with respect to single-and multi-loop failure are still unknown.

2.3 Decomposed Interaction Analysis

To investigate the interactions between an arbitrary loop y_i–u_j and all other loops of the multivariable processes in their open and closed states, we use the relative interaction (RI) [60, 61, 62, 63] to measure the interactions among loops because it can directly reflect the increment of process gain and the direction of interaction with its sign.

Definition 2.3 ([60,61,62,63]). *The RI for loop pairing y_i–u_j is defined as the ratio of two elements: the increment of the process gain after all other control loops are closed and the apparent gain in the same loop when all other control loops are open.*

$$\phi_{ij} = \frac{(\partial y_i/\partial u_j)_{y_{l \neq i}\text{constant}} - (\partial y_i/\partial u_j)_{u_{k \neq j}\text{constant}}}{(\partial y_i/\partial u_j)_{u_{k \neq j}\text{constant}}} = \frac{1}{\lambda_{ij}} - 1. \tag{2.7}$$

Equation (2.7) shows that, even though the interpretation of RI is different from the RGA, they are, nevertheless, equivalent through transformation of coordinates. Hence, the properties of RI can be easily derived from the RGA [60, 61, 62, 63]. Similarly to the RGA based loop pairing rule [32], one can obtain the following loop pairing rule in terms of the RI as:

$$\phi_{ij,n-1} \rightarrow 0, \quad \text{and} \quad \phi_{ij,n-1} > -1. \tag{2.8}$$

For an arbitrary loop y_i–u_j and the remaining subsystem block, there are four interaction scenarios corresponding to the combination of their open and closed states, which are listed in Table 2.2 and indicated in Fig. 2.3.

Table 2.2. Four Interaction scenarios for loop y_i–u_j

Case	Loop y_i–u_j	All Other Loops	Fig. 2.3
1	open loop	open loop	(a)
2	open loop	closed loop	(b)
3	closed loop	open loop	(c)
4	closed loop	closed loop	(d)

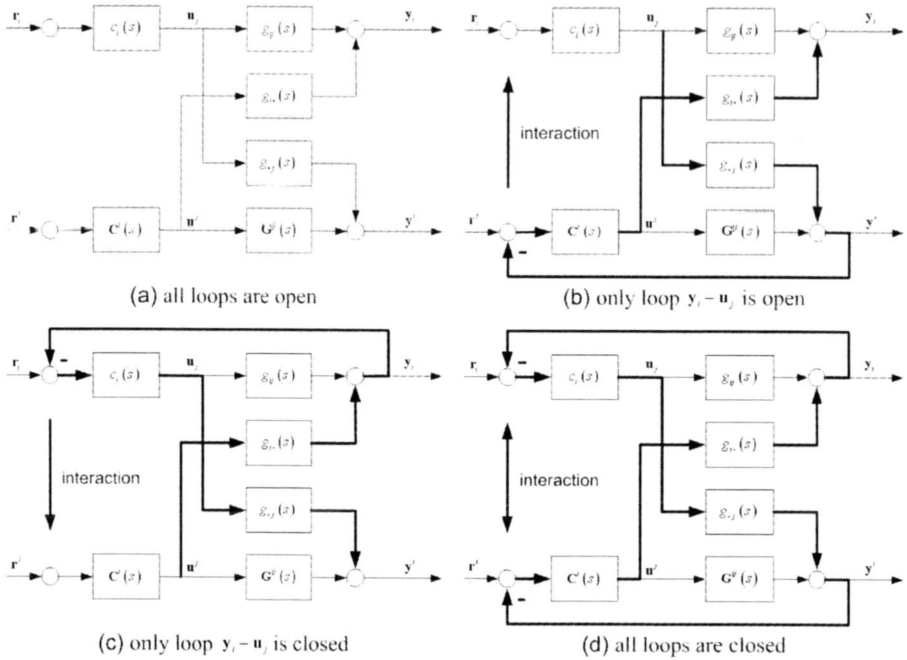

(a) all loops are open

(b) only loop $y_i - u_j$ is open

(c) only loop $y_i - u_j$ is closed

(d) all loops are closed

Fig. 2.3. Four interaction scenarios for loop y_i–u_j

Case 1

All loops are open, as shown in Fig. 2.3(a). Obviously, no interaction exists among loops; thus, the process gain of an arbitrary pairing y_i–u_j is the element g_{ij} in transfer function matrix G.

Case 2

Loop y_i–u_j is open and all other loops are closed, as shown in Fig. 2.3(b). This is the very condition for deriving the RGA, and the direction of interactions is from the subsystem block of closed loops to loop y_i–u_j. RI can be obtained by matrix operation as

$$\phi_{ij} = -\frac{1}{g_{ij}} g_{i*}^{ij} (G^{ij})^{-1} g_{*j}^{ij} \tag{2.9}$$

Case 3

Loop y_i–u_j is closed and all other loops are open, as shown in Fig. 2.3(c). In this case, the interaction is from closed loop y_i–u_j to all other open loops, and the RI can be derived from the 2×2 subsystem that includes loop y_i–u_j and an arbitrary loop y_k–u_l of subsystem G^{ij} ($k \neq i$ and $l \neq j$). This describes the 2×2 subsystem as

$$\begin{bmatrix} y_i \\ y_k \end{bmatrix} = \begin{bmatrix} g_{ij} & g_{il} \\ g_{kj} & g_{kl} \end{bmatrix} \begin{bmatrix} u_j \\ u_l \end{bmatrix}.$$

Hence, according to Definition 2.3, we obtain the RI of loop y_k–u_l from loop y_i–u_j as

$$\phi_{kl}^{ij} = \frac{g_{kl}|_{y_i-u_j,\text{CL}} - g_{kl}}{g_{kl}} = \frac{\triangle g_{kl}^{ij}}{g_{kl}}, \tag{2.10}$$

where

$$\triangle g_{kl}^{ij} = -g_{il}g_{kj}/g_{ij} \tag{2.11}$$

is the incremental process gain of loop y_k–u_l when loop y_i–u_j is closed. By extending (2.10) and (2.11) to all loops of the subsystem G^{ij}, we have

$$\Phi^{ij} = (G^{ij}|_{y_i-u_j,\text{CL}} - G^{ij}) \odot G^{ij} = \triangle G^{ij} \odot G^{ij}, \tag{2.12}$$

where "$\odot$" indicates element-by-element division, and

$$\triangle G^{ij} = -\frac{1}{g_{ij}} g_{*j}^{ij} g_{i*}^{ij} \tag{2.13}$$

is the incremental process gain matrix of subsystem G^{ij} when loop y_i–u_j is closed. Thus, Φ^{ij} is called the RIA to the open subsystem G^{ij} when loop y_i–u_j is closed.

Case 4

All loops are closed, as shown in Fig. 2.3(d). In this case, the interaction between the closed loop y_i–u_j and all other closed loops is bidirectional. Because the interaction to loop y_i–u_j from all other closed loops is the same as what has been studied in case 2, the interactions in this direction are not considered here. We now focus on the interaction from the isolated loop y_i–u_j to all other closed loops.

Suppose all loops of subsystem G^{ij} are closed and the following:

(i) If loop y_i–u_j is open, then the gain of an arbitrary loop y_k–u_l is affected by the RGA value of G^{ij}; therefore, according to the definition of RGA, it is

$$\hat{g}_{kl} = g_{kl}/\lambda_{kl}^{ij}, \tag{2.14}$$

where "arc" indicates the changed process gain in the closed subsystem G^{ij}. When (2.14) is extended to all loops of subsystem G^{ij}, the closed-loop transfer function matrix is obtained as

$$\hat{G}^{ij} = G^{ij} \odot \Lambda^{ij} \tag{2.15}$$

(ii) If loop y_i–u_j is closed, then the subsystem that includes loops y_i–u_j and y_k–u_l in the closed subsystem $\hat{G}^{ij}$ is given as

$$\begin{bmatrix} y_i \\ y_k \end{bmatrix} = \begin{bmatrix} g_{ij} & g_{il} \\ g_{kj} & \hat{g}_{kl} \end{bmatrix} \begin{bmatrix} u_j \\ u_l \end{bmatrix}.$$

Thus, the RI from loop y_i–u_j to loop y_k–u_l is obtained as

$$\psi_{kl}^{ij} = \frac{\hat{g}_{kl}|_{y_i-u_j,\,\text{CL}} - \hat{g}_{kl}}{\hat{g}_{kl}} = \frac{\triangle \hat{g}_{kl}^{ij}}{\hat{g}_{kl}}, \tag{2.16}$$

where $\triangle g_{kl}^{ij}$ is the increment of closed-loop process gain $\hat{g}_{kl}$ after loop y_i–u_j is closed and it is the same as that shown in (2.11). Extending (2.16) to all loops of the subsystem $\hat{G}^{ij}$, we obtain

$$\Psi^{ij} = (\hat{G}^{ij}|_{y_i-u_j,\ \mathrm{CL}} - \hat{G}^{ij}) \odot \hat{G}^{ij} = \triangle G^{ij} \odot \hat{G}^{ij}, \tag{2.17}$$

where $\triangle G^{ij}$ is the incremental matrix of subsystem $\hat{G}^{ij}$ after loop y_i–u_j is closed and it is the same as that shown in (2.13). Thus, Ψ^{ij} is called the RIA to the closed subsystem $\hat{G}^{ij}$, which reflects the interactions to $\hat{G}^{ij}$ when loop y_i–u_j is closed.

Remark 2.1. As an arbitrary element of G, g_{ij} may be zero. In such a case, the values of $\triangle g_{kl}^{ij}$ and ϕ_{kl}^{ij} in (2.10) and (2.11) are indefinite. Fortunately, those variables are only used during the course of derivation (where $\varepsilon \to 0$ may be used to replace the zero elements).

For practical calculation, Ψ^{ij} of the ijth element g_{ij} can be directly obtained by substituting (2.3) with (2.15) into (2.17) to result in

$$\Psi^{ij} = \triangle G^{ij} \otimes (G^{ij})^{-T} \tag{2.18}$$

To explore the inherent relationship between RGA and RI and on the basis of the definition of the RGA, rewrite (2.16) as

$$\psi_{kl}^{ij} = \frac{\triangle g_{kl}^{ij}}{\hat{g}_{kl}^{ij}} = \frac{\triangle g_{kl}^{ij}}{g_{kl}} \frac{g_{kl}}{\hat{g}_{kl}} = \phi_{kl}^{ij} \lambda_{kl}^{ij} \tag{2.19}$$

Extending (2.19) to all loops of subsystem $\hat{G}^{ij}$, we obtain another important matrix form for Ψ^{ij}

$$\Psi^{ij} = \Phi^{ij} \otimes \Lambda^{ij} \tag{2.20}$$

Furthermore, the relationship between ϕ_{ij} and Ψ^{ij} for an arbitrary nonzero element of system G is given by the following theorem.

Theorem 2.3. *For an arbitrary nonzero element g_{ij} of G, the corresponding ϕ_{ij} is the sum of all elements in Ψ^{ij}, i.e.,*

$$\phi_{ij} = \left\| \Psi^{ij} \right\|_{\Sigma} = \sum_{k=1,k \neq i}^{n} \sum_{l=1,l \neq j}^{n} \psi_{kl}^{ij} \tag{2.21}$$

where $\| \cdot \|_{\Sigma}$ is the summation of all matrix elements.

Proof. Using (2.9) and (2.12)–(2.13), we have

$$\begin{aligned}
\phi_{ij} &= -\frac{1}{g_{ij}} g_{i*}^{ij} (G^{ij})^{-1} g_{*j}^{ij} \\[2mm]
&= \left\| \left(-\frac{1}{g_{ij}} g_{*j}^{ij} g_{i*}^{ij} \right) \otimes (G^{ij})^{-T} \right\|_{\Sigma} \\[2mm]
&= \left\| (\triangle G^{ij}) \otimes (G^{ij})^{-T} \right\|_{\Sigma}
\end{aligned}$$

Then from (2.18), we obtain the result of (2.21).

Because Ψ^{ij} reveals the information on loop-by-loop interaction, we define it as DRIA. The significance of the development in this section is as follows:

(i) Equation (2.10) and (2.11) indicates that the pairing structure with a small value of ϕ_{kl}^{ij}, i.e., $\phi_{kl}^{ij} \to 0$, should be preferred.

(ii) Equation (2.16) indicates that a large value of ψ_{kl}^{ij} implies that the interaction from the closed loop $y_i\text{--}u_j$ to an arbitrary loop $y_k\text{--}u_l$ in the closed subsystem $\hat{G}^{ij}$ is large. Therefore, the corresponding pairing structure should be avoided.

(iii) Equation (2.19) indicates that ψ_{kl}^{ij}, because the RI from loop $y_i\text{--}u_j$ to loop $y_k\text{--}u_l$ in the closed subsystem $\hat{G}^{ij}$ is the product of λ_{kl}^{ij} and ϕ_{kl}^{ij} and because the ideal situation is $\lambda_{kl}^{ij} \to 1$ and $\phi_{kl}^{ij} \to 0$, ψ_{kl}^{ij} should be $\psi_{kl}^{ij} \to 0$ for minimal interactions.

(iv) Equation (2.20) indicates that, compared with Case 3, the RI from loop $y_i\text{--}u_j$ to an arbitrary loop $y_k\text{--}u_l$ of the subsystem G^{ij} in Case 4 is changed by a factor of λ_{kl}^{ij}, whereas this RGA is determined by the other loops of subsystem G^{ij}; thus, the element ψ_{kl}^{ij} of DRIA reflects more information on the interaction effect to $y_i\text{--}u_j$ from all of the other loops working together. Apparently, the best pairing structure should be $\lambda_{kl}^{ij} \to 1$, which is consistent with the conventional pairing rule.

(v) Equation (2.21) indicates that Ψ^{ij} given in a matrix form provides loop-by-loop information of the RIs between $y_i\text{--}u_j$ and all other loops as well as their distributions; therefore, it is more precise than ϕ_{ij} in measuring the loop interactions.

To illustrate the implication of DRIA in analyzing loop interactions, we continue with Example 2.1.

Example 2.2 (Example 2.1 Continued). For the two variable pairings derived from Example 2.1, the values of RGA, RI, and DRIA are listed in Table 2.3.

Table 2.3. RGA, RI and DRIA corresponding to different pairing structures

	Feasible Pairing Structure	
	1–1/2–2/3–3	1–2/2–3/3–1
RGA	1/1/1	5/5/5
NI	0/0/0	$-0.8/-0.8/-0.8$
DRIA	$\Psi^{11} = \begin{bmatrix} 0.9620 & -5.9604 \\ 4.0346 & 0.9629 \end{bmatrix}$	$\Psi^{12} = \begin{bmatrix} 0.0074 & 0.1927 \\ 0.1927 & -1.1929 \end{bmatrix}$
	$\Psi^{22} = \begin{bmatrix} 0.9620 & 4.0346 \\ -5.9604 & 0.9629 \end{bmatrix}$	$\Psi^{23} = \begin{bmatrix} -1.1927 & 0.1927 \\ 0.1925 & 0.0074 \end{bmatrix}$
	$\Psi^{33} = \begin{bmatrix} 0.9638 & -5.9657 \\ 4.0382 & 0.9638 \end{bmatrix}$	$\Psi^{31} = \begin{bmatrix} 0.1927 & 0.0074 \\ -1.1927 & 0.1925 \end{bmatrix}$

Table 2.3 clearly shows the following:

(i) $\lambda_{11} = 1$ and $\phi_{11} = 0$ indicate that there are no interactions to loop y_1–u_1 when all other loops are closed.

(ii) ϕ_{11} is the summation of all elements in corresponding Ψ^{11}, as given in (2.21), but $\phi_{11} = 0$ does not necessarily mean that all elements of Ψ^{11} are zeros; some of them can even be large values.

(iii) From (2.19), ψ_{32}^{11} in Ψ^{11} is the product of ϕ_{32}^{11} and λ_{32}^{11}. The large value of ψ_{32}^{11} – 4.0346 indicates that either ϕ_{32}^{11} or λ_{32}^{11} is large, which points out that the interaction between loops y_1–u_1 and y_3–u_2 is large (a large ϕ_{32}^{11} value implies severe interaction between loops y_1–u_1 and y_3–u_2 when the subsystem G^{11} is open, while a large λ_{32}^{11} value implies that the gain of loop y_3–u_2 will undergo a big change after all other loops in subsystem G^{11} are closed).

(iv) For the 1–2/2–3/3–1 pairing, even though $\lambda_{ii} = 5$, $\phi_{ii} = -0.8$ is not as good as those of the 1–1/2–2/3–3 pairing based on the RGA loop-pairing criterion; the smaller elements in DRIA imply smaller interaction between the considered loop and the other loops.

The analysis above suggests that the RGA and RI may not be able to reflect the interactions among loops accurately, while through DRIA analysis, loop interactions in the matrix form can be revealed categorically.

2.4   Control Structure Selection

Based on the DRIA, we can now define loop-by-loop interaction energy as below.

Definition 2.4. *The GI ω_{ij} is the interaction "energy" (2-norm) of matrix Ψ^{ij}, and correspondingly ω_{ij} is the ijth element of the GI array (GIA) Ω*

$$\Omega \triangleq \left\{ \omega_{ij} \mid \omega_{ij} = \left\| \Psi^{ij} \right\|_2, \quad i, j = 1, 2, \cdots, \right\}, \tag{2.22}$$

where $\| \cdot \|_2$ denotes the 2-norm of the matrix defined as

$$\left\| \Psi^{ij} \right\|_2 \triangleq \sigma_{\max}\left(\Psi^{ij}\right).$$

Equation (2.22) represents the interaction "energy" to all closed loops of the subsystem G^{ij} from loop y_i–u_j. Moreover, it also reflects the interaction "energy" to loop y_i–u_j from all remaining $n - 1$ closed loops in G. Therefore, GI reflects the intensity of the interactions among all loops.

In analogy to RGA and NI, we here provide some important properties of the GI:

(i) The GI only depends on the steady-state gain of the multivariable system.

(ii) The GI is not affected by any permutation of G.

(iii) The GI is scaling-independent (e.g., independent of units chosen for $\mathbf{u}$ and $\mathbf{y}$); this property is easily proven from the property of RGA, (2.10), (2.11), (2.12), (2.13), (2.20), and (2.22).

(iv) If the transfer function matrix G is diagonal or triangular, the corresponding GI is equal to zero.

The new loop-pairing criterion based on Definition 2.4 is given as follows:

2.4.1 New Loop-Pairing Criterion

Manipulated and controlled variables in a decentralized control system should be paired in such a way that (i) all paired RGA elements are positive, (ii) NI is positive, and (iii) the pairings have the smallest ω_{ij} value.

Example 2.3 (Example 2.1 Continued). For the two possible pairings, the corresponding RGA, RI, and ω_{ij} values are shown in Table 2.4.

Table 2.4. ω_{ij} for different pairing structures

	Feasible Pairing Structure	
	1–1/2–2/3–3	1–2/2–3/3–1
RGA	1/1/1	5/5/5
NI	0/0/0	$-0.8/-0.8/-0.8$
GI	6.0/6.0/6.0	1.2/1.2/1.2

Thus, on the basis of the new loop-pairing criterion, the second pairing is preferred, which draws the same conclusion as that in ref [41].

Remark 2.2. For two or more pairing structures that have passed all three variable-pairing steps but with similar GI values, the pairing structure that has the smallest product of all GIs are preferred because the interactions are transferable through interactive loops, i.e., selecting

$$\min\left\{\prod_{i=1}^{n} \omega_i\right\}, \tag{2.23}$$

where ω_i is the GI corresponding to the ith output.

On the basis of the proposed loop-pairing criterion, an algorithm to select the best control structure can be summarized as follows:

2.4.1.1 Algorithm 2.1

Step 1.   For a given transfer function $G(s)$, obtain steady-state gain matrix $G(0)$.
Step 2.   Calculate RGA and NI by (2.3) and (2.4), respectively.
Step 3.   Eliminate pairs having negative RGA and NI values.
Step 4.   Calculate $\triangle G^{ij}$ and Ψ^{ij} by (2.13) and (2.18), respectively.
Step 5.   Calculate ω_{ij} and form Ω by (2.22).
Step 6.   Select the pairing that has the smallest value of ω_{ij} in Ω.
Step 7.   Use (2.23) to select the best one if two or more pairings have similar GI values.
Step 8.   End

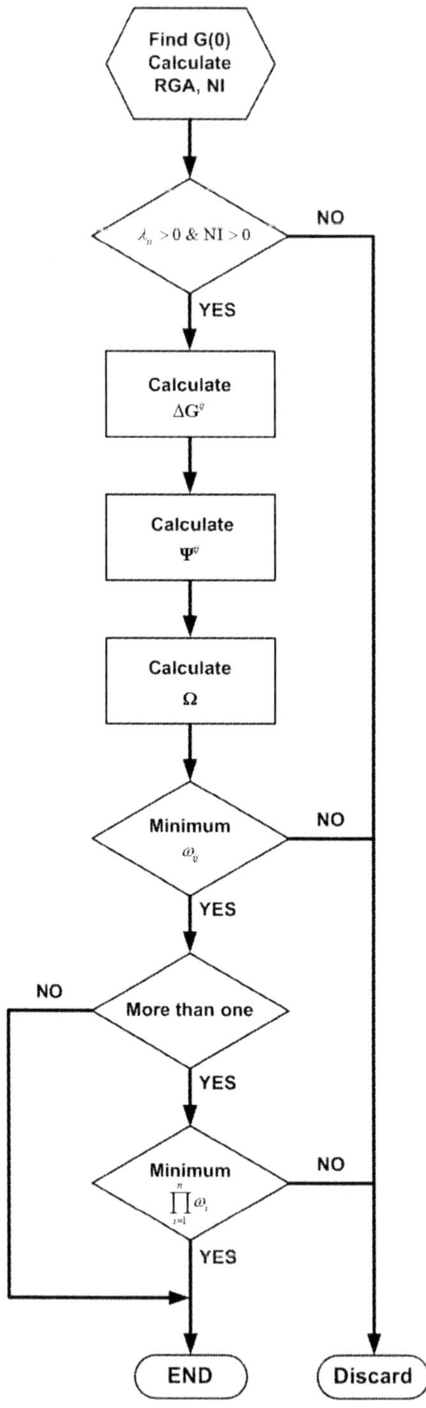

Fig. 2.4. Flow chart of variable pairing selection procedure

The procedure for the variable-pairing selection is demonstrated by the flowchart as shown in Fig. 2.4.

Remark 2.3. For the system that has the pure integral element, GIA and DRIA can be calculated by using a method similar to that proposed by Arkun and Downs [64].

Remark 2.4. For 2×2 system, from (2.9), (2.10), (2.11), (2.16), (2.19), and (2.22), we obtain the GI value of element g_{11} as

$$\omega_{11} = |\phi_{11}| = |\phi_{22}| = \left|\psi_{22}^{11}\right| = \left|\psi_{11}^{22}\right|.$$

Furthermore, from properties of RGA and RI, we can obtain an equation as

$$\lambda_{11} + \lambda_{12} = 1 \Rightarrow \phi_{11}\phi_{12} = 1,$$

which indicates ϕ_{11} and ϕ_{12} must have the same sign, and a smaller $|\phi_{11}|$ means less interaction. Therefore, selecting the pairings that have the smallest GIA elements is equivalent to the RGA-based loop-pairing criterion to select the RGA value closest to unity.

Remark 2.5. Even though the procedure to derive the new loop-pairing criterion is tedious, the calculation for the control structure configuration can be achieved automatically and can easily be programmed into a computer.

2.4.2 Case Study

In this section, we give two more examples to demonstrate the effectiveness of the proposed control-loop configuration criterion.

Example 2.4. Consider the process with its steady state gain matrix [40]

$$G(0) = \begin{bmatrix} 1.0 & 1.0 & -0.1 \\ 1.0 & -3.0 & 1.0 \\ 0.1 & 2.0 & -1.0 \end{bmatrix}$$

and the RGA is obtained as

$$\Lambda[G(0)] = \begin{bmatrix} 0.5348 & 0.5882 & -0.1230 \\ 0.4278 & 1.5882 & -1.0160 \\ 0.0374 & -1.1765 & 2.1390 \end{bmatrix}.$$

Obviously, pairings of 1–1/2–2/3–3 and 1–2/2–1/3–3 contain all positive RGA elements, and it is easy to verify that they all have positive NI (0.62 and 1.87, respectively), but because the RGA values of these two feasible pairings have similar deviations from unity, it is difficult to use RGA values to determine which pairing has fewer interactions. In such a case, the GIA-based criterion can make an effective selection, which is calculated as

$$\Omega = \begin{bmatrix} 1.0251 & 3.2787 & \infty \\ 4.5081 & 0.6811 & \infty \\ 53.2591 & \infty & 0.5031 \end{bmatrix},$$

where the infinity elements ∞ of Ω indicate that the elements of G have been filtered out by the RGA criteria (this representation is also used in the next example).

It can be easily seen from GIA that the pairing of 1–1/2–2/3–3 has smaller values compared with the second pairing of 1–2/2–1/3–3, which indicates less loop interaction from the first pairing structure than from the second.

Example 2.5. A three-product (Petlyuk) distillation column was studied by Wolff and Skogestad [33]. The process gain at steady-state operating conditions is given by

$$G(0) = \begin{bmatrix} 153.45 & -179.34 & 0.23 & 0.03 \\ -157.67 & 184.75 & -0.10 & 21.63 \\ 24.63 & -28.97 & -0.23 & -0.10 \\ -4.80 & 6.09 & 0.13 & -2.41 \end{bmatrix}$$

and the RGA is obtained as

$$\Lambda = \begin{bmatrix} 24.5230 & -23.6378 & 0.1136 & 0.0012 \\ -48.9968 & 49.0778 & 0.0200 & 0.8990 \\ 38.5591 & -38.6327 & 1.0736 & 0.0000 \\ -13.0852 & 14.1927 & -0.2072 & 0.0998 \end{bmatrix}.$$

It is easy to verify that there are six configurations that give positive values for both RGA and NI. According to the RGA-based loop-pairing criterion, the preferred control structure sequence is 1–1/2–4/3–3/4–2 > 1–3/2–4/3–1/4–2 > 1–1/2–2/3–3/4–4 > 1–3/2–2/3–1/4–4 > 1–1/2–3/3–4/4–2 > 1–4/2–3/3–1/4–2; i.e., the first pairing structure 1–1/2–4/3–3/4–2 is the most preferred. However, the pairing criterion based on GIA makes a very different selection. From Algorithm 2.1, the GIA of that process is obtained as

$$\Omega = \begin{bmatrix} 2.2032 & \infty & 771.3599 & 5.5671 \times 10^4 \\ \infty & 1.0259 & 2.9624 \times 10^3 & 75.4987 \\ 1.8562 & \infty & 44.8766 & 9.9018 \times 10^6 \\ \infty & 4.9251 & \infty & 193.7161 \end{bmatrix}. \tag{2.24}$$

It can be easily seen that if loop pairings y_1–u_4, y_2–u_3, or y_3–u_4 are selected, the interactions between loops will be very large. Therefore, these pairing structures should also be filtered out; i.e., these elements in (2.24) can be replaced by ∞, reducing the GIA to

$$
\Omega = \begin{bmatrix}
2.2032 & \infty & 771.3599 & \infty \\
\infty & 1.0259 & \infty & 75.4987 \\
1.8562 & \infty & 44.8766 & \infty \\
\infty & 4.9251 & \infty & 193.7161
\end{bmatrix}.
$$

For the remaining four possible pairings, their corresponding products of GIs are calculated and listed in Table 2.5.

Table 2.5. Four possible pairings and products of GI for Example 2.5

	Feasible Pairing			
	1–1/2–4	1–3/2–4	1–1/2–2	1–3/2–2
	/3–3/4–2	/3–1/4–2	/3–3/4–4	/3–1/4–4
Product of GIs	36 764.5	532 397.9	19 649.2	284 546.1

Apparently, the third pairing of 1–1/2–2/3–3/4–4 is the best one, with the smallest interactions between loops, and should be selected, which gives the same result as that in [33].

2.5 DRIS

We first reveal the relationship between DRIA and RGA, which is fundamental for the remaining developments.

Lemma 2.1. *For an arbitrary loop y_i–u_i in system G, the relationship between elements of $\Psi_{ii,n-1}$ and elements of Λ satisfies,*

$$
\frac{\lambda_{il}}{\lambda_{ii}} = \sum_{k=1,k \neq i}^{n} \psi_{ii,kl}, \qquad \forall i,l = 1,\cdots,n \text{ and } l \neq i, \tag{2.25}
$$

and

$$
\frac{\lambda_{ki}}{\lambda_{ii}} = \sum_{l=1,l \neq i}^{n} \psi_{ii,kl}, \qquad \forall i,k = 1,\cdots,n \text{ and } k \neq i. \tag{2.26}
$$

Because the relationship provided by (2.26) is similar to that provided by (2.25), only the relationship given by (2.25) is proved here.

Proof. Because

$$
\begin{aligned}
\frac{\lambda_{il}}{\lambda_{ii}} &= \frac{(-1)^{i+l} g_{il} \det G^{il} / \det G}{(-1)^{i+i} g_{ii} \det G^{ii} / \det G} = \frac{(-1)^{i+l} g_{il} \det G^{il}}{(-1)^{i+i} g_{ii} \det G^{ii}} \\
&= (-1)^{l-i} \frac{g_{il}}{g_{ii}} \times \frac{\sum_{k=1,k\neq i}^{n} (-1)^{k+i-1} g_{ki} \det \left(G^{ii}\right)^{kl}}{\det G^{ii}} \\
&= \sum_{k=1,k\neq i}^{n} -\frac{g_{il} g_{ki}}{g_{ii}} \times \frac{(-1)^{k+l} \det \left(G^{ii}\right)^{kl}}{\det G^{ii}},
\end{aligned}
$$

where $\left(G^{ii}\right)^{kl}$ is the transfer function matrix G with its ith, kth rows and jth, lth columns removed, using (2.19), we obtain,

$$
\frac{\lambda_{il}}{\lambda_{ii}} = \sum_{k=1,k\neq i}^{n} \left(\phi_{ii,kl} \lambda_{kl}^{ii}\right) = \sum_{k=1,k\neq i}^{n} \psi_{ii,kl},
$$

which completes the proof. $\square$

Remark 2.6. Lemma 2.1 presents an important relationship between the elements of DRIA and those of RGA. By the definition of RGA-number [48],

$$
\text{RGA number} = \|\Lambda - I\|_{\Sigma} = \sum_{i=1}^{n} \left[\lambda_{ii} |1 - 1/\lambda_{ii}| + \sum_{l=1,l\neq i}^{n} |\lambda_{il}/\lambda_{ii}| \right].
$$

It is obvious that both $\lambda_{ii} \to 1$ and $|\lambda_{il}/\lambda_{ii}| \to 0$ are desired. As indicated by Theorem 2.3 and Lemma 2.1, this is consistent with the expectation that RI, ϕ_{ii}, and all elements of DRIA have smaller values. Furthermore, a smaller element $\psi_{ii,kl}$ means less interaction either between loop y_i–u_i and loop y_k–u_l or between loop y_k–u_l and all the other loops in subsystem G^{ii}. Therefore, the DRIA provides more information than RGA, and to select loop pairings that have smaller elements of DRIA is more effective than the RGA based loop pairing rules.

Using the LR matrix factorization method [65] to DRIA, $\Psi_{ii,n-1}$ can be factorized as

$$
\Psi_{ii,n-1} = L_{ii,n-1} \times R_{ii,n-1}, \tag{2.27}
$$

where $L_{ii,n-1}$ is a $(n-1) \times (n-1)$ lower triangular matrix with its diagonal elements equal to unity and $R_{ii,n-1}$ is a $n-1 \times n-1$ upper triangular matrix. Then, we have the following lemma:

Lemma 2.2. *Given a subsystem G^{ii} of G, if its first $n-m-1$ loops are removed, then the relative interaction to loop y_i–u_i from the remaining m loops is the sum of all elements of the matrix that produced by the submatrices $L_{ii,m}$ and $R_{ii,m}$,*

$$
\phi_{ii,m} = \|\Psi_{ii,m}\|_{\Sigma} = \|L_{ii,m} \times R_{ii,m}\|_{\Sigma}. \tag{2.28}
$$

Proof. According to the LR factorization algorithm, the DRIA $\Psi_{ii,n-1}$ can be factorized step-by-step. The first step is given as:

$$
\Psi_{ii,n-1} = \begin{bmatrix}
\psi_{ii,11} & \psi_{ii,12} & \cdots & \psi_{ii,1n} \\
\psi_{ii,21} & \psi_{ii,22} & \cdots & \psi_{ii,2n} \\
\vdots & \vdots & \ddots & \vdots \\
\psi_{ii,n1} & \psi_{ii,n2} & \cdots & \psi_{ii,nn}
\end{bmatrix}_{n \neq i}
$$

$$
= \begin{bmatrix}
1 & 0 & \cdots & 0 \\
\psi_{ii,21}/\psi_{ii,11} & 1 & & 0 \\
\vdots & & \ddots & \\
\psi_{ii,n1}/\psi_{ii,11} & 0 & & 1
\end{bmatrix}_{n \neq i}
$$

$$
\times \begin{bmatrix}
\psi_{ii,11} & \psi_{ii,12} & \cdots & \psi_{ii,1n} \\
0 & & & \\
\vdots & & \tilde{\Psi}_{ii,n-2} & \\
0 & & &
\end{bmatrix}_{n \neq i}, \tag{2.29}
$$

where,

$$
\tilde{\Psi}_{ii,n-2} = \begin{bmatrix}
\psi_{ii,22} - \psi_{ii,12}\psi_{ii,21}/\psi_{ii,11} & \cdots & \psi_{ii,2n} - \psi_{ii,1n}\psi_{ii,21}/\psi_{ii,11} \\
\vdots & \ddots & \vdots \\
\psi_{ii,n2} - \psi_{ii,12}\psi_{ii,n1}/\psi_{ii,11} & \cdots & \psi_{ii,nn} - \psi_{ii,1n}\psi_{ii,n1}/\psi_{ii,11}
\end{bmatrix}_{n \neq i}.
$$

In (2.29), the vectors $[1,\cdots,\psi_{ii,n1}/\psi_{ii,11}]^T$ and $[\psi_{ii,11},\cdots,\psi_{ii,1n}]$ are the first column and the first row of triangular matrices $L_{ii,n-1}$ and $R_{ii,n-1}$, respectively. On the basis of (2.19), the klth element $\tilde{\psi}_{ii,kl}$ of $\tilde{\Psi}_{ii,n-2}$ can be simplified as,

$$
\begin{aligned}
\tilde{\psi}_{ii,kl} &= \psi_{ii,kl} - \psi_{ii,1l}\psi_{ii,k1}/\psi_{ii,11} \\
&= -\frac{g_{il}g_{ki}}{g_{ii}} \times \frac{\det\left(G^{ii}\right)^{11}\det\left(G^{ii}\right)^{kl} - \det\left(G^{ii}\right)^{1l}\det\left(G^{ii}\right)^{i1}}{\det G^{ii}} \\
&= -\frac{g_{il}g_{ki}}{g_{ii}} \times \frac{\det\left(\left(G^{ii}\right)^{11}\right)^{kl}}{\det G^{ii}} = \phi_{ii,kl}\lambda_{kl}^{ii(11)} = \psi_{ii,kl}^{11},
\end{aligned}
$$

where the superscript "11" means that loop y_1–u_1 is removed. Obviously,

$$
\begin{aligned}
\tilde{\Psi}_{ii,n-2} &= \left[\Psi_{ii,n-1} - [\Psi_{ii,n-1}]_{*1} \times [\Psi_{ii,n-1}]_{1*}/[\Psi_{ii,n-1}]_{11}\right]^{11} \\
&= \Psi_{ii,n-2}^{11}
\end{aligned} \tag{2.30}
$$

is the DRIA of loop y_i–u_i in subsystem G^{11}. Therefore based on Theorem 2.3,

$$\phi_{ii,n-2} = \|\Psi_{ii,n-2}\|_\Sigma = \|L_{ii,n-2} \times R_{ii,n-2}\|_\Sigma,$$

which completes the proof. $\qquad\qquad\qquad\qquad\qquad\qquad\qquad\qquad\qquad\qquad\qquad\square$

Above relationship can be applied straightforward to all loops of subsystem G^{ii}. Consequently, the result given by (2.27) is obtained. If the top-left corner element of the matrix is not equal to zero, the similar factorization step can be continued loop-by-loop till m to result (2.28).

Remark 2.7. According to (2.29), since the top-left corner elements of DRIA such as $\psi_{ii,11}$ are applied in denominator in every step of factorization, they must not be equal to zero, which requires all elements of G are not equal to zero. In practice, this problem can be easily solved by setting those zero elements to a very small value, say 10^{-9}. Our simulation results show that by using very small values to replace those zero elements do not avert the outcomes of failure tolerance property.

Theorem 2.4. *Suppose the control configuration of system G has been selected, for an arbitrary failure index M, the RI to loop y_i–u_i from the other $n-1$ loops can be represented by the summation of $n-1$ elements,*

$$\phi_{ii,n-1}^{M} = \sum_{p=1}^{n-1} s_{ii,p}^{M}, \tag{2.31}$$

and

$$s_{ii,p}^{M} = \sum \left[L_{ii,n-1}^{M}\right]_{*p} \times \sum \left[R_{ii,n-1}^{M}\right]_{p*}, \tag{2.32}$$

where "$*p$" and "$p*$" indicate the pth column and the pth row of matrix respectively.

Proof. Equation (2.31) can be derived straightforward from (2.28), since

$$
\begin{aligned}
\phi_{ii,n-1}^{M} &= \|\Psi_{ii,n-1}^{M}\|_\Sigma = \|L_{ii,n-1}^{M} \times R_{ii,n-1}^{M}\|_\Sigma \\[2mm]
&= \underbrace{\begin{bmatrix} 1 & \cdots & 1 \end{bmatrix}}_{n-1} \times L_{ii,n-1}^{M} \times R_{ii,n-1}^{M} \times \underbrace{\begin{bmatrix} 1 & \cdots & 1 \end{bmatrix}^{T}}_{n-1} \\[2mm]
&= \underbrace{\begin{bmatrix} \sum\left[L_{ii,n-1}^{M}\right]_{*1} & \cdots & \sum\left[L_{ii,n-1}^{M}\right]_{*n-1} \end{bmatrix}}_{n-1} \\[2mm]
&\quad \times \underbrace{\begin{bmatrix} \sum\left[R_{ii,n-1}^{M}\right]_{1*} & \cdots & \sum\left[R_{ii,n-1}^{M}\right]_{n-1*} \end{bmatrix}^{T}}_{n-1} \\[2mm]
&= \sum_{p=1}^{n-1} \left(\sum \left[L_{ii,n-1}^{M}\right]_{*p}\right) \times \left(\sum \left[R_{ii,n-1}^{M}\right]_{p*}\right) \\[2mm]
&= \sum_{p=1}^{n-1} s_{ii,p}^{M}.
\end{aligned}
$$

Definition 2.5. *For individual loop y_i–u_i in system G,*

$$S_{ii}^M = \left\{ s_{ii,1}^M, \cdots s_{ii,p}^M, \cdots, s_{ii,n-1}^M \right\}, \tag{2.33}$$

and its individual element $s_{ii,p}^M$ are defined as DRIS and DRIF to failure index M, respectively.

To explain the physical meaning of DRIF, we analyze an arbitrary element in DRIS, say $s_{11,1}^M$, as an example. If all control loops of subsystem $\left(G^{11}\right)^{22}$ are closed, the subsystem including loop y_1–u_1 and loop y_2–u_2 are given as.

$$\tilde{G}_{11-22} = \begin{bmatrix} \tilde{g}_{11} & \tilde{g}_{12} \\ \tilde{g}_{21} & \tilde{g}_{22} \end{bmatrix},$$

where "~" indicates subsystem $\left(G^{11}\right)^{22}$ is closed and

$$
\begin{aligned}
\tilde{g}_{11} &= g_{11}/\lambda_{11}^{22} = g_{11}/\left[g_{11}\det\left(G^{22}\right)^{11}/\det G^{22}\right] = \det G^{22}/\det\left(G^{22}\right)^{11}, \\
\tilde{g}_{12} &= g_{12}/\lambda_{12}^{21} = g_{12}/\left[g_{12}\det\left(G^{21}\right)^{12}/\det G^{21}\right] = \det G^{21}/\det\left(G^{22}\right)^{11}, \\
\tilde{g}_{21} &= g_{21}/\lambda_{21}^{12} = g_{21}/\left[g_{21}\det\left(G^{12}\right)^{21}/\det G^{12}\right] = \det G^{12}/\det\left(G^{22}\right)^{11}, \\
\tilde{g}_{22} &= g_{22}/\lambda_{22}^{11} = g_{22}/\left[g_{22}\det\left(G^{11}\right)^{22}/\det G^{11}\right] = \det G^{11}/\det\left(G^{22}\right)^{11}.
\end{aligned}
$$

Now, if loop y_2–u_2 is also closed, the incremental RI to loop y_1–u_1 can be obtained as,

$$
\begin{aligned}
\tilde{\phi}_{11,22} &= -\frac{\tilde{g}_{12}\tilde{g}_{21}}{g_{11}\tilde{g}_{22}} \\
&= -\frac{1}{g_{11}}\frac{\det G^{12}\det G^{21}/\left(\det\left(G^{22}\right)^{11}\right)^2}{\det G^{11}/\det\left(G^{22}\right)^{11}} \\
&= -\frac{1}{g_{11}}\frac{\det G^{12}\det G^{21}}{\det G^{11}\det\left(G^{22}\right)^{11}} \\
&= \frac{g_{12}\det G^{12}/\det G g_{21}\det G^{21}/\det G}{-g_{12}g_{21}/(g_{11}g_{22})g_{22}\det\left(G^{22}\right)^{11}/\det G^{11}\left(g_{11}\det G^{11}/\det G\right)^2} \\
&= \frac{1}{\psi_{22}^{11}}\frac{\lambda_{12}}{\lambda_{11}}\frac{\lambda_{21}}{\lambda_{11}}.
\end{aligned}
$$

Then, from Lemma 2.1,

$$
\begin{aligned}
\tilde{\phi}_{11,22} &= \frac{\left(\psi_{11,22} + \psi_{11,23} + \cdots + \psi_{11,2n}\right)\left(\psi_{11,22} + \psi_{11,32} + \ldots + \psi_{11,n2}\right)}{\psi_{11,22}} \\
&= s_{11,1}^M,
\end{aligned}
$$

which suggests the following:

(i) If subsystem $\left(G^{11}\right)^{22}$ is closed, closing loop y_2–u_2 will make the RI to loop y_1–u_1 increase by a value of $s^M_{11,1}$;

(ii) Reversely, if loop y_2–u_2 is taken out of service, the RI to loop y_1–u_1 will decrease by a value of $s^M_{11,1}$.

To generalize the above explanation, we conclude that, for an arbitrary loop y_i–u_i, if the first $p-1$ loops of subsystem G^{ii} has already been taken out of service, the removal of the pth loop will decrease $\phi^M_{ii,n-p}$ by a value of $s^M_{ii,p}$.

The significance of the development in this section are as follows:

(i) The RI, $\phi^M_{ii,n-1}$, to individual control loop y_i–u_i from all other $n-1$ control loops is decomposed as the DRIS, S^M_{ii}, according to the failure index M, such that the interaction from an arbitrary loop of the remaining closed loops is represented by the DRIF;

(ii) The DRIF, $s^M_{ii,p}$, indicates the interaction to individual control loop y_i–u_i from the pth control loop of the remaining $n-p$ closed loops, which means when the pth control loop is put in or taken out of service, the corresponding DRIF should be added to or subtracted from the overall interaction RI;

(iii) In terms of DRIS, not only the interactions between individual loop and the remaining loops but also the interactions to this individual loop from any combination of loops taken out of service can be reflected precisely.

2.6 Tolerance to Loops Failures

From (2.7) and (2.8), both large values and values close to -1 of RI imply significant interaction among individual loops. Because we are investigating the property of loop failure tolerance, only the lower boundary (-1) is considered. By selecting the maximum DRIF from all possible values, we can determine a failure index $\bar{M}$ corresponding DRIS $S^{\bar{M}}_{ii}$ of individual loop y_i–u_i as,

$$S^{\bar{M}}_{ii} = \left\{ s^{\bar{M}}_{ii,p} \,\middle|\, s^{\bar{M}}_{ii,p} = \max \left\{ s^M_{ii,p} \right\}, p = 1,2,\cdots,n-1 \right\}, \tag{2.34}$$

Therefore, taking the pth loop out of service according to failure index $\bar{M}$ will result

$$\phi^{\bar{M}}_{ii,n-p-1} = \min \left\{ \phi^M_{ii,n-p-1} \right\}. \tag{2.35}$$

The value of RI is closest to -1, implying that the particular combination of loop failures has the most significant effect on the DCLI. On the basis of (2.8), (2.35), and Theorem 2 of [49], we now provide the necessary and sufficient conditions if individual loop y_i–u_i is DCLI to single-loop failure.

Theorem 2.5. *For decentralize controlled multivariable process G, individual loop y_i–u_i is DCLI to single-loop failure, if and only if*

$$\phi^{\bar{M}}_{ii,n-2} > -1. \tag{2.36}$$

or

$$s^{\bar{M}}_{ii,1} < 1/\lambda_{ii}. \tag{2.37}$$

Proof. Sufficient: In the case of an arbitrary loop failure, (2.31) and (2.34) gives the RI to individual loop $y_i\text{--}u_i$ as,

$$\phi^M_{ii,n-2} = \sum_{p=2}^{n-1} s^M_{ii,p} = \phi^M_{ii,n-1} - s^M_{ii,1} \geq \phi^{\bar{M}}_{ii,n-1} - s^{\bar{M}}_{ii,1} = \phi^{\bar{M}}_{ii,n-2}.$$

Obviously, when (2.36) holds, inequality $\phi^M_{ii,n-2} > -1$ holds. Therefore, the sign of steady-state gain for individual loop $y_i\text{--}u_i$ does not change in the face of single-loop failure.

Necessary: Because individual loop $y_i\text{--}u_i$ possesses single-loop failure tolerance, the sign of its steady-state loop gain does not change in the face of any single-loop failure,

$$\phi^M_{ii,n-2} > -1, \quad \forall M, \quad \Rightarrow \quad \phi^{\bar{M}}_{ii,n-2} > -1.$$

Then, according to (2.7) and (2.31), (2.37) can be obtained. $\square$

Similar as single-loop failure, on the basis of (2.8), (2.35), and Theorem 2 of [49], the necessary and sufficient condition for individual loop $y_i\text{--}u_i$ is DCLI for multiple-loop failures are given as follows.

Theorem 2.6. *For decentralized control multivariable process G, individual loop $y_i\text{--}u_i$ is DCLI to multiple-loop failures if and only if*

$$\phi^{\bar{M}}_{ii,m_{min}} > -1, \tag{2.38}$$

where,

$$\phi^{\bar{M}}_{ii,m_{\min}} = \min\left\{ \sum_{p=n-m_{\min}}^{n-1} s^{\bar{M}}_{ii,p} \,\middle|\, m_{\min} = 1,\cdots,n-1 \right\}. \tag{2.39}$$

Proof. Sufficient: In the case of $n-m-1$ loops failure, (2.31) and (2.34) show that the RI to individual loop $y_i\text{--}u_i$ from the remaining m loops is,

$$\phi^M_{ii,m} = \sum_{p=n-m}^{n-1} s^M_{ii,p} = \phi^M_{ii,n-1} - \sum_{p=1}^{n-m-1} s^M_{ii,p}$$

$$\geq \phi^{\bar{M}}_{ii,n-1} - \sum_{p=1}^{n-m_{\min}-1} s^{\bar{M}}_{ii,p} = \phi^{\bar{M}}_{ii,m_{\min}}.$$

Obviously, when (2.38) holds, $\phi^M_{ii,m} > -1$ always holds. Therefore, the sign of the steady-state gain for individual loop $y_i\text{--}u_i$ does not change in the face of multiple-loop failure.

Necessary: Because individual loop $y_i\text{--}u_i$ possesses multiple-loop failure tolerance, the sign of its steady-state loop gain does not change in the face of any single-loop failure,

$$\phi^M_{ii,m} > -1, \quad \forall M, \quad \Rightarrow \quad \phi^{\bar{M}}_{ii,m_{\min}} > -1.$$

Remark 2.8. The significance of Theorems 2.5 and 2.6 are as follows:

(i) The necessary and sufficient conditions for both single- and multiple-loop failure tolerance are provided.

(ii) In the case where two or more control structures are DCLI, the one with, $\phi^{\bar{M}}_{ii,m_{\min}} \to 0$, should be preferred.

(iii) Single-loop failure is a special case of multiple-loop failure.

2.7 DCLI Evaluation for Loop Pairings

2.7.1 Algorithm

In subsystem G^{ii}, the DRIF $s^M_{ii,p}$ may have as many as $n - p$ possible values according to different failure sequence of the remaining $n - p$ loops. Therefore, to find either $\phi^{\bar{M}}_{ii,m_{\min}}$ or $\phi^{\bar{M}}_{ii,n-2}$, one is required first to determine the index $\bar{M}$ and to calculate the DRIS $S^{\bar{M}}_{ii}$, where $s^M_{ii,p}$ can be determined by the first row and column of $\Psi^M_{ii,n-p}$ (equations (2.29) and (2.32)),

$$s^M_{ii,p} = \sum_{k=1}^{n-p} \left[\Psi^M_{ii,n-p}\right]_{1k} \times \sum_{k=1}^{n-p} \left[\Psi^M_{ii,n-p}\right]_{k1} / \left[\Psi^M_{ii,n-p}\right]_{11}.$$

However, there is no need to arrange elements of DRIA $\Psi^M_{ii,n-p}$ $n - p$ times to calculate $s^M_{ii,p}$, because the elements of DRIA are permutation- independent (equations (2.18) and (2.13)). In fact, once the DRIA $\Psi_{ii,n-p}$ for $p = 1$ has been obtained (equation (2.18)), the DRIF $s^{\bar{M}}_{ii,p}$ can be directly calculated from,

$$\left[s^{\bar{M}}_{ii,p}, l\right] = \max\left\{ \text{diag}\left(\left(\sum_{k=1}^{n-p}[\Psi_{ii,n-p}]_{*k} \times \sum_{k=1}^{n-p}[\Psi_{ii,n-p}]_{k*} \right) \odot \Psi_{ii,n-p} \right) \right\}, \qquad (2.40)$$

where, function $\max\{A\}$ finds the maximum diagonal element of matrix A and provides its row number l in matrix $\Psi_{ii,n-p}$, $\text{diag}\{A\}$ is a diagonal matrix contains the diagonal elements of matrix A. "$\odot$" indicates element-by-element division.

For checking the DCLI of individual loop y_i–u_i against failure of $p + 1$ loops, the DRIA $\Psi_{ii,n-p-1}$ can be recursively calculated as (equations (2.29) and (2.30)),

$$\Psi_{ii,n-p-1} = \left[\Psi_{ii,n-p} - [\Psi_{ii,n-p}]_{*l} \times [\Psi_{ii,n-p}]_{l*}/[\Psi_{ii,n-p}]_{ll}\right]^{ll}, \qquad (2.41)$$

and the DRIF $s^{\bar{M}}_{ii,p+1}$ can be calculated by applying DRIA $\Psi_{ii,n-p-1}$ to (2.40).

Therefore, for individual loop y_i–u_i of $n \times n$ system G, after $\Psi_{ii,n-1}$ is obtained, its DRIS $S^{\bar{M}}_{ii}$ can be calculated by using iterative (2.40) and (2.41) $n - 2$ times which requires only one matrix inverse of $n - 1$ order to calculated the DRIA as shown by (2.18), the computational load is much reduced compared with that of permutation methods.

For a given multivariable process $G(s)$, its control configuration can be obtained based on its steady-state transfer function matrix $G(0)$ by using the loop pairing criterion such as the one developed in [66]. After all elements in matrix $G(0)$ have been rearranged to place the gains of control loops in the diagonal position, the proposed method can be used to verify DCLI of the selected control configuration and an algorithm is given as follow.

2.7.1.1 Algorithm 2.2

Step 1.   Calculate $\Psi_{ii,n-1}$ of loop y_i–u_i by (2.18) and (2.13);
Step 2.   Obtain $s^{\bar{M}}_{ii,p}$ and $S^{\bar{M}}_{ii}$ of loop y_i–u_i by (2.40) and (2.41);
Step 3.   Verify single loop failure tolerance by (2.36);
Step 4.   Obtain $\phi^{\bar{M}}_{ii,m_{\min}}$ to loop y_i–u_i form the other loops by (2.39);
Step 5.   Verify multiple loop failure tolerance by referring to (2.38);
Step 6.   Repeat the previous 5 steps loop-by-loop until any one loop fails or all loops pass;
Step 7.   End.

The procedure for the determination of DCLI for a decentralized control system is illustrated by the flowchart shown in Fig. 2.5.

2.7.2 Case Study

Example 2.6. Consider the following 4×4 process [55] with the process steady-state transfer function matrix given by

$$G(0) = \begin{bmatrix} 8.72 & -15.80 & 2.98 & 2.81 \\ 6.54 & -20.79 & 2.50 & -2.92 \\ -5.82 & -7.51 & -1.48 & 0.99 \\ -7.23 & 7.86 & 3.11 & 2.92 \end{bmatrix}.$$

To verify DCLI to single-loop failure of the first loop y_1–u_1 by using RGA based criterion, three alternatives have to be tested, namely, calculation of RGAs of subsystems G^{22}, G^{33} and G^{44} for single loop failure of y_2–u_2, y_3–u_3, and y_4–u_4, respectively. Furthermore, to verify DCLI to multiple-loop failures, an additional three RGAs need to be calculated. Consequently, six inverse matrices have to be performed.

Table 2.6. DCLI verification of control loop y_1–u_1

Loop	RI	DRIS	Failed Loop	DCLI-SLF[a]	DCLI-MLF[b]
	$\phi^{\bar{M}}_{11,3} = 1.4142$				
	$\downarrow$	$s^{\bar{M}}_{11,1} = 2.4095$	y_4–u_4		
	$\phi^{\bar{M}}_{11,2} = -0.9953$				
y_1–u_1	$\downarrow$	$s^{\bar{M}}_{11,2} = 0.3486$	y_2–u_2	YES	NO
	$\phi^{\bar{M}}_{11,1} = -1.3439$				
	$\downarrow$	$s^{\bar{M}}_{11,3} = -1.3439$	y_3–u_3		
	0				

[a] SLF: single-loop failure.
[b] MLF: multi-loop failure.

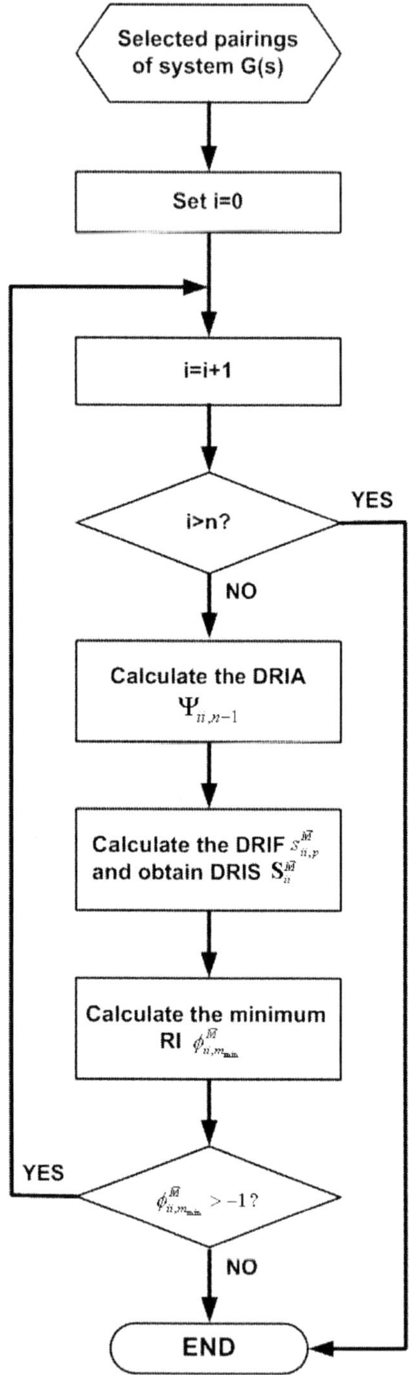

Fig. 2.5. Flowchart for determining DCLI

From application of Algorithm 2 and with one matrix inverse, the DRIA of control loop y_1–u_1 is obtained, and DRIS is calculated through a series of vector operations. The results for DCLI to single- and multiple-loop failures are listed in Table 2.6.

Initially, when all control loops in subsystem G^{11} are closed, the RI $\phi_{11,3}^{\bar{M}} = 1.4142 > -1$, implying that there is no sign change before and after subsystem G^{11} has been closed. Following Table 2.6, DCLI information of loop y_1–u_1 can be obtained as follows:

(i) Loop y_4–u_4 provides the maximum interaction, and if it fails, the RI of loop y_1–u_1 will decrease in value of $s_{11,1}^{\bar{M}} = 2.4095$ and is $\phi_{11,2}^{\bar{M}} = -0.9953 > -1$, and for loop y_1–u_1, the sign of its loop gain does not change for any single-loop failure. Therefore, loop y_1–u_1 is DCLI for single loop failure.

(ii) Loop y_2–u_2 provides the maximum interaction among the two remaining loops after loop y_4–u_4 has already been taken out of service. If loop y_2–u_2 fails, the RI of loop y_1–u_1 will decrease in a value of $s_{11,2}^{\bar{M}} = 0.3486$ and is $\phi_{11,1}^{\bar{M}} = -1.3439 < -1$. Hence, the process gain of loop y_1–u_1 will change its sign and it is not DCLI when both y_4–u_4 and y_2–u_2 fail (implying that G is not DCLI for multiple loop failure).

To show how DCLI is pairing dependant, reconfigure the control structure using the loop pairing criterion proposed in [66] as follows

$$
G(0) = \begin{bmatrix}
2.81 & -15.80 & 8.72 & 2.98 \\
-2.92 & -20.79 & 6.54 & 2.50 \\
0.99 & -7.51 & -5.82 & -1.48 \\
2.92 & 7.86 & -7.23 & 3.11
\end{bmatrix}.
$$

The DCLI results of y_1–u_1 are listed in Table 2.7.

Table 2.7. DCLI verification of control loop y_1–u_1

Loop	RI	DRIS	Failed Loop	DCLI-SLF[a]	DCLI-MLF[b]
	$\phi_{11,3}^{\bar{M}} = 1.1237$				
	$\downarrow$	$s_{11,1}^{\bar{M}} = 0.6885$	y_2–u_2		
	$\phi_{11,2}^{\bar{M}} = 0.4352$				
y_1–u_1	$\downarrow$	$s_{11,2}^{\bar{M}} = 1.4309$	y_3–u_3	YES	YES
	$\phi_{11,1}^{\bar{M}} = -0.9957$				
	$\downarrow$	$s_{11,3}^{\bar{M}} = -0.9957$	y_4–u_4		
	0				

[a] SLF: single-loop failure.
[b] MLF: multi-loop failure.

From Table 2.7, we observe the following:

(i) When all control loops in subsystem G^{11} are closed, the RI $\phi^{\bar{M}}_{11,3} = 1.1237 > -1$, implying that there is no sign change before and after subsystem G^{11} has been closed.

(ii) Loop y_1–u_1 can tolerate any single-loop failure because the minimal $\phi^{\bar{M}}_{11,2} = 0.4352 > -1$

(iii) Loop y_1–u_1 can tolerate any double-loop failure since the minimal $\phi^{\bar{M}}_{11,1} = -0.9957 > -1$

(iv) Since $\phi^{\bar{M}}_{11,3} > \phi^{\bar{M}}_{11,2} > 0$, if loop y_2–u_2 fails, interaction between loop y_1–u_1 and the remaining loops will be smaller.

(v) If loop loop y_3–u_3 also fails, interaction between loop y_1–u_1 and loop y_4–u_4 becomes significant for $\phi^{\bar{M}}_{11,1} = -0.9957 \rightarrow -1$, implying the equivalent process gain of loop y_1–u_1 will undergo a big change in the case of where either loop y_4–u_4 is closed first in system G or loop y_2–u_2 and loop y_3–u_3 fail first in closed subsystem G^{11}.

Using Algorithm 2.2, DRIS of the other three control loops can be obtained and list in Table 2.8, all control loops are DCLI to both single-loop and multiple-loop failures.

Table 2.8. DCLI verification of other three control loops

Loop	RI	DRIS	Failed Loop	DCLI-SLF[a]	DCLI-MLF[b]
	$\phi^{\bar{M}}_{11,3} = 1.2873$	$s^{\bar{M}}_{22,1} = 0.7416$	y_1–u_1		
y_2–u_2	$\phi^{\bar{M}}_{11,2} = 0.5458$	$s^{\bar{M}}_{22,2} = 0.2418$	y_3–u_3	YES	YES
	$\phi^{\bar{M}}_{11,1} = 0.3039$	$s^{\bar{M}}_{22,3} = 0.3039$	y_4–u_4		
	$\phi^{\bar{M}}_{11,3} = 1.4765$	$s^{\bar{M}}_{33,1} = 0.8086$	y_4–u_4		
y_3–u_3	$\phi^{\bar{M}}_{11,2} = 0.6679$	$s^{\bar{M}}_{33,2} = 0.2620$	y_1–u_1	YES	YES
	$\phi^{\bar{M}}_{11,1} = 0.4059$	$s^{\bar{M}}_{33,3} = 0.4059$	y_2–u_2		
	$\phi^{\bar{M}}_{11,3} = 0.7498$	$s^{\bar{M}}_{44,1} = 0.5713$	y_3–u_3		
y_4–u_4	$\phi^{\bar{M}}_{11,2} = 0.1785$	$s^{\bar{M}}_{44,2} = 1.1742$	y_2–u_2	YES	YES
	$\phi^{\bar{M}}_{11,1} = -0.9957$	$s^{\bar{M}}_{44,3} = -0.9957$	y_1–u_1		

[a] SLF: single-loop failure.
[b] MLF: multi-loop failure.

Example 2.7. Consider the 4×4 distillation column studied by Chiang and Luyben (CL column) [67]. The steady state transfer function matrix is given as follow

$$
G(0) = \begin{bmatrix}
4.45 & -7.4 & 0 & 0.35 \\
17.3 & -41 & 0 & 9.2 \\
0.22 & -4.66 & 3.6 & 0.042 \\
1.82 & -34.5 & 12.2 & -6.92
\end{bmatrix}.
$$

When the zero elements in $G(0)$ are set to 10^{-9} to make the zero interaction a micro-interaction, the maximum DRIFs and DRIS of all control loops are obtained as listed in Table 2.9.

Table 2.9. DCLI verification of CL column

Loop	RI	DRIS	Failed Loop	DCLI-SLF[a]	DCLI-MLF[b]
$y_1\text{-}u_1$	$\phi^{\bar{M}}_{11,3} = -0.5233$	$s^{\bar{M}}_{11,1} = 0.1783$	$y_4\text{-}u_4$	YES	YES
	$\phi^{\bar{M}}_{11,2} = -0.7017$	$s^{\bar{M}}_{11,2} = 0.0000$	$y_3\text{-}u_3$		
	$\phi^{\bar{M}}_{11,1} = -0.7017$	$s^{\bar{M}}_{11,3} = -0.7017$	$y_2\text{-}u_2$		
$y_2\text{-}u_2$	$\phi^{\bar{M}}_{11,3} = -0.2490$	$s^{\bar{M}}_{22,1} = 0.4527$	$y_4\text{-}u_4$	YES	YES
	$\phi^{\bar{M}}_{11,2} = -0.7017$	$s^{\bar{M}}_{22,2} = 0.0000$	$y_3\text{-}u_3$		
	$\phi^{\bar{M}}_{11,1} = -0.7017$	$s^{\bar{M}}_{22,3} = -0.7017$	$y_1\text{-}u_1$		
$y_3\text{-}u_3$	$\phi^{\bar{M}}_{11,3} = -0.3394$	$s^{\bar{M}}_{33,1} = -0.1074$	$y_1\text{-}u_1$	YES	YES
	$\phi^{\bar{M}}_{11,2} = -0.2320$	$s^{\bar{M}}_{33,2} = -0.2320$	$y_4\text{-}u_4$		
	$\phi^{\bar{M}}_{11,1} = 0.0000$	$s^{\bar{M}}_{33,3} = 0.0000$	$y_2\text{-}u_2$		
$y_4\text{-}u_4$	$\phi^{\bar{M}}_{11,3} = 1.6000$	$s^{\bar{M}}_{44,1} = 1.5672$	$y_2\text{-}u_2$	YES	YES
	$\phi^{\bar{M}}_{11,2} = 0.0328$	$s^{\bar{M}}_{44,2} = 0.0122$	$y_1\text{-}u_1$		
	$\phi^{\bar{M}}_{11,1} = 0.0206$	$s^{\bar{M}}_{44,3} = 0.0206$	$y_3\text{-}u_3$		

[a] SLF: single-loop failure.
[b] MLF: multi-loop failure.

Obliviously, as Table 2.9 indicates, all four control loops are DCLI to multiple-loop failures.

2.8 Conclusion

In this chapter, a new loop-pairing criterion based on a new interaction measurements for the control structure configuration of the multivariable process was proposed. DRIA was defined to evaluate all possible interactions among loops, and GI based on the concept of interaction energy and DRIA was introduced for control-loop interaction measurement. An effective algorithm that combines RGA, NI, and GI rules was developed that can accurately and systematically solve the loop configuration problem. Several examples were used to demonstrate the effectiveness of the new loop-pairing criterion.

Compared with the RGA-based loop-pairing criterion, the new loop-pairing criterion only provides measurements for loop interaction, while RGA not only gives the interaction measurement but also provides a guide for controller tuning. Therefore, how to design decentralized controllers based on the proposed interaction measurement to achieve the best performance and at the same time guarantee integrity of the overall process will be investigated in our future work.

Moreover, a novel approach for evaluating DCLI for multivariable control systems was also proposed. The DRIS was introduced to represent the RI to a particular loop from other loops. The maximum DRIF was used to find the maximum interaction from the remaining loops among all possible failure indexes. Consequently, the necessary and sufficient conditions for DCLI of an individual loop under both single- and multiple-loop failures were provided. A simple and effective algorithm for verifying DCLI for multivariable control systems was developed. Two classical examples were used to illustrate the effectiveness of the proposed approach. Because DRIS provides more detailed information of interactions among loops, it can be used to design robust multiloop controllers for multivariable processes.

3 Loop Gain Margins and Stabilizing PID Ranges

In the previous chapter, a new loop paring criterion is presented to achieve the minimum loop interaction under the decentralized multivariable control. When a decentralized PID controller is applied according to the paring criterion, it should stabilize the multivariable system at first. More importantly, it is desirable to know the stabilizing ranges, not just values, of PID parameters so that the stability of the closed-loop MIMO system may not be violated in case of modeling error. This chapter addresses such a problem whose solution also leads to the concept of loop gain margin. It should be noted that the simple and effective methods to determine the gain margin of SISO systems are not applicable for MIMO systems due to the loop interactions. Hence, the new theory and/or algorithm for loop gain margins needs to be found.

3.1 Introduction

Stability is a fundamental requirement for all control systems. Nevertheless, it is not enough for the system analysis if only stability is guaranteed. Besides, one must know further how stable the system is because system stability may be deteriorated in case of noise or disturbance. That is the reason why the concept of the stability margins, gain and phase margins, are introduced to measure the relative stability/stability robustness of a system.

PID controllers have dominated industrial applications for more than fifty years because of their simplicity in controller structure, robustness to modeling errors and disturbances, and the availability of numerous tuning methods [2, 17]. Stability analysis of SISO PID systems is straightforward. Usually, the Nyquist stability theorem is utilized with help of the Nyquist curve of the open loop transfer function. For MIMO systems, the generalized Nyquist stability theorem was addressed by Rosenbrock [19], MacFarlane [15], Nwokah [68] and Morari [49]; and effectively unified by Nwokah et al. [69]. The relevant tools such as characteristic loci, Nyquist arrays and Gershgorin bands are developed to help MIMO system analysis and design in frequency domain, which is similar to the SISO case in nature but not as convenient as their counterparts in SISO case, owing to their complexity.

Q.-G. Wang et al.: PID Control for Multivariable Processes, LNCIS 373, pp. 39–70, 2008.
springerlink.com   © Springer-Verlag Berlin Heidelberg 2008

Although great progress on PID control has been made recently, some fundamental issues remain to be addressed for better understanding and applications of PID controllers, especially for MIMO case. The very first task at the outset of PID control is to get a stabilizing PID controller for a given process; If possible, it would be desirable to find their parameter regions for stabilizing a given process. This problem is of great importance, both theoretically and practically, and also related to the problem of stability margins or robustness. Unfortunately, most of existing methods mentioned above can only determine some values (but not ranges) of stabilizing PID parameters. For SISO systems, the gain and phase margins are well defined and can easily be determined graphically or numerically. The characterization of the set of all stabilizing PID controllers is developed for a SISO delay-free linear time invariant (LTI) plant in [70], and for a SISO LTI plant with time-delay in [71], based on the Hermite-Biehler Theorem, its extensions, and some optimization techniques. However, it is pointed out [72, 73] that their methods are unlikely to be extended to the MIMO case. In the context of MIMO PID systems, not much work has been done. Safonov and Athans [74] proposed a singular value approach to multiloop stability analysis, where the sufficient condition of stability and some characterization of frequency-dependent gain and phase margins for multiloop systems is developed. But their criterion is conservative. Morari [49] introduced the concept of integral controllability, that is, for any k such that $0 < k \leq k^*$ ($k^* > 0$), the feedback system with the open loop as $G(s)k/s$ is stable. He also gave the necessary and sufficient conditions of integral controllability for MIMO systems but did not tell how to determine k^*. On the other hand, Yaniv [75] developed a control method to meet some stability margins which are defined loop by loop like a single variable system. Li and Lee [76] showed that the H_∞ norm of a sensitivity function matrix for a stable multivariable closed-loop system is related to some common gain and phase margins for all the loops. Ho et al. [77] defined gain and phase margins and use them for multivariable control system design assuming that the process is diagonally dominant or made so. Such definitions of gain and phase margins based on Gershgorin bands or other frequency domain techniques are more or less conservative, which brings some limitation of their applications. Doyle [78] developed the μ-analysis, which is utilized as an effective tool for robust stabilizing analysis in multivariable feedback control [48]. As a method in frequency domain, the μ-analysis treats system uncertainties as complex valued. But the parameters of PID controllers are all real. Thus, when μ-analysis is used to determine the stabilizing ranges of PID controllers, conservativeness is inevitable, and a detailed analysis on this is shown in the example of Section 6. In summary, it can been concluded that there seems neither satisfactory definition for MIMO gain and phase margins, nor effective technique for determining them so far. To our best knowledge, no results are available to find the stabilizing PID ranges for MIMO processes.

It should be noted that recent developments in the time domain approaches to MIMO PID control is appealing [79,80,81,82,83]. The basic idea in such approaches is to transform MIMO PID control system to an equivalent static output feedback (SOF) system and then to solve a convex optimization problem through iterative algorithms based on linear matrix inequalities (LMI) [79, 80]. Though the static output feedback stabilizability is still hard to solve, Lyapunov-like conditions [84] and the solution of some

Linear Quadratic (LQ) control problem [85, 86] have been developed to enable stability analysis and stabilization. For stochastic systems, a generalized PI control strategy in discrete-time context is presented in [81, 82] for solving the constrained tracking problem. Wang et al. [87] developed a nonlinear PI controller for a class of nonlinear systems based on singular perturbation theory. Crusius et al. [88] showed how to convert an LQ problem in a new parameter space such that the resulting equivalent problem is convex. Boyd et al. [89] showed how to convert control design problems to a class of convex programming problems with linear objective function and constraints expressed in terms of LMI. Cao et al. [90] proposed an iterative LMI approach for static output feedback stabilization, and sufficient LMI conditions for such a control problem were given by Crusius and Trofino [91]. It seems that time domain approach with help of the LMI-like tools opens an new direction to analysis and design of MIMO PID control systems and makes it possible to give better results than classical frequency domain methods mentioned above.

In this chapter, we investigate a linear MIMO plant under a diagonal (or block-diagonal) PID control and fully-coupled PI control structure using time-domain approach to determine the PID stabilizing ranges as well as the gain margins. We will modify our recent descriptor model approach to transform the problem into a robust stability problem for a linear polytopic system. In this way, a detailed scheme in descriptor version is provided for the robust stability test and an effective procedure is given to find the parameter ranges of PID controllers by improving the methods in [92, 93, 94]. The scheme incorporates a relaxed LMI technique which not only effectively solves the considered PID problem, but also leads to better results than the existing methods for special cases of standard polytopic systems [92, 94, 95, 93]. The present procedure is a kind of quasi-LMI based convex computation which can be fulfilled through LMI-Toolbox [89, 96].

Notation: $\mathbb{R}^n$ denotes the n-dimensional real Euclidean space; the superscript "T" stands for the matrix transpose; $W > 0$ ($W \geq 0$) means that W is real, symmetric and positive-definite (positive-semidefinite).

3.2 Problem Formulation

To illustrate mutual dependence of loop gains which stabilizes a coupled multivariable system under decentralized control, let us consider a 2×2 system with the transfer function matrix:

$$G(s) = \begin{bmatrix} \dfrac{1}{s+1} & \dfrac{2}{s+1} \\ \dfrac{3}{s+1} & \dfrac{4}{s+1} \end{bmatrix}$$

A decentralized proportional controller $K(s) = \text{diag}\{k_1, k_2\}$ is applied to it in the unity negative feedback configuration, as shown in Fig. 3.1. It follows [19, 97] that the characteristic equation of the closed-loop system is

$$\begin{aligned} P_c(s) &= P_G(s)P_K(s)\det[I + G(s)K(s)] \\ &= s^2 + (k_1 + 4k_2 + 2)s + (k_1 + 4k_2 + 1 - 2k_1k_2) = 0, \end{aligned} \qquad (3.1)$$

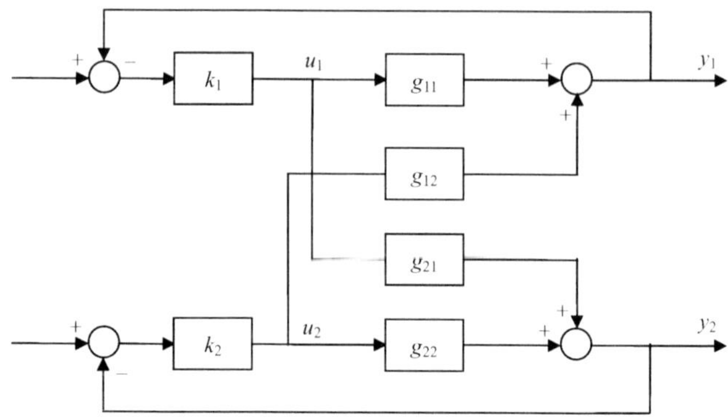

Fig. 3.1. Block diagram of TITO system

where $P_G(s)$ and $P_K(s)$ are the pole polynomials of $G(s)$ and $K(s)$, respectively. The closed-loop system is stable if and only if all the roots of $P_c(s)$ have negative real parts, or

$$\begin{cases} k_1 + 4k_2 + 2 > 0, \\ k_1 + 4k_2 + 1 - 2k_1k_2 > 0. \end{cases} \tag{3.2}$$

The solution of (3.2) is

$$\begin{cases} k_2 > -\dfrac{1}{4}k_1 - \dfrac{1}{2}, \\ k_2 < \dfrac{k_1+1}{2(k_1-2)}, & \text{if } k_1 > 2, \\ k_2 > \dfrac{k_1+1}{2(k_1-2)}, & \text{if } k_1 < 2. \end{cases} \tag{3.3}$$

This stabilizing range is drawn as the shaded region in Fig. 3.2. For example, when $k_1 = 1$, we get from the figure or (3.3) that $k_2 > -3/4$.

Alternately, we may look at the system loop by loop. From Fig. 3.1, it is straightforward [98] to see that

$$\begin{aligned} y_2 &= g_{22}u_2 + g_{21}u_1, \\ &= \left(g_{22} - \frac{k_1g_{12}g_{21}}{1+k_1g_{11}} \right) u_2. \end{aligned} \tag{3.4}$$

Thus, the equivalent open-loop transfer function which k_2 stabilizes when the first loop is closed with gain k_1 is

$$\widetilde{g}_{22}(s) = \frac{y_2(s)}{u_2(s)} = g_{22} - \frac{k_1g_{12}g_{21}}{1+k_1g_{11}} = \frac{4s+4-2k_1}{(s+1)(s+1+k_1)}.$$

Then, the Nyquist stability theorem for SISO systems can be applied to determine the stabilizing range of k_2, if k_1 is specified. For example, when $k_1 = 1$, the Nyquist curve of $\widetilde{g}_{22}(j\omega)$ is depicted as Fig. 3.3. Since the open-loop system $\widetilde{g}_{22}$ has no poles in the

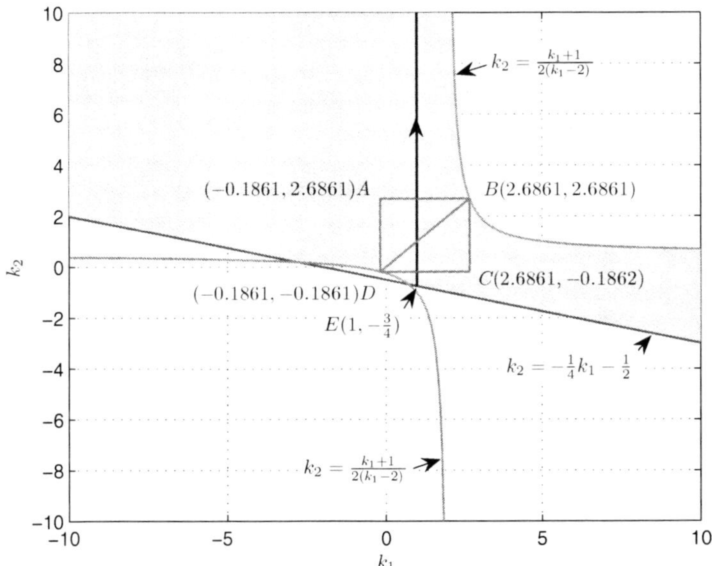

Fig. 3.2. Stabilization region of (k_1, k_2)

RHP, the closed-loop system is stable if and only if the Nyquist curve of $\widetilde{g}_{22}$ does not encircle the point $(-1/k_2, 0)$, or $k_2 > -3/4$, which is the same as before. In general, the characteristic equation for the second equivalent loop is $1 + k_2\widetilde{g}_{22}(s) = 0$, which gives exactly (3.1). Similarly, we may work with the first equivalent open-loop transfer function and lead to the same results.

It is seen from this example that the stabilizing range of one loop gain depends on the value of other loop's gain. This range can be computed with the SISO method for the equivalent SISO plant derived from the given MIMO system with all other loops closed with the fixed loop gains k_j, $j \neq i$. If k_1 is fixed at some value, the stabilizing range for k_2 is uniquely determined. For instance, $k_1 = 1$ yields $k_2 > -3/4$, and graphically such a stabilizing range for k_2 is between two intersection points of line $k_1 = 1$ with the lower and upper boundaries of the shaded (stabilizing) region of Fig. 3.2. Note that loop 1 may have some uncertainties on its parameters and/or k_1 needs to be tuned or de-tuned separately. When k_1 or loop 1 has some change, the previous stabilizing range for k_2 may not be stabilizing any more. Such results are not very useful in the context of MIMO gain margins and their applications as they are sensitive to other loops' gains. Therefore, it is more practical and useful to prescribe a range for k_1 when determining the stabilizing range for k_2. In general, if k_1 varies in some range, the stabilizing range for k_2 can be uniquely determined. For instance, $k_1 \in [1,2]$ yields $k_2 \in [-3/4, +\infty)$. Graphically such a stabilizing region for both k_1 and k_2 is a rectangle with length k_1 from 1 to 2 and width k_2 from $-3/4$ to $+\infty$. When the range of k_1 changes, so does the stabilizing range of k_2. For instance, $\{(k_1, k_2) | k_1 \in [3,4], k_2 \in [-5/4, 5/4]\}$ gives another stabilizing rectangle for k_1 and k_2. In view of the above observations, we are motivated to find such stabilizing rectangles and formulate the problem as follows.

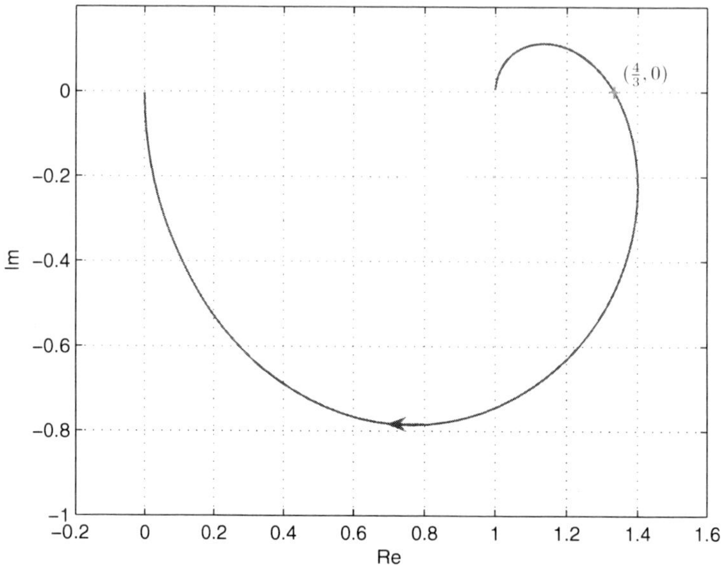

Fig. 3.3. Nyquist curve of $\tilde{g}_{22}$ for $k_1 = 1$

Consider an $m \times m$ square plant $G(s)$ with n-dimensional state-space realization:

$$\begin{aligned}
\dot{x}(t) &= Ax(t) + Bu(t), \\
y(t) &= Cx(t),
\end{aligned} \tag{3.5}$$

where $x \in \mathbb{R}^n$ is the state, $y \in \mathbb{R}^m$ is the output, B and C are real constant matrices with appropriate dimensions. This chapter focuses on the following form of PID controllers: $U(s) = K(s)E(s)$, where $e(t) = r(t) - y(t)$, $r(t)$ is the set point and

$$\begin{aligned}
K(s) &= K_1 + \frac{K_2}{s} + K_3 s \\
&= \text{diag}\{k_{11}I_{11}, \ldots, k_{1r_1}I_{1r_1}\} \\
&\quad + \frac{1}{s}\text{diag}\{k_{21}I_{21}, \ldots, k_{2r_2}I_{2r_2}\} \\
&\quad + s\,\text{diag}\{k_{31}I_{31}, \ldots, k_{3r_3}I_{3r_3}\},
\end{aligned} \tag{3.6}$$

where k_{1i}, k_{2j} and k_{3l} are scalars to be determined, I_{1i}, I_{2j} and I_{3l} are identity matrices with dimensions m_{1i}, m_{2j} and m_{3l}, respectively, and $\sum_{i=1}^{r_1} m_{1i} = \sum_{j=1}^{r_2} m_{2j} = \sum_{l=1}^{r_3} m_{3l} = m$. Since our concern in this chapter is stabilization, $r(t)$ has no effect and can be ignored. The controller in (3.6) can be re-written in time domain as

$$\begin{aligned}
u(t) &= -K_1 y(t) - K_2 \int_0^t y(\theta)d\theta - K_3 \dot{y}(t) \\
&:= -\sum_{i=1}^{r_1} k_{1i}\bar{I}_{1i}y(t) - \sum_{i=1}^{r_2} k_{2i}\bar{I}_{2i}\int_0^t y(\theta)d\theta - \sum_{i=1}^{r_3} k_{3i}\bar{I}_{3i}\dot{y}(t),
\end{aligned} \tag{3.7}$$

where

$$\bar{I}_{vi} = \text{diag}\{0,\ldots,0,I_{vi},0,\ldots,0\} \in \mathbb{R}^{m \times m},$$
$$v = 1,2,3, \quad i = 1,2,\ldots,r_v.$$

The problem considered in this chapter is as follows.

Problem 3.1. For a plant (3.5) under the controller (3.7), find the ranges of scalars k_{1i}, k_{2j} and k_{3l}, $i = 1,\ldots,r_1$, $j = 1,\ldots,r_2$, $l = 1,\ldots,r_3$, such that the closed-loop system is stable for all allowable k_{1i}, k_{2j} and k_{3l} in these ranges.

If the controller is full-coupled and the plant is non-square, Problem 3.1 becomes

Problem 3.2. Consider an $m \times l$ plant $G(s)$. Find the maximum ranges of $k_{1,ij}$, $k_{2,ij}$ and $k_{3,ij}$, $i = 1,\cdots,m$, $j = 1,\cdots,l$, such that the closed-loop system is stable under PID controller

$$K(s) = K_1 + \frac{1}{s}K_2 + sK_3 = [k_{1ij}] + \frac{1}{s}[k_{2ij}] + s[k_{3,ij}], \tag{3.8}$$

for all allowable $k_{1,ij}$, $k_{2,ij}$ and $k_{3,ij}$ in these ranges.

Note that in Problem 3.1, the controller of the form $K(s) = (k_1 + \frac{k_2}{s} + k_3 s)I_m$ corresponds to the specially chosen $r_1 = r_2 = r_3 = 1$; the controller of the form $K(s) = \text{diag}\{k_{1i}\} + \frac{1}{s}\text{diag}\{k_{2i}\} + s\text{diag}\{k_{3i}\}$ corresponds to the specially chosen $r_1 = r_2 = r_3 = m$ (or $m_{1i} = m_{2j} = m_{3l} = 1$).

It is worth mentioning that the gain margins for MIMO systems can readily be defined and obtained as by-products of solutions to Problem 3.1. Consider the example again in the special case where $K(s) = kI_2$, or k_1 and k_2 are equal to each other. Then, it follows from (3.3) that the stabilizing range is $k \in [(5 - \sqrt{33})/4, (5 + \sqrt{33})/4]$. Graphically, such a stabilizing range is obtained as the straight line, $\overline{BD}$, in Fig. 3.2, where B and D are two intersection points of line $k_1 = k_2$ with the boundary of shaded region. $\overline{BD}$ is uniquely determined. In general, for an $m \times m$ square plant in (3.5) under the decentralized proportional controller form in (3.7) with the common gain for all loops, $K(s) = kI_m$, suppose that the solution to Problem 1 is

$$k \in [\underline{k},\bar{k}]. \tag{3.9}$$

This stabilizing range is uniquely determined and called as the common gain margin of the system. Graphically, such a stabilizing range (3.9) is the largest line segment of $k_1 = k_2 = \cdots = k_m$ available in the stabilizing region for k_i, $i = 1,2,\cdots,m$.

Consider now the decentralized proportional controller, $K(s) = \text{diag}\{k_1, k_2,\cdots,k_m\}$, with probably different gains for different loops. Suppose that the solution to Problem 3.1 is

$$k_i \in [\underline{k}_i,\bar{k}_i], \quad i = 1,2,\cdots,m. \tag{3.10}$$

Then, the closed-loop remains stable even when the gain for the i-th loop, k_i, varies between $\underline{k}_i$ and $\bar{k}_i$, provided that other loop gains, k_j, $j = 1,2,\cdots,m$, $j \neq i$, are (arbitrary but) also within their respective ranges. $[\underline{k}_i,\bar{k}_i]$ is called the gain margin for the i-th loop, subject to other loops' gain margins within $[\underline{k}_j,\bar{k}_j]$, $j = 1,2,\cdots,m$, $j \neq i$. Note that the margins so defined allow variations/uncertainties of other loops' gains within $[\underline{k}_j, \bar{k}_j]$, which facilitates their use in loop tunings. Actually, such ranges are usually quite large for normal stable plants. The formulation here truly reflects the distinct feature of a MIMO system from the SISO case that the stabilizing range for a loop depends on other loops' gains in general.

3.3 The Proposed Approach

We will transform the closed-loop system into a descriptor form analogous to that of Lin et al. [80, 99]. It should be pointed out that the descriptor model in Lin et al. [80] is obtained by introducing an augmented state $\bar{x}(t) = [x^T(t), \int_0^t x^T(\theta)d\theta, \dot{x}^T(t)]^T$. This brings conservatism since the resulting design is only applicable to a narrow class of systems with matrix C being of full column rank. To overcome such a drawback, we replace the augmented state by $\bar{x}(t) = [x^T(t), \int_0^t y^T(\theta)d\theta, \dot{y}^T(t)]^T$ in this chapter. The new output remains the same as in Lin et al. [80], i.e., $\bar{y}(t) = [y^T(t), \int_0^t y^T(\theta)d\theta, \dot{y}^T(t)]^T$. Noticing the fact that $\dot{y}(t) = CAx(t) + CBu(t)$, system (3.5) with (3.7) is then transformed into the following descriptor control system:

$$
\begin{aligned}
\bar{E}\dot{\bar{x}}(t) &= \bar{A}\bar{x}(t) + \bar{B}u(t), \\
\bar{y}(t) &= \bar{C}\bar{x}(t), \\
u(t) &= -\bar{K}\bar{y}(t),
\end{aligned}
\tag{3.11}
$$

where

$$
\bar{E} = \begin{bmatrix} I_n & 0 & 0 \\ 0 & I_m & 0 \\ 0 & 0 & 0 \end{bmatrix}, \quad
\bar{A} = \begin{bmatrix} A & 0 & 0 \\ C & 0 & 0 \\ CA & 0 & -I_m \end{bmatrix}, \quad
\bar{B} = \begin{bmatrix} B \\ 0 \\ CB \end{bmatrix},
$$

$$
\bar{C} = \begin{bmatrix} C & 0 & 0 \\ 0 & I_m & 0 \\ 0 & 0 & I_m \end{bmatrix}, \quad
\bar{K} = \begin{bmatrix} K_1 & K_2 & K_3 \end{bmatrix}.
$$

In view of the diagonal structure of PID controller (3.7), the closed-loop system of (3.11) is rewritten as

$$
\begin{aligned}
\bar{E}\dot{\bar{x}}(t) &= (\bar{A} - \bar{B}\bar{K}\bar{C})\bar{x}(t) \\
&= (\bar{A} - \sum_{i=1}^{r_1} k_{1i}\bar{A}_{1i} - \sum_{j=1}^{r_2} k_{2j}\bar{A}_{2j} - \sum_{l=1}^{r_3} k_{3l}\bar{A}_{3l})\bar{x}(t) \\
&:= \bar{A}_{cl}\bar{x}(t),
\end{aligned}
\tag{3.12}
$$

where

$$
\bar{A}_{1i} = \begin{bmatrix} B\bar{I}_{1i}C & 0 & 0 \\ 0 & 0 & 0 \\ CB\bar{I}_{1i}C & 0 & 0 \end{bmatrix}, \quad
\bar{A}_{2j} = \begin{bmatrix} 0 & B\bar{I}_{2j} & 0 \\ 0 & 0 & 0 \\ 0 & CB\bar{I}_{2j} & 0 \end{bmatrix}, \quad
\bar{A}_{3l} = \begin{bmatrix} 0 & 0 & B\bar{I}_{3l} \\ 0 & 0 & 0 \\ 0 & 0 & CB\bar{I}_{3l} \end{bmatrix},
$$

for $i = 1, 2, \cdots, r_1$, $j = 1, 2, \cdots, r_2$ and $l = 1, 2, \cdots, r_3$. A descriptor system of the form (3.12) is called admissible if the system, or say, the pair $(\bar{E}, \bar{A}_{cl})$, is regular, impulse-free and stable. Please refer to Dai [100] and Masubuchi et al. [101] for detailed definitions.

So far, Problem 3.1 has been converted to the following problem.

Problem 3.3. Find the ranges of scalars k_{1i}, k_{2j} and k_{3l}, $i = 1, \ldots, r_1$, $j = 1, \ldots, r_2$, $l = 1, \ldots, r_3$, such that the closed-loop system (3.12) is admissible for all allowable k_{1i}, k_{2j} and k_{3l} in these ranges.

To solve Problem 3.3, one could adopt the structured singular value method (i.e., μ-analysis) as presented in Lin et al. [102, 103]. Noticing the fact that the μ-analysis may produce conservative results due to the requirement of common perturbation bounds, we next suggest an alternative method in the polytopic context.

To address Problem 3.3, the first step is to find k^0_{vi} are such that $(\bar{E}, \bar{A}^0_{cl})$ with $\bar{A}^0_{cl} = \bar{A} - \sum_{i=1}^{r_1} k^0_{1i}\bar{A}_{1i} - \sum_{i=1}^{r_2} k^0_{2i}\bar{A}_{2i} - \sum_{i=1}^{r_3} k^0_{3i}\bar{A}_{3i}$ is admissible. This step can be done by standard techniques available [90, 79, 80]. A specific procedure is provided in Sect. 3.5 to fulfill this step. Next, set $\bar{k}_{vi} = k_{vi} - k^0_{vi}$, $v = 1, 2, 3$, $i = 1, 2, \cdots, r_v$. Then,

$$\bar{A}_{cl} = \bar{A}^0_{cl} - \sum_{i=1}^{r_1} \bar{k}_{1i}\bar{A}_{1i} - \sum_{i=1}^{r_2} \bar{k}_{2i}\bar{A}_{2i} - \sum_{i=1}^{r_3} \bar{k}_{3i}\bar{A}_{3i}. \tag{3.13}$$

The task now is to compute the perturbation ranges for scalars $\bar{k}_{vi}$ such that $(\bar{E}, \bar{A}_{cl})$ remains admissible. To this end, we specify the lower and upper bounds for $\bar{k}_{vi}$ as β^{low}_{vi} and β^{upp}_{vi}, respectively, i.e.,

$$\bar{k}_{vi} \in [\beta^{\text{low}}_{vi}, \beta^{\text{upp}}_{vi}], \quad v = 1, 2, 3, \quad i = 1, 2, \ldots, r_v. \tag{3.14}$$

For brevity, relabel them as β^{low}_i and β^{upp}_i with $i = 1, 2, \cdots, r_1 + r_2 + r_3$. Let $r_0 = r_1 + r_2 + r_3$ and $\beta = [\beta^{\text{low}}_1, \beta^{\text{upp}}_1, \cdots, \beta^{\text{low}}_{r_0}, \beta^{\text{upp}}_{r_0}]$. Then, $\bar{A}_{cl}$ is equivalently recast as a matrix polytope with $\bar{r} = 2^{r_0}$ vertices denoted by $\bar{A}_j(\beta) \in \mathbb{R}^{(n+2m) \times (n+2m)}$,

$$\bar{A}_{cl} \in \left\{ \bar{A}(\alpha) : \bar{A}(\alpha) = \sum_{j=1}^{\bar{r}} \alpha_j \bar{A}_j(\beta); \sum_{j=1}^{\bar{r}} \alpha_j = 1; \alpha_j \geq 0; j = 1, 2, \cdots, \bar{r} \right\}. \tag{3.15}$$

By the work of Masubuchi et al. [101], it is known that a nominal pair (E, A) is admissible if and only if there exists a matrix P such that $P^T A + A^T P < 0$ with $P^T E = E^T P \geq 0$. Therefore, the pair $(\bar{E}, \bar{A}_{cl})$ is robustly admissible if and only if there exists a parameter-dependent Lyapunov matrix $P(\alpha)$ such that

$$P(\alpha)^T \bar{E} = \bar{E}^T P(\alpha) \geq 0, \tag{3.16}$$

$$P(\alpha)^T \bar{A}(\alpha) + \bar{A}(\alpha)^T P(\alpha) < 0. \tag{3.17}$$

An alternative result which is equivalent to the above criterion is easy to be established as follows.

Lemma 3.1. *The pair $(\bar{E},\bar{A}(\alpha))$ is robustly admissible if and only if there exist parameter-dependent matrices $P(\alpha)$, $F(\alpha)$ and $H(\alpha)$ such that*

$$P(\alpha)^T \bar{E} = \bar{E}^T P(\alpha) \geq 0, \tag{3.18}$$

$$\begin{bmatrix} F(\alpha)\bar{A}(\alpha) + \bar{A}(\alpha)^T F(\alpha)^T & \star \\ P(\alpha) - F(\alpha)^T + H(\alpha)^T \bar{A}(\alpha) & -H(\alpha) - H(\alpha)^T \end{bmatrix} < 0. \tag{3.19}$$

Here and in the sequel, an ellipsis $\star$ denotes a block induced by symmetry.

Proof. The proof is parallel to that for standard systems in Geromel et al. [92] and Peaucelle et al. [94]. $\qquad\square$

Using Lemma 3.1, we have the following LMI-based result.

Proposition 3.1. *The pair $(\bar{E},\bar{A}_{cl})$ is robustly admissible if there exist matrices P_j, F_j, H_j and X_{jl} with $X_{jj} = X_{jj}^T$, $l \leq j$, $j,l = 1,2,\ldots,\bar{r}$, such that*

$$P_j^T \bar{E} = \bar{E}^T P_j \geq 0, \tag{3.20}$$

$$\bar{\Theta}_{jl} + \bar{\Theta}_{lj} < X_{jl} + X_{jl}^T, \quad j = 1,2,\ldots,\bar{r}, \ l \leq j, \tag{3.21}$$

$$\begin{bmatrix} X_{11} & \star & \cdots & \star \\ X_{21} & X_{22} & \cdots & \star \\ \vdots & \vdots & \ddots & \vdots \\ X_{\bar{r}1} & X_{\bar{r}2} & \cdots & X_{\bar{r}\bar{r}} \end{bmatrix} \leq 0, \tag{3.22}$$

where

$$\bar{\Theta}_{jl} = \begin{bmatrix} F_j \bar{A}_l(\beta) + \bar{A}_l(\beta)^T F_j^T & \star \\ P_j - F_j^T + H_j^T \bar{A}_l(\beta) & -H_j - H_j^T \end{bmatrix}.$$

Proof. Let the parameter-dependent matrices $P(\alpha)$, $F(\alpha)$ and $H(\alpha)$ be

$$P(\alpha) = \sum_{j=1}^{\bar{r}} \alpha_j P_j, \quad F(\alpha) = \sum_{j=1}^{\bar{r}} \alpha_j F_j, \quad H(\alpha) = \sum_{j=1}^{\bar{r}} \alpha_j H_j. \tag{3.23}$$

If conditions (3.21)–(3.22) are true, substituting (3.23) into the matrix of (3.19), yields

$$\begin{bmatrix} F(\alpha)\bar{A}(\alpha) + \bar{A}(\alpha)^T F(\alpha)^T & \star \\ P(\alpha) - F(\alpha)^T + H(\alpha)^T \bar{A}(\alpha) & -H(\alpha) - H(\alpha)^T \end{bmatrix}$$

$$\begin{aligned}
&= \sum_{j=1}^{\bar{r}} \sum_{l=1}^{\bar{r}} \alpha_j \alpha_l \bar{\Theta}_{jl} \\
&= \sum_{j=1}^{\bar{r}} \alpha_j^2 \bar{\Theta}_{jj} + \sum_{l<j}^{\bar{r}} \alpha_j \alpha_l (\bar{\Theta}_{jl} + \bar{\Theta}_{lj}) \\
&< \sum_{j=1}^{\bar{r}} \alpha_j^2 X_{jj} + \sum_{l<j}^{\bar{r}} \alpha_j \alpha_l (X_{jl} + X_{jl}^T) \\
&= [\alpha_1 I, \cdots, \alpha_{\bar{r}} I] \begin{bmatrix} X_{11} & \cdots & \star \\ \vdots & \ddots & \vdots \\ X_{\bar{r}1} & \cdots & X_{\bar{r}\bar{r}} \end{bmatrix} \begin{bmatrix} \alpha_1 I \\ \vdots \\ \alpha_{\bar{r}} I \end{bmatrix} \\
&\leq 0.
\end{aligned}$$

This completes the proof from Lemma 3.1. $\qquad\square$

We remark that, when $\bar{E} = I$ (i.e., for regular systems), Proposition 3.1 reduces to the robust stability test for $\bar{A}_{cl}$ to be Hurwitz. In this case, the present method is less conservative than those given in Geromel et al. [92], Peaucelle et al. [94], Ramos and Peres [95] because the conditions in (3.20)–(3.22) reduce to those in these papers when setting $X_{jl} = 0$.

Proposition 3.1 provides a quasi-LMI condition to search for β. Based on Proposition 3.1, we present an LMI-based algorithm to compute ranges of PID controller gains. Note that (3.20)-(3.21) can be combined to a single LMI. Let $L = [0,0,I_n]^T \in \mathbb{R}^{(2n+m)\times n}$. Then, similar to Lin et al. [99], (3.21) with (3.20) is equivalent to the following LMI for additional matrices $Z_j > 0$ and $Y_j \in \mathbb{R}^{n\times(2n+m)}$:

$$\Omega_{jl} + \Omega_{lj} < X_{jl} + X_{jl}^T, \quad j = 1,2,\cdots,\bar{r}, \ l \leq j, \tag{3.24}$$

where

$$\Omega_{jl} = \begin{bmatrix} F_j \bar{A}_l(\beta) + \bar{A}_l(\beta)^T F_j^T & \star \\ Z_j \bar{E} + L Y_j - F_j^T + H_j^T \bar{A}_l(\beta) & -H_j - H_j^T \end{bmatrix}.$$

Procedure 3.1

Step 1. Find a set of scalars k_{vi}^0 such that $(\bar{E},\bar{A}_{cl}^0)$ with $\bar{A}_{cl}^0 = \bar{A} - \sum_{i=1}^{r_1} k_{1i}^0 \bar{A}_{1i} - \sum_{i=1}^{r_2} k_{2i}^0 \bar{A}_{2i} - \sum_{i=1}^{r_3} k_{3i}^0 \bar{A}_{3i}$ is admissible.

Step 2. Find the maximum $\beta_0 \geq 0$ such that LMIs (3.22) and (3.24) are feasible for $\beta = [-\beta_0, \beta_0, \ldots, -\beta_0, \beta_0]$.

Step 3. Find $\beta_1^{\text{low}} \leq -\beta_0$ such that LMIs (3.22) and (3.24) are feasible for $\beta = [\beta_1^{\text{low}}, \beta_0, \ldots, -\beta_0, \beta_0]$.

Step 4. Find $\beta_1^{\text{upp}} \geq \beta_0$ such that LMIs (3.22) and (3.24) are feasible for $\beta = [\beta_1^{\text{low}}, \beta_1^{\text{upp}}, -\beta_0, \beta_0, \ldots, -\beta_0, \beta_0]$.

Step 5. Repeat Steps 2 and 3 such that LMIs (3.22) and (3.24) are feasible for $\beta = [\beta_1^{\text{low}}, \beta_1^{\text{upp}}, \ldots, \beta_{r_0}^{\text{low}}, \beta_{r_0}^{\text{upp}}]$.

Step 6. Calculate the range of k_{vi} from (3.14) by $k_{vi} = k_{vi}^0 + [\beta_{vi}^{\text{low}}, \beta_{vi}^{\text{upp}}]$.

The above procedure is a theoretic summary on determining stabilizing ranges of PID parameters. Yet, it leaves for many practical implementation problems, such as how to find the initial stabilizing parameter k_{vi}^0 and how to make the stabilizing ranges as large as possible. Also, it should be pointed out that different solutions may be obtained if the parameters of PID controller are reordered. To obtain a reasonable stabilizing ranges as large as possible, we add some modifications to Procedure 3.1 from the practical point of view, which will be described in detail in Sect. 3.5 and summarized as Algorithm 3.1.

3.4 Special Cases

For three special cases of PID control, namely, P, PD and PI control, their transformed state-space representations are different, which leads to different LMI conditions. Hence, in this section, we would like to give such representations and conditions for these three special cases for easy reference and applications.

3.4.1 P Control

In this special case, $K_2 = 0$ and $K_3 = 0$ in (3.6). Then,

$$
\begin{aligned}
\dot{x}(t) &= Ax(t) + Bu(t), \\
y(t) &= Cx(t), \\
u(t) &= -K_1 y(t).
\end{aligned}
\tag{3.25}
$$

or, rewritten as

$$
\dot{x}(t) = (A - \sum_{i=1}^{r_1} k_{1i} A_{1i}) x(t) := A_{cl} x(t),
\tag{3.26}
$$

where

$$
A_{1i} = B \bar{I}_{1i} C, \quad i = 1, 2, \ldots, r_1.
\tag{3.27}
$$

As processed before, equivalently recast A_{cl} as a matrix polytope with $r = 2^{r_1}$ vertices denoted by $A_j(\beta) \in \mathbb{R}^{n \times n}$,

$$
A_{cl} \in \left\{ A(\alpha) : A(\alpha) = \sum_{j=1}^{r} \alpha_j A_j(\beta); \ \sum_{j=1}^{r} \alpha_j = 1; \ \alpha_j \geq 0; \ j = 1, 2, \cdots, r \right\}.
\tag{3.28}
$$

Therefore, in the proportional control case, we have the following result.

Proposition 3.2. *The polytope A_{cl} is robustly stable if there exist matrices $P_j > 0$, F_j, H_j and X_{jl} with $X_{jj} = X_{jj}^T$, $l \leq j$, $j, l = 1, 2, \cdots, r$, such that*

$$
\Theta_{jl} + \Theta_{lj} < X_{jl} + X_{jl}^T, \quad j = 1, 2, \ldots, r, \ l \leq j,
\tag{3.29}
$$

$$
\begin{bmatrix}
X_{11} & \cdots & \star \\
\vdots & \ddots & \vdots \\
X_{r1} & \cdots & X_{rr}
\end{bmatrix} \leq 0,
\tag{3.30}
$$

where

$$\Theta_{jl} = \begin{bmatrix} F_j A_l(\beta) + A_l(\beta)^T F_j^T & \star \\ P_j - F_j^T + H_j^T A_l(\beta) & -H_j - H_j^T \end{bmatrix}.$$

3.4.2 PD Control

In this special case, $K_2 = 0$ in (3.6). Let $\widehat{x}(t) = [x^T(t), \dot{y}^T(t)]^T$ and $\widehat{y}(t) = [y^T(t), \dot{y}^T(t)]^T$. Then we have the following control system:

$$\begin{bmatrix} I_n & 0 \\ 0 & 0 \end{bmatrix} \dot{\widehat{x}}(t) = \begin{bmatrix} A & 0 \\ CA & -I_m \end{bmatrix} \widehat{x}(t) + \begin{bmatrix} B \\ CB \end{bmatrix} u(t),$$

$$\widehat{y}(t) = \begin{bmatrix} C & 0 \\ 0 & I_m \end{bmatrix} \widehat{x}(t), \tag{3.31}$$

$$u(t) = -\begin{bmatrix} K_1 & K_3 \end{bmatrix} \widehat{y}(t),$$

or, rewritten as

$$\widehat{E}\dot{\widehat{x}}(t) = (\widehat{A} - \sum_{i=1}^{r_1} k_{1i}\widehat{A}_{1i} - \sum_{i=1}^{r_3} k_{3i}\widehat{A}_{3i})\widehat{x}(t)$$

$$:= \widehat{A}_{cl}\widehat{x}(t), \tag{3.32}$$

where

$$\widehat{E} = \begin{bmatrix} I_n & 0 \\ 0 & 0 \end{bmatrix}, \ \widehat{A} = \begin{bmatrix} A & 0 \\ CA & -I_m \end{bmatrix}, \ \widehat{A}_{1i} = \begin{bmatrix} B\bar{I}_{1i}C & 0 \\ CB\bar{I}_{1i}C & 0 \end{bmatrix}, \ \widehat{A}_{3l} = \begin{bmatrix} 0 & B\bar{I}_{3l} \\ 0 & CB\bar{I}_{3l} \end{bmatrix},$$

for $i = 1, 2, \cdots, r_1$ and $l = 1, 2, \cdots, r_3$. As proceeded in PID and PI cases, $\widehat{A}_{cl}$ can be equivalently recast as a matrix polytope with $\widehat{r} = 2^{r_1+r_3}$ vertices denoted by $\widehat{A}_j(\beta) \in \mathbb{R}^{(n+m)\times(n+m)}$,

$$\widehat{A}_{cl} \in \left\{ \widehat{A}(\alpha) : \widehat{A}(\alpha) = \sum_{j=1}^{\widehat{r}} \alpha_j \widehat{A}_j(\beta); \ \sum_{j=1}^{\widehat{r}} \alpha_j = 1; \ \alpha_j \geq 0; \ j = 1, 2, \ldots, \widehat{r} \right\}. \tag{3.33}$$

Therefore, in the PD control case, we have the following result.

Proposition 3.3. *The pair* $(\widehat{E}, \widehat{A}_{cl})$ *is robustly admissible if there exist matrices* P_j, F_j, H_j *and* X_{jl} *with* $X_{jj} = X_{jj}^T$, $l \leq j$, $j, l = 1, 2, \cdots, \widehat{r}$, *such that*

$$P_j^T \bar{E} = \bar{E}^T P_j \geq 0, \tag{3.34}$$

$$\widehat{\Theta}_{jl} + \widehat{\Theta}_{lj} < X_{jl} + X_{jl}^T, \quad j = 1, 2, \cdots, \widehat{r}, \ l \leq j, \tag{3.35}$$

$$\begin{bmatrix} X_{11} & \cdots & \star \\ \vdots & \ddots & \vdots \\ X_{\widehat{r}1} & \cdots & X_{\widehat{r}\widehat{r}} \end{bmatrix} \leq 0, \tag{3.36}$$

where

$$\widehat{\Theta}_{jl} = \begin{bmatrix} F_j \widehat{A}_l(\beta) + \widehat{A}_l(\beta)^T F_j^T & \star \\ P_j - F_j^T + H_j^T \widehat{A}_l(\beta) & -H_j - H_j^T \end{bmatrix}.$$

3.4.3 PI Control

In this special case, $K_3 = 0$ in (3.6). Let $\widetilde{x}(t) = [x^T(t), \ \int_0^t y^T(\theta)d\theta]^T$ and $\widetilde{y}(t) = [y^T(t), \int_0^t y^T(\theta)d\theta]^T$. Then we have the following control system:

$$\begin{aligned} \dot{\widetilde{x}}(t) &= \begin{bmatrix} A & 0 \\ C & 0 \end{bmatrix} \widetilde{x}(t) + \begin{bmatrix} B \\ 0 \end{bmatrix} u(t), \\ \widetilde{y}(t) &= \begin{bmatrix} C & 0 \\ 0 & I_m \end{bmatrix} \widetilde{x}(t), \\ u(t) &= -\begin{bmatrix} K_1 & K_2 \end{bmatrix} \widetilde{y}(t), \end{aligned} \tag{3.37}$$

or, rewritten as

$$\begin{aligned} \dot{\widetilde{x}}(t) &= (\widetilde{A} - \sum_{i=1}^{r_1} k_{1i}\widetilde{A}_{1i} - \sum_{i=1}^{r_2} k_{2i}\widetilde{A}_{2i})\widetilde{x}(t) \\ &:= \widetilde{A}_{cl}\widetilde{x}(t), \end{aligned} \tag{3.38}$$

where

$$\widetilde{A} = \begin{bmatrix} A & 0 \\ C & 0 \end{bmatrix}, \quad \widetilde{A}_{1i} = \begin{bmatrix} B\bar{I}_{1i}C & 0 \\ 0 & 0 \end{bmatrix}, \quad \widetilde{A}_{2j} = \begin{bmatrix} 0 & B\bar{I}_{2j} \\ 0 & 0 \end{bmatrix}, \tag{3.39}$$

for $i = 1, 2, \cdots, r_1$ and $j = 1, 2, \ldots, r_2$. Assume that a set of scalars k_{vi}^0 are such that $\widetilde{A}_{cl}^0 = \widetilde{A} - \sum_{i=1}^{r_1} k_{1i}^0 \widetilde{A}_{1i} - \sum_{i=1}^{r_2} k_{2i}^0 \widetilde{A}_{2i}$ is Hurwitz stable. Set $\widetilde{k}_{vi} = k_{vi} - k_{vi}^0$, $v = 1, 2$, $i = 1, 2, \cdots, r_v$. Then,

$$\widetilde{A}_{cl} = \widetilde{A}_{cl}^0 - \sum_{i=1}^{r_1} \widetilde{k}_{1i}\widetilde{A}_{1i} - \sum_{i=1}^{r_2} \widetilde{k}_{2i}\widetilde{A}_{2i}. \tag{3.40}$$

The following task is to compute the perturbation ranges for scalars $\widetilde{k}_{vi}$ such that $\widetilde{A}_{cl}$ remains stable. Now, we specify the lower and upper bounds for $\widetilde{k}_{vi}$ as β_{vi}^{low} and β_{vi}^{upp}, respectively, and relabel them as β_i^{low} and β_i^{upp} with $i = 1, 2, \cdots, r_1 + r_2$. Then, $\widetilde{A}_{cl}$ is equivalently recast as a matrix polytope with $\widetilde{r} = 2^{r_1 + r_2}$ vertices denoted by $\widetilde{A}_j(\beta) \in \mathbb{R}^{(n+m)\times(n+m)}$,

$$\widetilde{A}_{cl} \in \left\{ \widetilde{A}(\alpha) : \ \widetilde{A}(\alpha) = \sum_{j=1}^{\widetilde{r}} \alpha_j \widetilde{A}_j(\beta); \ \sum_{j=1}^{\widetilde{r}} \alpha_j = 1; \ \alpha_j \geq 0; \ j = 1, 2, \ldots, \widetilde{r} \right\}.$$

$$(3.41)$$

Therefore, in the PI control case, we have the following result.

Proposition 3.4. *The polytope $\widetilde{A}_{cl}$ is robustly stable if there exist matrices $P_j > 0$, F_j, H_j and X_{jl} with $X_{jj} = X_{jj}^T$, $l \leq j$, $j,l = 1, 2, \ldots, \widetilde{r}$, such that*

$$\widetilde{\Theta}_{jl} + \widetilde{\Theta}_{lj} < X_{jl} + X_{jl}^T, \quad j = 1, 2, \ldots, \widetilde{r}, \ l \leq j, \tag{3.42}$$

$$\begin{bmatrix} X_{11} & \cdots & \star \\ \vdots & \ddots & \vdots \\ X_{\widetilde{r}1} & \cdots & X_{\widetilde{r}\widetilde{r}} \end{bmatrix} \leq 0, \tag{3.43}$$

where

$$\widetilde{\Theta}_{jl} = \begin{bmatrix} F_j \widetilde{A}_l(\beta) + \widetilde{A}_l(\beta)^T F_j^T & \star \\ P_j - F_j^T + H_j^T \widetilde{A}_l(\beta) & -H_j - H_j^T \end{bmatrix}.$$

3.5 Computational Algorithm

In Step 1 of Procedure 3.1, scalars k_{vi}^0, $v = 1, 2, 3$, $i = 1, 2, \cdots, r_v$, are determined such that the closed-loop system is stable. The selection of k_{vi}^0 will determine the location of the stabilizing range of k_{vi} obtained in later steps. From a practical point of view, a control engineer would like the origin to be contained in the stabilizing range of k_{vi} to facilitate control tuning if possible (this is the case if the plant is stable). The reason is that the open-loop corresponds to a zero gain controller, or the origin in the parameter space of k_{vi}. To have a closed-loop control, a practising engineer will typically gradually increase loop gains from zero, and will have great choice of such gains and beneficial loop performance if he or she is given a sufficiently large stabilizing range containing the origin. In this section, we modify Procedure 3.1 such that the initial settings and the subsequent search for the desired stabilizing ranges are carried out in a systematic way, and the largest stabilizing range containing the origin and other ranges of interests are obtained if they are not empty.

To illustrate our ideas and resulting modifications, consider again the example in Sect. 3.2, where a decentralized proportional controller, $K(s) = \mathrm{diag}\{k_1, k_2\}$ is applied to stabilize the plant with the transfer function matrix:

$$G(s) = \begin{bmatrix} \dfrac{1}{s+1} & \dfrac{2}{s+1} \\ \dfrac{3}{s+1} & \dfrac{4}{s+1} \end{bmatrix}$$

in the unity negative feedback configuration. We proceed as follows.

(i) Start from the simplest common gain controller, $K(s) = kI_2$ (or $k_1 = k_2 = k$) to stabilize $G(s)$, where k is a scalar. Let k^0 be a stabilizing point. Since this plant, $G(s)$, is stable, the origin ($k^0 = 0$) is already a stabilizing point.

(ii) Let $\bar{k} = k - k^0$. By Barmish [104] (see Proposition 3.5 below), the stabilizing range of $\bar{k}$ is calculated as $\bar{k} \in (-0.1861, 2.6861)$. Thus, the stabilizing range in terms of k is $k = k^0 + \bar{k} \in (-0.1861, 2.6861)$. Graphically, such a stabilizing range is the straight line, $\overline{BD}$, shown in Fig. 3.2.

(iii) The mid-point of the above stabilizing range of k is $(2.6861 - 0.1861)/2 = 1.25$. One may reset $k_1^0 = k_2^0 = 1.25$ and calculate $\beta_0 = 1.4361$ by Step 2 of Procedure 3.1 such that LMIs (3.29) and (3.30) are feasible for $\beta = [-\beta_0, \beta_0, \cdots, -\beta_0, \beta_0]$. Thus, the initial stabilizing range of the independent gain controller, $K(s) = \mathrm{diag}\{k_1, k_2\}$, is $k_i \in [k_i^0 - \beta_0, k_i^0 + \beta_0] = [-0.1861, 2.6861]$, $i = 1, 2$, which contains the origin, $k_1 = k_2 = 0$. Note that so calculated k_1 and k_2 can now vary mutually independently within this range while the closed-loop remains stable. Graphically, this range is the square, $ABCD$, shown in Fig. 3.2. One may wish to shift the range to reflect different scaling and/or importance of different gains.

(iv) Before Step 3 of Procedure 3.1 is applied, we arrange k_1 and k_2 in decreasing order of their importance, i.e., if k_1 needs to be as large as possible (most important), then resize its stabilizing range firstly, and so on. As we pointed out in Sect. 3.2 that the stabilizing range of k_1 usually has effects on the stabilizing range of k_2, our ordering of k_i helps to obtain a stabilizing range as large as possible at the desired location. Suppose that k_1 is more important than k_2, one wishes to shift the initial stabilizing range, $k_i \in [-0.1861, 2.6861]$, $i = 1, 2$, to the new location as desired. Note that B (or D) lies on the boundary of the stabilizing region (see Fig. 3.2), the upper bound (or the lower bound) of k_i can not be extended. Hence, to obtain a stabilizing range as large as possible, we shift the initial stabilizing ranges of k_i from the point A or C, that is, choose initial values $(\bar{k}_1^0, \bar{k}_2^0) = [k_1^0 - \beta_0, k_2^0 + \beta_0] = (-0.1861, 2.6861)$ or $(\bar{k}_1^0, \bar{k}_2^0) = [k_1^0 + \beta_0, k_2^0 - \beta_0] = (2.6861, -0.1861)$. Thus, LMIs (3.29) and (3.30) are feasible for $\beta = [k_1^0 - \beta_0, k_1^0 + \beta_0, k_2^0 - \beta_0, k_2^0 + \beta_0] - [\bar{k}_1^0, \bar{k}_1^0, \bar{k}_2^0, \bar{k}_2^0] = [0, 2.8722, -2.8722, 0]$ or $[-2.8722, 0, 0, 2.8722]$.

(v) Since our range shift starts from the point A or C, the corner of the square $ABCD$, some "0" items will appear in β, which implies that the upper bound or the lower bound of k_i may be possibly extended so that they should be resized a priori. Moreover, β should also be relaxed as $\beta^* = \alpha\beta$, where $\alpha \in (0, 1)$ with $\alpha = 0.5$ by default. Then, the stabilizing range of k_1 and k_2 at such a shifted position can be calculated by Steps 3 to 6 of Procedure 3.1. For example, suppose that $(\bar{k}_1^0, \bar{k}_2^0) = (-0.1861, 2.6861)$, k_1 is more important than k_2. Let $\alpha = 0.5$ and $\beta^* = \alpha\beta = [0, 1.4361, -1.4361, 0]$. We firstly find the lower bound of k_1, secondly the upper bound of k_2, thirdly the upper bound of k_1 and finally the lower bound of k_2. The resulting stabilizing range of k_1 and k_2 is

$$k_1 \in [-7, 1.2561], \quad k_2 \in [1.25, 202.6861].$$

Similarly, if the alternative $(\bar{k}_1^0, \bar{k}_2^0) = (2.6861, -0.1861)$ is used, the stabilizing ranges become

$$k_1 \in [1.2468, 4], \quad k_2 \in [-0.8117, 1.25].$$

Such a range shifting to the different location will lead to a larger stabilizing range if it exists there.

One sees from the above example that our modifications to Procedure 3.1 consist of four steps. Firstly, find a common gain controller which can stabilize the plant. In the case of stable plants, the default gain is zero; Secondly, determine, by some formula, the stabilizing range of the common gain controller containing the above stabilizing point; Thirdly, compute the initial stabilizing range for independent gain controller by Step 2 of Procedure 3.1; Finally, shift it to the desired location, resize it, and compute the new range with Steps 3 to 6 of Procedure 3.1. The same technique is now applied to general cases as follows.

P Control

Start with the common gain controller, $K(s) = kI_m$, or $u(t) = -ky(t)$, where k is a scalar. By (3.26), the closed-loop system becomes

$$\dot{x}(t) = (A - kBC)x(t) = A_{cl}x(t),$$

which is a regular system. Let $\bar{k} = k - k^0$. Then

$$A_{cl} = A_{cl}^0 - \bar{k}BC, \tag{3.44}$$

where $A_{cl}^0 = A - k^0BC$, and k^0 is to stabilize the plant. If the plant is stable, take $k^0 = 0$; otherwise, find such a stabilizing non-zero k^0 [85]. If $G(s)$ can not be stabilized by any common gain controller, we have to find a controller in the general form of (3.6), i.e., k_{vi}^0, $v = 1, 2, 3$, $i = 1, 2, \cdots, r_v$. This is the stabilization problem by static output feedback, which can be solved by a few standard techniques [79, 80, 90].

Denote by λ_{min}^- and λ_{max}^+, respectively, the minimum negative eigenvalue and the maximum positive eigenvalue of a square matrix (set as zero if none). For regular stable systems, the stabilizing range of $\bar{k}$ is calculated from the following formula.

Proposition 3.5 (Barmish [104]). *The matrix A_{cl} given in (3.44) with a stable A_{cl}^0 and an uncertain $\bar{k}$ remains robustly stable if*

$$\bar{k} \in (\bar{k}_{min}, \bar{k}_{max}), \tag{3.45}$$

where

$$\bar{k}_{\min} = \frac{1}{\lambda_{\min}^{-}\left(-(A_{cl}^0 \otimes I_n + I_n \otimes A_{cl}^0)^{-1}((-BC) \otimes I_n + I_n \otimes (-BC))\right)},$$

$$\bar{k}_{\max} = \frac{1}{\lambda_{\max}^{+}\left(-(A_{cl}^0 \otimes I_n + I_n \otimes A_{cl}^0)^{-1}((-BC) \otimes I_n + I_n \otimes (-BC))\right)}.$$

Here, '$\otimes$' denotes the Kronecker product.

Consider now the independent gain controller. Reset $k_{vi}^0 \rightarrow k_{vi}^0 + (\bar{k}_{\min} + \bar{k}_{\max})/2$ and calculate the β_0 with Step 2 of Procedure 3.1, i.e., find the maximum $\beta_0 \geq 0$ such that LMIs (3.29) and (3.30) are feasible for $\beta = [-\beta_0, \beta_0, \cdots, -\beta_0, \beta_0]$. This yields the mutually independent stabilizing range of k_{vi} as $k_{vi} \in [k_{vi}^0 - \beta_0, k_{vi}^0 + \beta_0]$, $v = 1, 2, 3$, $i = 1, 2, \cdots, m$. Graphically, these initial stabilizing ranges form an m-dimension super cube.

Next, to get the stabilizing range at any desired location, start the range shift from the corners of the above cube, i.e., choose $\bar{k}_{vi}^0 = k_{vi}^0 - \beta_0$ or $\bar{k}_{vi}^0 = k_{vi}^0 + \beta_0$. As the sequence of resizing the stabilizing range for each loop is important, arrange k_{vi} in decreasing order of their importance, that is, if k_{11} needs to be as large as possible (most important), take it at the first place in the sequence of k_{vi}, and so on. Suppose that k_{vi} is arranged in decreasing order of their importance as $[k_{11}, \cdots, k_{3m}]$, then, $\beta = [\underline{\beta}_{11}^0, \overline{\beta}_{11}^0, \cdots, \underline{\beta}_{3m}^0, \overline{\beta}_{3m}^0]$ with $\underline{\beta}_{vi}^0 = k_{vi}^0 - \beta_0 - \bar{k}_{vi}^0$ and $\overline{\beta}_{vi}^0 = k_{vi}^0 + \beta_0 - \bar{k}_{vi}^0$.

Since we start from the corner of the super cube, some "0" items may appear in β and they should be tuned a priori. Moreover, β should also be relaxed as $\beta^+ = \alpha\beta$, where $\alpha \in (0, 1)$ with $\alpha = 0.5$ by default. All these measures guarantee that our search will succeed in getting the stabilizing range of k_{vi} at the desired location. Finally, following Steps 3 to 6 of Procedure 3.1, the stabilizing range of k_{vi} is actually determined.

PI control

Start with the common gain controller, $K(s) = k(1 + 1/s)I_m$, or $u(t) = -k[I_m, I_m]\tilde{y}(t)$, where k is a scalar. By (3.38), the closed-loop system becomes

$$\dot{\tilde{x}}(t) = (\tilde{A} - k\tilde{H})\tilde{x}(t) = \tilde{A}_{cl}\tilde{x}(t), \tag{3.46}$$

which is a regular system, where

$$\tilde{H} = \sum_{i=1}^{r_1} \tilde{A}_{1i} + \sum_{i=1}^{r_2} \tilde{A}_{2i} = \begin{bmatrix} BC & B \\ 0 & 0 \end{bmatrix}.$$

Let $\tilde{k} = k - k^0$. Then

$$\tilde{A}_{cl} = \tilde{A}_{cl}^0 - \tilde{k}\tilde{H},$$

where $\tilde{A}_{cl}^0 = \tilde{A} - k^0\tilde{H}$, and k^0 is to stabilize the plant. Note that unlike the P control case, even if the plant is stable, we can not take k^0 as zero because an integrator is present here. This is the problem of so called *integral controllability*, and the following proposition gives the criterion for the existence of such a k^0.

Proposition 3.6. *Consider the plant $G(s)$ under the PI control $K(s) = k^0(1 + 1/s)I_m$. Suppose that $G(s)$ is strictly proper, then*

(i) $\widetilde{A}^0_{cl} = \widetilde{A} - k^0\widetilde{H}$ *is stable for some k^0 if:*

    (a) $k^0 > 0$ *and all the eigenvalues of $G(0)$ lie in the open right half complex plane;*

    or,

    (b) $k^0 < 0$ *and all the eigenvalues of $G(0)$ lie in the open left half complex plane.*

(ii) $\widetilde{A}^0_{cl} = \widetilde{A} - k^0\widetilde{H}$ *is unstable for any k^0 if:*

    (a) $k^0 > 0$ *and the number of eigenvalues of $G(0)$ in open left half complex plane is odd; or,*

    (b) $k^0 < 0$ *and the number of eigenvalues of $G(0)$ in open right half complex plane is odd.*

Proof. Let $G^*(s) = (s+1)G(s)$. The results follow directly from Morari [49]. $\square$

If $G(s)$ can not be stabilized by any common gain controller, we have to find k^0_{vi}, $v = 1, 2, 3$, $i = 1, 2, \cdots, r_v$, to stabilize the plant instead. The same techniques mentioned in the case of P control can be still applied to find such k^0_{vi}. Once k^0 (or k^0_{vi}) is determined, Barmish's formula becomes applicable to calculate the stabilizing range of $\widetilde{k}$ by replacing BC with $\widetilde{H}$. After that, follow the same procedure as in the case of P control.

PD control

Start with the common gain controller, $K(s) = k(1 + s)I_m$, or $u(t) = -k[I_m, I_m]\widehat{y}(t)$, where k is a scalar. By (3.32), the closed-loop system becomes

$$\widehat{E}\dot{\widehat{x}}(t) = (\widehat{A} - k\widehat{H})\widehat{x}(t) = \widehat{A}_{cl}\widehat{x}(t),$$

which is a singular system, where

$$\widehat{H} = \sum_{i=1}^{r_1}\widehat{A}_{1i} + \sum_{i=1}^{r_3}\widehat{A}_{3i} = \begin{bmatrix} BC & B \\ CBC & CB \end{bmatrix}.$$

Let $\widehat{k} = k - k^0$. Then

$$\widehat{A}_{cl} = \widehat{A}^0_{cl} - \widehat{k}\widehat{H}. \tag{3.47}$$

where

$$\widehat{A}^0_{cl} = \widehat{A} - k^0\widehat{H} = \begin{bmatrix} A - k^0BC & -k^0B \\ CA - k^0CBC & -I_m - k^0CB \end{bmatrix} := \begin{bmatrix} A_1 & A_2 \\ A_3 & A_4 \end{bmatrix},$$

and k^0 is such that the pair $(\widehat{E}, \widehat{A}^0_{cl})$ is admissible. If the plant is stable, take $k^0 = 0$; otherwise, use the same method as in the case of P control to determine a non-zero k^0 or k^0_{vi}, $v = 1, 2, 3$, $i = 1, 2, \cdots, m$, such that the pair $(\widehat{E}, \widehat{A}^0_{cl})$ is admissible.

For singular stable systems, the stabilizing range of $\widehat{k}$ can be calculated from the following formula.

Proposition 3.7 (Lee et al. [105]). *The largest interval of $\widehat{k}$ such that the pair $\{\widehat{E}, \widehat{A}^0_{cl} - \widehat{k}\widehat{H}\}$ remains admissible is given by*

$$\widehat{k} \in (\widehat{k}_{\min}, \widehat{k}_{\max}) = (k_1, k_2) \cap (k_3, k_4), \tag{3.48}$$

where

$$k_1 = \frac{1}{\lambda^-_{\min}(-CB)}, \quad k_2 = \frac{1}{\lambda^+_{\max}(-CB)}, \quad k_3 = \frac{1}{\lambda^-_{\min}(-T_1^{-1}T_2)},$$

$$k_4 = \frac{1}{\lambda^+_{\max}(-T_1^{-1}T_2)},$$

$$T_1 = \begin{bmatrix} A_1 \otimes I_n + I_n \otimes A_1 & I_n \otimes A_2 & A_2 \otimes I_n \\ I_n \otimes A_3 & I_n \otimes A_4 & 0 \\ A_3 \otimes I_n & 0 & A_4 \otimes I_n \end{bmatrix},$$

$$T_2 = \begin{bmatrix} (-BC) \otimes I_n + I_n \otimes (-BC) & I_n \otimes (-B) & (-B) \otimes I_n \\ I_n \otimes (-CBC) & I_n \otimes (-CB) & 0 \\ (-CBC) \otimes I_n & 0 & (-CB) \otimes I_n \end{bmatrix}.$$

Here, '$\otimes$' denotes the Kronecker product.

Once $(\widehat{k}_{\min}, \widehat{k}_{\max})$ is found as above, let $k_{vi} = k^0_{vi} + \widehat{k} \in (k^0_{vi} + \widehat{k}_{\min}, k^0_{vi} + \widehat{k}_{\max})$. Then, follow the same procedure as in the case of P control.

PID control

Start with the common gain controller, $K(s) = k(1 + 1/s + s)I_m$, or the control signal $u(t) = -k[I_m, I_m, I_m]\bar{y}(t)$, where k is a scalar. By (3.12), the closed-loop system becomes

$$\bar{E}\dot{\bar{x}}(t) = (\bar{A} - k\bar{H})\bar{x}(t) = \bar{A}_{cl}\bar{x}(t), \tag{3.49}$$

which is a singular system, where

$$\bar{H} = \sum_{i=1}^{r_1} \bar{A}_{1i} + \sum_{i=1}^{r_2} \bar{A}_{2i} + \sum_{i=1}^{r_3} \bar{A}_{3i} = \begin{bmatrix} BC & B & | & B \\ 0 & 0 & | & 0 \\ \hline CBC & CB & | & CB \end{bmatrix} = \begin{bmatrix} H_1 & | & H_2 \\ \hline H_3 & | & H_4 \end{bmatrix}.$$

Let $\bar{k} = k - k^0$. Then

$$\bar{A}_{cl} = \bar{A}^0_{cl} - \widehat{k}\bar{H}.$$

where

$$\bar{A}^0_{cl} = \bar{A} - k^0 \bar{H} = \left[\begin{array}{cc:c} A - k^0 BC & -k^0 B & -k^0 B \\ C & 0 & 0 \\ \hdashline CA - k^0 CBC & -k^0 CB & -I_m - k^0 CB \end{array} \right] = \left[\begin{array}{c:c} A_1 & A_2 \\ \hdashline A_3 & A_4 \end{array} \right],$$

and k^0 is such that the pair $(\bar{E}, \bar{A}^0_{cl})$ is admissible. Note that unlike the PD control case, even if the plant is stable, we can not take k^0 as zero because an integrator is present here. If $G(s)$ can not be stabilized by any common gain controller, we have to find k^0_{vi}, $v = 1,2,3$, $i = 1,2,\cdots,r_v$, to stabilize the plant instead. Once k^0 (or k^0_{vi}) is determined, Lee et al's formula becomes applicable to calculate the stabilizing range of $\bar{k}$ by replacing BC with H_1, B with H_2, CBC with H_3, and CB with H_4, respectively. After that, follow the same procedure as in the case of P control.

The above development is summarized as follows.

Algorithm 3.1

Step 1. Find a common gain controller, $K(s)$, to stabilize the plant, $G(s)$. If $K(s)G(s)$ is stable, take $k^0 = 0$; otherwise, use any standard technique [90, 79, 80] to find the scalar k^0. Let $k^0_{vi} = k^0$. If $G(s)$ can not be stabilized by any common gain controller, find a controller in the general form of (3.6), i.e., k^0_{vi}, $v = 1,2,3$, $i = 1,2,\cdots,r_v$.

Step 2. Let $\bar{k} = k - k^0$ or $\bar{k}_{vi} = k_{vi} - k^0_{vi}$. Calculate the stabilizing ranges of $\bar{k}$ or $\bar{k}_{vi}$ as $(\bar{k}_{\min}, \bar{k}_{\max})$ by formula of Barmish [104] for P/PI control or Lee et al. [105] for PD/PID control.

Step 3. Reset $k^0_{vi} \rightarrow k^0_{vi} + (\bar{k}_{\min} + \bar{k}_{\max})/2$ and find the maximum $\beta_0 \geq 0$ such that LMIs (3.22) and (3.24) are feasible for $\beta = [-\beta_0, \beta_0, \cdots, -\beta_0, \beta_0]$. Obtain the mutually independent stabilizing range of k_{vi} as $k_{vi} \in [k^0_{vi} - \beta_0, k^0_{vi} + \beta_0]$.

Step 4. Arrange k_{vi} in decreasing order of their importance and choose initial values $\bar{k}^0_{vi} = k^0_{vi} - \beta_0$ or $\bar{k}^0_{vi} = k^0_{vi} + \beta_0$. Thus, LMIs (3.22) and (3.24) are still feasible for $\beta = [\underline{\beta}^0_{v1}, \overline{\beta}^0_{v1}, \cdots, \underline{\beta}^0_{vm}, \overline{\beta}^0_{vm}]$, where $\underline{\beta}^0_{vi} = k^0_{vi} - \beta_0 - \bar{k}^0_{vi}$ and $\overline{\beta}^0_{vi} = k^0_{vi} + \beta_0 - \bar{k}^0_{vi}$.

Step 5. Relax β as $\beta^* = \alpha\beta$, $\alpha \in (0,1)$ with $\alpha = 0.5$ by default. If $\underline{\beta}^*_{vi} = 0$ (or $\overline{\beta}^*_{vi} = 0$), find $\beta^{low}_{vi} \leq 0$ (or $\beta^{upp}_{vi} \geq 0$) such that LMIs (3.22) and (3.24) are feasible for $i = 1,2,\cdots,m$.

Step 6. If $\underline{\beta}^*_{vi} \neq 0$ (or $\overline{\beta}^*_{vi} \neq 0$), find $\beta^{low}_{vi} \leq \alpha\underline{\beta}^0_{vi}$ (or $\beta^{upp}_{vi} \geq \alpha\overline{\beta}^0_{vi}$) such that LMIs (3.22) and (3.24) are still feasible for $i = 1,2,\cdots,m$.

Step 7. Calculate the range of k_{vi} from (3.14) by $k_{vi} = \bar{k}^0_{vi} + [\beta^{low}_{vi}, \beta^{upp}_{vi}]$.

3.6 An Example

Example 3.6.1. Consider a process with transfer function [17]

$$G(s) = \left[\begin{array}{cc} \dfrac{s-1}{(s+1)(s+3)} & \dfrac{4}{s+3} \\[2ex] \dfrac{1}{s+2} & \dfrac{3}{s+2} \end{array} \right].$$

Its state-space minimum realization is of the form (3.5) with

$$A = \begin{bmatrix} -1.255 & -0.2006 & -0.1523 \\ 1.452 & -0.465 & 0.8761 \\ -5.496 & -4.108 & -4.28 \end{bmatrix}, \quad B = \begin{bmatrix} 0.1458 & 0.03811 \\ -0.113 & 0.06613 \\ 0.4033 & 0.7228 \end{bmatrix},$$

$$C = \begin{bmatrix} -0.3773 & 7.907 & 4.831 \\ 5.273 & 8.879 & 3.06 \end{bmatrix}.$$

We now use the proposed algorithm to compute the stabilizing ranges of different types of controllers. Suppose the largest available range of parameters is ± 100.

Case 1: P control. Consider a common gain controller, $K(s) = kI_2$, to stabilize the plant, $G(s)$. Since $G(s)$ is stable, take $k^0 = 0$.

Let $\bar{k} = k - k^0$. By Barmish's formula in Proposition 3.5, we compute the stabilizing range of $\bar{k}$ as $\bar{k} \in [-0.5522, \ 1.5513]$. Thus, the stabilizing range of k is obtained as

$$k = k^0 + \bar{k} \in [-0.5522, \ 1.5513].$$

Reset $k_1^0 = k_2^0 = k^0 + (-0.5522 + 1.5513)/2 = 0.4995$ and calculate $\beta_0 = 1.0518$. Then, the stabilizing range with mutually independent gains of k_i is $k_i \in [-0.5522, \ 1.5513], i = 1,2$.

Suppose that k_1 is more important than k_2 and choose $\bar{k}_1^0 = -0.5522$ and $\bar{k}_2^0 = 1.5513$ as initial values. Then, LMIs (3.29) and (3.30) are still feasible for $\beta = [-0.5522, 1.5513, -0.5522, 1.5513] - [-0.5522, -0.5522, 1.5513, 1.5513] = [0, 2.1035, -2.1035, 0]$.

Let $\alpha = 0.5$ and relax β as $\beta^* = \alpha\beta$. The sequence of range shifting is as follows: firstly find the lower bound of k_1, secondly the upper bound of k_2, thirdly the upper bound of k_1 and finally the lower bound of k_2.

Fix the stabilizing range of k_1 as $k_1 \in [-1.6556, \ 1.2]$ and compute the stabilizing range of k_2 as

$$[\beta_2^{\text{low}}, \ \beta_2^{\text{upp}}] = [-2.00355, \ 100],$$

which yields the stabilizing proportional controller gain ranges as

$$k_1 \in [-1.6556, \ 1.2], \quad k_2 \in [-0.45225, \ 100]. \tag{3.50}$$

If the stabilizing range of k_2 is fixed to $k_2 \in [-0.3529, \ 4.0967]$ and the stabilizing range of k_1 is calculated as

$$[\beta_1^{\text{low}}, \ \beta_1^{\text{upp}}] = [-3.2436, \ 1.94905],$$

which yields the stabilizing proportional controller gain ranges as

$$k_1 \in [-3.7958, \ 1.39685], \quad k_2 \in [-0.3529, \ 4.0967]. \tag{3.51}$$

Comparison with Ho's method

Ho et al. [77] gave a definition for loop's gain margin of MIMO systems based on Gershgorin bands under the assumption that the plant is diagonally dominant. For this example, the gain margins of each loop are computed by Ho's method as

$$k_1 \in [-1.6556,\ 1.2], \quad k_2 \in [-0.3529,\ 4.0967].$$

One sees that when the stabilizing range for one loop is the same, the range for other loop is much more conservative by Ho's method than ours. Furthermore, the common gain margin by Ho's method is

$$k \in [-1.6556,\ 1.2] \cap [-0.3529,\ 4.0967] = [-0.3529,\ 1.2],$$

which is also more conservative than ours $k \in [-0.5522,\ 1.5513]$. Note that our method goes beyond P-control and finds the stabilizing parameter ranges for PI, PD and PID controlllers, which is not possible by Ho's method or Gershgorin's theorem.

Comparison with the μ-analysis

Let $K = \mathrm{diag}\{k_1, k_2\}$ and $k_i = k_i^0(1 + w_i\widetilde{\Delta}_i)$, where k_i^0 are the nominal stabilizing gains and $\widetilde{\Delta}_i$ are the parameter uncertainties scaled by weights w_i, $i = 1,2$. To get the maximum gain ranges from the μ-analysis, one may proceed as follows with knowledge of our result in (3.50). Set k_i^0 as the mid-point of the stabilizing range of k_i given by (3.50) as $k_1^0 = (-1.6556 + 1.2)/2 = -0.2278$ and $k_2^0 = (-0.45225 + 100)/2 = 49.7739$. If $w_1 = w_2$, the μ-analysis yields $\mu = 1.0179$, which results in the allowable perturbation $|\widetilde{\Delta}_i| < 1/\mu = 0.9825$ and the stabilizing ranges of k_i as

$$k_1 \in [-0.4516,\ -0.0040], \quad k_2 \in [0.8710,\ 98.6768]. \tag{3.52}$$

Adjusting w_i will lead to different stabilizing ranges of k_i from which the least conservative one is obtained as

$$k_1 \in [-1.2494,\ 0.7938], \quad k_2 \in [13.8371,\ 85.7107], \tag{3.53}$$

which is more conservative than those in (3.50). Similarly, if we set k_i^0 with knowledge of our ranges in (3.51), the least conservative stabilizing ranges of k_i are obtained as

$$k_1 \in [-1.9143,\ -0.4847], \quad k_2 \in [1.2594,\ 2.4844], \tag{3.54}$$

which is still conservative than ours.

It should be pointed out that all the above calculations with μ-analysis have made use of the known stabilizing ranges of k_i obtained by our method, and that the stabilizing ranges of k_i depend not only on the weights, but also on the nominal gains. If such prior knowledge about the stabilizing ranges of k_i is unknown, one has to start with some stabilizing gains determined by users, which are unlikely to be the mid-point of the actual (but unknown yet) stabilizing ranges, and the results from μ-analysis will certainly become more conservative. For example, If k_i^0 deviate from the mid-point of the ranges in (3.50), say, $k_1^0 = (-1.6556) \times 3/4 + 1.2/4 = -0.9417$ and $k_2^0 = (-0.45225) \times 3/4 + 100/4 = 24.6608$, then the least conservative stabilizing ranges of k_i in this case are

$$k_1 \in [-1.2494, -0.6340], \quad k_2 \in [13.8347, 35.4869], \tag{3.55}$$

which is even more conservative than those in (3.53) indeed.

It is concluded that the μ-analysis gives conservative results for computing stabilizing controller gains. One reason is that the parameters of PID controllers are all real, while the μ-analysis treats all systems uncertainties as complex valued. The other reason is that the actual stabilizing ranges are generally not symmetric with respect to the nominal value while the allowable perturbations in the μ-analysis is always symmetric with respect to the nominal stabilizing value. The proposed method do not suffer such disadvantages and thus produce much stronger stabilizing results. Besides, the computational complexity of computing μ has a combinatoric growth with the number of parameters involved. Although practical algorithms are possible in such a case, they are very time consuming.

Case 2: PI Control. Consider the common gain controller, $K(s) = k(1 + 1/s)I_2$, to stabilize the plant, $G(s)$. Since

$$G(0) = \begin{bmatrix} -\frac{1}{3} & \frac{4}{3} \\ \frac{1}{2} & \frac{3}{2} \end{bmatrix}$$

with eigenvalues $\lambda_1 = -0.6442 < 0$ and $\lambda_2 = 1.8109 > 0$, by Proposition 3.6, no k exists to stabilize the plant. Hence, let $K(s) = \text{diag}\{k_1, k_2\} + \frac{1}{s}\text{diag}\{k_3, k_4\}$. One sees that $k_1^0 = 1$, $k_2^0 = 5$, $k_3^0 = -1$ and $k_4^0 = 5$ can stabilize the plant. By Algorithm 5.1, we compute the stabilizing PI controller ranges as

$$k_1 \in [-15.4516, 1.0542], \quad k_2 \in [4.9458, 100],$$
$$k_3 \in [-1.0545, -0.0001], \quad k_4 \in [4.9451, 5.0550].$$

Additionally, if we choose another stabilizer as $k_1^0 = 1$, $k_2^0 = 1$, $k_3^0 = -0.1$ and $k_4^0 = 0.1$, the stabilizing PI ranges become

$$k_1 \in [-1.9482, 0.7263], \quad k_2 \in [-0.1479, 100],$$
$$k_3 \in [-0.1, -0.05], \quad k_4 \in [0.1, 1.1035].$$

Case 3: PD Control. Consider the common gain controller, $K(s) = k(1 + s)I_2$, to stabilize the plant, $G(s)$. Since $G(s)$ is stable, take $k^0 = 0$.

Let $\widehat{k} = k - k^0$. By Lee et al.'s formula in Proposition 3.7, we compute the stabilizing range of $\widehat{k}$ as $\widehat{k} \in (-0.2361, 4.2389) \cap (-0.2361, 1.5515) = (-0.2361, 1.5515)$. Thus, the stabilizing range of k is obtained as $k = k^0 + \widehat{k} \in (-0.2361, 1.5515)$.

Suppose that $K(s) = (k_1 + k_2 s)I_2$. Algorithm 3.1 then yields the stabilizing PD ranges as

$$k_1 \in [-0.5522, 0.6578], \quad k_2 \in [0.6577, 2.0822].$$

To find other possible stabilizing ranges of k_1 and k_2, take $k^0 = 5$ and $k^0 = -5$, respectively. It is easy to check that k^0 in both cases can stabilize $G(s)$. Then, we compute the stabilizing PD controller ranges as

$$k_1 \in [1.5515, 100], \qquad k_2 \in [4.2389, 100], \qquad k^0 = 5;$$
$$k_1 \in [-100, -0.5683], \quad k_2 \in [-100, -0.2361], \quad k^0 = -5.$$

Case 4: PID Control. Consider the common gain controller, $K(s) = k(1 + 1/s + s)I_2$, to stabilize the plant, $G(s)$. By the standard techniques [90, 79, 80], we obtain $k^0 = 5$.

Let $\bar{k} = k - k^0$. By Lee *et al.*'s formula in Proposition 3.7, we compute the stabilizing range of $\bar{k}$ as $\bar{k} \in (-0.7611, +\infty) \cap (-2.2222, 1.7203) = (-0.7611, 1.7203)$. Thus, the stabilizing range of k is obtained as $k = k^0 + \bar{k} \in (4.2389, 6.7203)$.

Suppose that $K(s) = (k_1 + k_2/s + k_3 s)I_2$. After descriptor transformation, we have the following closed-loop system of the form (3.12)

$$\bar{E}\dot{\bar{x}}(t) = (\bar{A} - k_1\bar{A}_1 - k_2\bar{A}_2 - k_3\bar{A}_3)\bar{x}(t) := \bar{A}_{cl}\bar{x}(t), \tag{3.56}$$

where

$$\bar{E} = \begin{bmatrix} I_5 & 0 \\ 0 & 0 \end{bmatrix}, \quad \bar{A} = \begin{bmatrix} A & 0 & 0 \\ C & 0 & 0 \\ CA & 0 & -I_2 \end{bmatrix},$$

$$\bar{A}_1 = \begin{bmatrix} BC & 0 & 0 \\ 0 & 0 & 0 \\ CBC & 0 & 0 \end{bmatrix}, \quad \bar{A}_2 = \begin{bmatrix} 0 & B & 0 \\ 0 & 0 & 0 \\ 0 & CB & 0 \end{bmatrix}, \quad \bar{A}_3 = \begin{bmatrix} 0 & 0 & B \\ 0 & 0 & 0 \\ 0 & 0 & CB \end{bmatrix}.$$

For $k_1^0 = k_2^0 = k_3^0 = k^0$, the pair $(\bar{E}, \bar{A}_{cl}^0)$ is admissible where $\bar{A}_{cl}^0 = \bar{A} - k_1^0\bar{A}_1 - k_2^0\bar{A}_2 - k_3^0\bar{A}_3$. To find the ranges of $\bar{k}_i = k_i - k_i^0$, $i = 1, 2, 3$, such that

$$\bar{E}\dot{\bar{x}}(t) = \bar{A}_{cl}\bar{x}(t) = (\bar{A}_{cl}^0 - \bar{k}_1\bar{A}_1 - \bar{k}_2\bar{A}_2 - \bar{k}_3\bar{A}_3)\bar{x}(t), \tag{3.57}$$

is robustly admissible, we let $\bar{k}_i \in [\beta_i^{\text{low}}, \beta_i^{\text{upp}}]$. Then, $\bar{A}_{cl}$ is equivalently recast as a matrix polytope with 8 vertices

$$A_1(\beta) = \bar{A}_{cl}^0 - \beta_1^{\text{low}}\bar{A}_1 - \beta_2^{\text{low}}\bar{A}_2 - \beta_3^{\text{low}}\bar{A}_3,$$
$$A_2(\beta) = \bar{A}_{cl}^0 - \beta_1^{\text{upp}}\bar{A}_1 - \beta_2^{\text{low}}\bar{A}_2 - \beta_3^{\text{low}}\bar{A}_3,$$
$$A_3(\beta) = \bar{A}_{cl}^0 - \beta_1^{\text{low}}\bar{A}_1 - \beta_2^{\text{upp}}\bar{A}_2 - \beta_3^{\text{low}}\bar{A}_3,$$
$$A_4(\beta) = \bar{A}_{cl}^0 - \beta_1^{\text{low}}\bar{A}_1 - \beta_2^{\text{low}}\bar{A}_2 - \beta_3^{\text{upp}}\bar{A}_3,$$
$$A_5(\beta) = \bar{A}_{cl}^0 - \beta_1^{\text{low}}\bar{A}_1 - \beta_2^{\text{upp}}\bar{A}_2 - \beta_3^{\text{upp}}\bar{A}_3,$$
$$A_6(\beta) = \bar{A}_{cl}^0 - \beta_1^{\text{upp}}\bar{A}_1 - \beta_2^{\text{upp}}\bar{A}_2 - \beta_3^{\text{low}}\bar{A}_3,$$
$$A_7(\beta) = \bar{A}_{cl}^0 - \beta_1^{\text{upp}}\bar{A}_1 - \beta_2^{\text{low}}\bar{A}_2 - \beta_3^{\text{upp}}\bar{A}_3,$$
$$A_8(\beta) = \bar{A}_{cl}^0 - \beta_1^{\text{upp}}\bar{A}_1 - \beta_2^{\text{upp}}\bar{A}_2 - \beta_3^{\text{upp}}\bar{A}_3.$$

Reset $k_1^0 = k_2^0 = k_3^0 = k^0 + (-0.7611 + 1.7302)/2 = 5.4796$ and calculate $\beta_0 = 1.1342$, the stabilizing range with the mutually independent gains of k_i is $k_i \in [4.3454, 6.6138]$, $i = 1, 2, 3$.

Suppose that the importance of k_i is in order of k_1, k_2 and k_3. Let $\alpha = 0.5$ and choose initial values $\bar{k}_1^0 = \bar{k}_2^0 = \bar{k}_3^0 = 6.6138$. By Algorithm 5.1, we compute

$$[\beta_1^{\text{low}}, \beta_1^{\text{upp}}] = [-5.0794, 100],$$
$$[\beta_2^{\text{low}}, \beta_2^{\text{upp}}] = [-1.1535, 100],$$
$$[\beta_3^{\text{low}}, \beta_3^{\text{upp}}] = [-1.3316, 100],$$

which yields the stabilizing PID ranges as

$$k_1 \in [1.5344, 100], \quad k_2 \in [5.4603, 100], \quad k_3 \in [5.2821, 100].$$

3.7 Extension to the Centralized Controller

Consider an n-dimensional state-space realization of $G(s)$:

$$\begin{aligned} \dot{x}(t) &= Ax(t) + Bu(t), \\ y(t) &= Cx(t), \end{aligned} \tag{3.58}$$

where $x \in \mathbb{R}^n$ is the state variable, $y \in \mathbb{R}^l$ is the output variable, $u \in \mathbb{R}^m$ is the control input, $B \in \mathbb{R}^{n \times m}$ and $C \in \mathbb{R}^{l \times n}$ are real constant matrices. The centralized PI controller in (3.8) corresponds to

$$u(t) = K_1 y(t) + K_2 \int_0^t y(\theta)\mathrm{d}\theta. \tag{3.59}$$

After a standard system transformation by introducing new state variable and output variable as

$$\bar{x}(t) = \left[x^T(t), \int_0^t y^T(\theta)\mathrm{d}\theta \right]^T \quad \text{and} \quad \bar{y}(t) = \left[y^T(t), \int_0^t y^T(\theta)\mathrm{d}\theta \right]^T,$$

system (3.58) with (3.59) is transformed into the following augmented control system:

$$\begin{aligned} \dot{\bar{x}}(t) &= \bar{A}\bar{x}(t) + \bar{B}u(t), \\ \bar{y}(t) &= \bar{C}\bar{x}(t), \\ u(t) &= \bar{K}\bar{y}(t), \end{aligned} \tag{3.60}$$

where

$$\bar{A} = \begin{bmatrix} A & 0 \\ C & 0 \end{bmatrix}, \quad \bar{B} = \begin{bmatrix} B \\ 0 \end{bmatrix}, \quad \bar{C} = \begin{bmatrix} C & 0 \\ 0 & I_l \end{bmatrix}, \quad \bar{K} = \begin{bmatrix} K_1 & K_2 \end{bmatrix}.$$

Assume that with $\bar{K}_0 = [K_1^0, K_2^0]$ the resultant matrix $\bar{A}_0 := \bar{A} + \bar{B}\bar{K}_0\bar{C}$ is Hurwitz. Let $\bar{K}_1 = K_1 - K_1^0 = [\bar{k}_{1ij}]$ and $\bar{K}_2 = K_2 - K_2^0 = [\bar{k}_{2ij}]$, $i = 1, \cdots, m$, $j = 1, \cdots, l$. Then, the close-loop system of (3.60) is rewritten as

$$\dot{\bar{x}}(t) = (\bar{A}_0 + \bar{B}[\bar{K}_1, \bar{K}_2]\bar{C})\bar{x}(t) := \bar{A}_{cl}\bar{x}(t).$$

Denote $\bar{I}_{ij}$ an $m \times 2l$ matrix with the (i,j) element being 1 and other elements being 0, $i = 1, 2, \cdots, m$, $j = 1, 2, \cdots, 2l$. We obtain

$$\bar{A}_{cl} = \bar{A}_0 + \sum_{i=1}^{m} \sum_{j=1}^{l} (\bar{k}_{1ij} \bar{B} \bar{I}_{ij} \bar{C} + \bar{k}_{2ij} \bar{B} \bar{I}_{i,j+l} \bar{C}).$$

The following task is to compute the maximum perturbation ranges for scalars $\bar{k}_{1ij}$ and $\bar{k}_{2ij}$, $i = 1, 2, \cdots, m$, $j = 1, 2, \cdots, l$, such that $\bar{A}_{cl}$ remains Hurwitz. To achieve this purpose, one could adopt the structured singular value method (i.e., μ-analysis) as presented in [102, 103]. Noticing the fact that the μ-analysis may produce conservative results due to the common perturbation bound, we next suggest an alternative method in the polytopic context. Let β_{1ij}^{low} and β_{1ij}^{upp} (and, β_{2ij}^{low} and β_{2ij}^{upp}) refer to the lower and upper bounds for $\bar{k}_{1ij}$ (and, $\bar{k}_{2ij}$), respectively, i.e.,

$$\bar{k}_{\sigma ij} \in [\beta_{\sigma ij}^{\text{low}}, \beta_{\sigma ij}^{\text{upp}}],$$

where $\sigma = 1, 2$, $i = 1, \cdots, m$, $j = 1, \cdots, l$. For brevity, rename them as β_i^{low} and β_i^{upp} with $i = 1, 2, \cdots, 2ml$. Let $\beta = [\beta_1^{\text{low}}, \beta_1^{\text{upp}}, \cdots, \beta_{2ml}^{\text{low}}, \beta_{2ml}^{\text{upp}}]$. Then, $\bar{A}_{cl}$ is equivalently recast as a matrix polytope with $r = 2^{2ml}$ vertices denoted by $A_j(\beta) \in \mathbb{R}^{(n+l) \times (n+l)}$,

$$\bar{A}_{cl} \in \{A(\alpha) : A(\alpha) = \sum_{j=1}^{r} \alpha_j A_j(\beta); \ \sum_{j=1}^{r} \alpha_j = 1; \ \alpha_j \geq 0; \ j = 1, 2, \cdots, r\}.$$

It is well-known that matrix A is Hurwitz stable if and only if there exists a matrix $P > 0$ such that

$$PA + A^T P < 0. \tag{3.61}$$

An alternative criterion that is equivalent to (3.61) is as follows.

Lemma 3.2 ([87,93]). *Matrix A is Hurwitz stable if and only if there exist matrices F, H and $P > 0$ such that*

$$\begin{bmatrix} FA + A^T F^T & \star \\ P - F^T + H^T A & -H - H^T \end{bmatrix} < 0. \tag{3.62}$$

Here and in the sequel, an ellipsis $\star$ denotes a block induced by symmetry.

It is seen from Lemma 3.2 that the matrix P is decoupled from A by introducing slack variables F and H. Based on Lemma 3.2, a robust stability test is given in [92, 94]. In the following, we give another LMI-based method by introducing additional slack variables which make it more flexible to solve LMIs than the method in [92, 94].

Lemma 3.3. *Matrix A is Hurwitz stable if and only if there exist matrices F, H, S, T, U, W and $P > 0$ such that*

$$\begin{bmatrix} FA + A^T F^T & \star & \star \\ P - F^T + H^T A - A^T S^T & -H - H^T - TA - A^T T^T & \star \\ S^T + W^T A & T^T - W^T + U^T A & -U - U^T \end{bmatrix} < 0. \tag{3.63}$$

Proof. Necessity. Suppose that matrix A is Hurwitz stable. We need only to show that (3.62) implies (3.63). Note that (3.62) implies that

$$\begin{bmatrix} FA + A^T F^T & \star & \star \\ P - F^T + H^T A & -H - H^T & \star \\ 0 & c\Lambda & -2\varepsilon I \end{bmatrix} < 0, \tag{3.64}$$

holds for a sufficiently small scalar $\varepsilon > 0$. From (3.64), one sees that (3.63) holds by choosing $S = 0$, $T = 0$, $W = 0$ and $U = \varepsilon I$.

Sufficiency. Suppose that (3.63) holds. Multiplying (3.63) from the left and right, respectively, by

$$\begin{bmatrix} I & A^T & (A^T)^2 \end{bmatrix}$$

and its transpose, yields

$$PA + A^T P < 0,$$

which is exactly (3.61). This completes the proof. $\qquad\square$

Based on Lemma 3.3, we have the following result.

Proposition 3.8. *Matrix $\bar{A}_{cl}$ is robustly stable if there exist matrices F_j, H_j, S_j, T_j, U_j, W_j and $P_j > 0$, $j = 1, 2, \cdots, r$, such that*

$$\Theta_{ij} + \Theta_{ji} < 0, \quad i, j = 1, 2, \cdots, r, \quad i \le j, \tag{3.65}$$

where

$$\Theta_{ij} = \begin{bmatrix} F_j A_i + A_i^T F_j^T & \star & \star \\ P_j - F_j^T + H_j^T A_i - A_i^T S_j^T & -H_j - H_j^T - T_j A_i - A_i^T T_j^T & \star \\ S_j^T + W_j^T A_i & T_j^T - W_j^T + U_j^T A_i & -U_j - U_j^T \end{bmatrix}.$$

Proof. Let the parameter-dependent Lyapunov matrix $P(\alpha) > 0$ and other parameter-dependent matrices be

$$P(\alpha) = \sum_{j=1}^{r} \alpha_j P_j, \quad F(\alpha) = \sum_{j=1}^{r} \alpha_j F_j, \quad H(\alpha) = \sum_{j=1}^{r} \alpha_j H_j, \quad S(\alpha) = \sum_{j=1}^{r} \alpha_j S_j,$$

$$T(\alpha) = \sum_{j=1}^{r} \alpha_j T_j, \quad U(\alpha) = \sum_{j=1}^{r} \alpha_j U_j, \quad W(\alpha) = \sum_{j=1}^{r} \alpha_j W_j. \tag{3.66}$$

If condition (3.65) is true, substituting (3.66) into the matrix of (3.63) with A replaced by $A(\alpha)$, yields

$$
\begin{bmatrix}
F(\alpha)A(\alpha)+A(\alpha)^T F(\alpha)^T & \star & \star \\
\begin{array}{c} P(\alpha)-F(\alpha)^T + \\ H(\alpha)^T A(\alpha)-A(\alpha)^T S(\alpha)^T \end{array} & \begin{array}{c} -H(\alpha)-H(\alpha)^T - \\ T(\alpha)A(\alpha)-A(\alpha)^T T(\alpha)^T \end{array} & \star \\
S(\alpha)^T + W(\alpha)^T A(\alpha) & \begin{array}{c} T(\alpha)^T - W(\alpha)^T \\ +U(\alpha)^T A(\alpha) \end{array} & -U(\alpha)-U(\alpha)^T
\end{bmatrix}
$$

$$
= \sum_{i=1}^{r}\sum_{j=1}^{r}\alpha_i\alpha_j\Theta_{ij} = \sum_{i=1}^{r}\alpha_i^2\Theta_{ii} + \sum_{i=1}^{r}\sum_{j>i}^{r}\alpha_i\alpha_j(\Theta_{ij}+\Theta_{ji}) < 0.
$$

The proof follows immediately by Lemma 3.3. $\qquad\square$

Proposition 3.8 provides a quasi-LMI condition to search β. Based on Proposition 3.8, we present the following LMI-based algorithm to compute maximum ranges of PI controller parameters.

Algorithm 3.2

Step 1. Find the maximum $\beta_0 \geq 0$ such that LMI (3.65) is feasible for

$$
\beta = [-\beta_0, \beta_0, \cdots, -\beta_0, \beta_0].
$$

Step 2. Find $\beta_1^{\text{low}} \leq -\beta_0$ such that LMI (3.65) is feasible for

$$
\beta = [\beta_1^{\text{low}}, \beta_0, \cdots, -\beta_0, \beta_0].
$$

Step 3. Find $\beta_1^{\text{upp}} \geq \beta_0$ such that LMI (3.65) is feasible for

$$
\beta = [\beta_1^{\text{low}}, \beta_1^{\text{upp}}, -\beta_0, \beta_0, \cdots, -\beta_0, \beta_0].
$$

Step 4. Repeat Steps 2 and 3 such that LMI (3.65) is feasible for

$$
\beta = [\beta_1^{\text{low}}, \beta_1^{\text{upp}}, \cdots, \beta_{2ml}^{\text{low}}, \beta_{2ml}^{\text{upp}}].
$$

Example 3.7.1. Consider a process with transfer function

$$
G(s) = \begin{bmatrix} -\dfrac{2s+1}{s^2(s+1)} & \dfrac{1}{s} \end{bmatrix}.
$$

It is obvious that the system is unstable. Now we use the proposed PI control scheme to compute the maximum stabilizing controller ranges (if any). Its state-space realization is of the form (3.58) with

$$
A = \begin{bmatrix} -1 & 1 & 0 \\ 0 & 0 & 0 \\ 1 & 2 & 0 \end{bmatrix}, \quad B = \begin{bmatrix} 0 & 0 \\ 1 & 0 \\ 0 & 1 \end{bmatrix}, \quad C = \begin{bmatrix} 0 & 0 & 1 \end{bmatrix}.
$$

After system transformation, we have the augmented control system of the form (3.60) with

$$\bar{A} = \begin{bmatrix} -1 & 1 & 0 & 0 \\ 0 & 0 & 0 & 0 \\ 1 & 2 & 0 & 0 \\ 0 & 0 & 1 & 0 \end{bmatrix}, \quad \bar{B} = \begin{bmatrix} 0 & 0 \\ 1 & 0 \\ 0 & 1 \\ 0 & 0 \end{bmatrix}, \quad \bar{C} = \begin{bmatrix} 0 & 0 & 1 & 0 \\ 0 & 0 & 0 & 1 \end{bmatrix},$$

$$\bar{K} = \begin{bmatrix} K_1 & K_2 \end{bmatrix} = \begin{bmatrix} k_{11} & k_{12} \\ k_{21} & k_{22} \end{bmatrix}.$$

Let $K_1^0 = [-5, -6]^T$ and $K_2^0 = [-7, -4]^T$. Then the matrix $\bar{A}_0 := \bar{A} + \bar{B}\bar{K}_0\bar{C}$ with $\bar{K}_0 = [K_1^0, k_2^0]$ is Hurwitz. Denote $\bar{K}_1 = K_1 - K_1^0 = [\bar{k}_{11}, \bar{k}_{21}]^T$, $\bar{K}_2 = K_2 - K_2^0 = [\bar{k}_{12}, \bar{k}_{22}]^T$ and $\bar{I}_{ij}$ a 2×2 matrix with the (i, j) element being 1 and other elements being 0, $i, j = 1, 2$. Then, the closed-loop system of (3.60) is rewritten as

$$\dot{\bar{x}}(t) = \left(\bar{A}_0 + \sum_{i=1}^{2} \sum_{j=1}^{2} \bar{k}_{ij} \bar{B} \bar{I}_{ij} \bar{C} \right) \bar{x}(t) := \left(\bar{A}_0 + \sum_{i=1}^{4} \bar{k}_i \bar{A}_i \right) \bar{x}(t),$$

where

$$\bar{k}_1 = \bar{k}_{11}, \quad \bar{k}_2 = \bar{k}_{21}, \quad \bar{k}_3 = \bar{k}_{12}, \quad \bar{k}_4 = \bar{k}_{22},$$
$$\bar{A}_1 = \bar{B}\bar{I}_{11}\bar{C}, \quad \bar{A}_2 = \bar{B}\bar{I}_{21}\bar{C}, \quad \bar{A}_3 = \bar{B}\bar{I}_{12}\bar{C}, \quad \bar{A}_4 = \bar{B}\bar{I}_{22}\bar{C}.$$

Next, we use the proposed algorithm to find the maximum ranges of $\bar{k}_i$, $i = 1, 2, 3, 4$, such that $\bar{A}_0 + \sum_{i=1}^{4} \bar{k}_i \bar{A}_i$ is Hurwitz. Let $\bar{k}_i \in [\beta_i^{\text{low}}, \beta_i^{\text{upp}}]$, $i = 1, 2, 3, 4$. Then, $\bar{A}_0 + \sum_{i=1}^{4} \bar{k}_i \bar{A}_i$ is equivalently recast as a matrix polytope with 16 vertices

$$
\begin{aligned}
A_1(\beta) &= \bar{A}_0 + \beta_1^{\text{low}}\bar{A}_1 + \beta_2^{\text{low}}\bar{A}_2 + \beta_3^{\text{low}}\bar{A}_3 + \beta_4^{\text{low}}\bar{A}_4, \\
A_2(\beta) &= \bar{A}_0 + \beta_1^{\text{upp}}\bar{A}_1 + \beta_2^{\text{low}}\bar{A}_2 + \beta_3^{\text{low}}\bar{A}_3 + \beta_4^{\text{low}}\bar{A}_4, \\
A_3(\beta) &= \bar{A}_0 + \beta_1^{\text{low}}\bar{A}_1 + \beta_2^{\text{upp}}\bar{A}_2 + \beta_3^{\text{low}}\bar{A}_3 + \beta_4^{\text{low}}\bar{A}_4, \\
A_4(\beta) &= \bar{A}_0 + \beta_1^{\text{low}}\bar{A}_1 + \beta_2^{\text{low}}\bar{A}_2 + \beta_3^{\text{upp}}\bar{A}_3 + \beta_4^{\text{low}}\bar{A}_4, \\
A_5(\beta) &= \bar{A}_0 + \beta_1^{\text{low}}\bar{A}_1 + \beta_2^{\text{low}}\bar{A}_2 + \beta_3^{\text{low}}\bar{A}_3 + \beta_4^{\text{upp}}\bar{A}_4, \\
A_6(\beta) &= \bar{A}_0 + \beta_1^{\text{upp}}\bar{A}_1 + \beta_2^{\text{upp}}\bar{A}_2 + \beta_3^{\text{low}}\bar{A}_3 + \beta_4^{\text{low}}\bar{A}_4, \\
A_7(\beta) &= \bar{A}_0 + \beta_1^{\text{upp}}\bar{A}_1 + \beta_2^{\text{low}}\bar{A}_2 + \beta_3^{\text{upp}}\bar{A}_3 + \beta_4^{\text{low}}\bar{A}_4, \\
A_8(\beta) &= \bar{A}_0 + \beta_1^{\text{upp}}\bar{A}_1 + \beta_2^{\text{low}}\bar{A}_2 + \beta_3^{\text{low}}\bar{A}_3 + \beta_4^{\text{upp}}\bar{A}_4, \\
A_9(\beta) &= \bar{A}_0 + \beta_1^{\text{low}}\bar{A}_1 + \beta_2^{\text{upp}}\bar{A}_2 + \beta_3^{\text{upp}}\bar{A}_3 + \beta_4^{\text{low}}\bar{A}_4, \\
A_{10}(\beta) &= \bar{A}_0 + \beta_1^{\text{low}}\bar{A}_1 + \beta_2^{\text{upp}}\bar{A}_2 + \beta_3^{\text{low}}\bar{A}_3 + \beta_4^{\text{upp}}\bar{A}_4, \\
A_{11}(\beta) &= \bar{A}_0 + \beta_1^{\text{low}}\bar{A}_1 + \beta_2^{\text{low}}\bar{A}_2 + \beta_3^{\text{upp}}\bar{A}_3 + \beta_4^{\text{upp}}\bar{A}_4,
\end{aligned}
$$

$$
\begin{aligned}
A_{12}(\beta) &= \bar{A}_0 + \beta_1^{\text{upp}}\bar{A}_1 + \beta_2^{\text{upp}}\bar{A}_2 + \beta_3^{\text{upp}}\bar{A}_3 + \beta_4^{\text{low}}\bar{A}_4, \\
A_{13}(\beta) &= \bar{A}_0 + \beta_1^{\text{upp}}\bar{A}_1 + \beta_2^{\text{upp}}\bar{A}_2 + \beta_3^{\text{low}}\bar{A}_3 + \beta_4^{\text{upp}}\bar{A}_4, \\
A_{14}(\beta) &= \bar{A}_0 + \beta_1^{\text{upp}}\bar{A}_1 + \beta_2^{\text{low}}\bar{A}_2 + \beta_3^{\text{upp}}\bar{A}_3 + \beta_4^{\text{upp}}\bar{A}_4, \\
A_{15}(\beta) &= \bar{A}_0 + \beta_1^{\text{low}}\bar{A}_1 + \beta_2^{\text{upp}}\bar{A}_2 + \beta_3^{\text{upp}}\bar{A}_3 + \beta_4^{\text{upp}}\bar{A}_4, \\
A_{16}(\beta) &= \bar{A}_0 + \beta_1^{\text{upp}}\bar{A}_1 + \beta_2^{\text{upp}}\bar{A}_2 + \beta_3^{\text{upp}}\bar{A}_3 + \beta_4^{\text{upp}}\bar{A}_4.
\end{aligned}
$$

Suppose the largest available range is limited to be ± 50. By Algorithm, we compute that $\beta_0 = 2.3$ and

$$
\begin{aligned}
&[\beta_1^{\text{low}}, \beta_1^{\text{upp}}] = [-50, 2.3], \quad [\beta_2^{\text{low}}, \beta_2^{\text{upp}}] = [-6.9, 2.3], \\
&[\beta_3^{\text{low}}, \beta_3^{\text{upp}}] = [-2.3, 4.8], \quad [\beta_4^{\text{low}}, \beta_4^{\text{upp}}] = [-2.3, 2.3],
\end{aligned}
$$

which yield the stabilizing PI controller range as

$$
\begin{aligned}
&k_{11} \in [-50, -2.7], \quad k_{12} \in [-9.3, -2.2], \\
&k_{21} \in [-12.9, -3.7] \quad k_{22} \in [-6.3, -1.7].
\end{aligned}
$$

3.8 Conclusions

The problem of determining the parameter ranges of stabilizing multi-loop PID controllers has been investigated in this chapter. A detailed scheme has been proposed using the descriptor model approach. Linearly parameter-dependent technique and convex optimization method have been employed to establish basic criteria for computing the controller parameter ranges. Numerical examples have been given to illustrate the use of the present procedure. It has been seen that the stabilizing ranges obtainable from our procedure is large and sufficient for practical tuning purpose.

It should be pointed out that our algorithm provides stabilizing ranges in "sufficiency" sense only. In other words, the ranges are not necessary since there might be some PID settings which are not in the computed ranges but yet stabilize the plant. Besides, if the plant has time delay, one may use Pade approximation for time delay so as to apply our procedure. For instance, consider the example in section 6, and suppose that a time delay $L = 0.4$ is present at the second loop, that is

$$
G(s) = \begin{bmatrix} \dfrac{s-1}{(s+1)(s+3)} & \dfrac{4}{s+3}e^{-0.4s} \\ \dfrac{1}{s+2} & \dfrac{3}{s+2}e^{-0.4s} \end{bmatrix}.
$$

Using the Pade approximation $e^{-Ls} \approx (1 - Ls/2)/(1 + Ls/2)$, $G(s)$ is approximated as

$$
\hat{G}(s) = \begin{bmatrix} \dfrac{s-1}{(s+1)(s+3)} & \dfrac{4(1-0.2s)}{(s+3)(1+0.2s)} \\ \dfrac{1}{s+2} & \dfrac{3(1-0.2s)}{(s+2)(1+0.2s)} \end{bmatrix}.
$$

For a proportional control $K(s) = \text{diag}\{k_1, k_2\}$, the stabilizing region for k_1 and k_2 are calculated as: $k_1 \in [-1.7174, -0.1079]$ and $k_2 \in [0.5580, 0.8910]$. The time-delay case for our problem will lead to a different system description. The feedback of delay output gives rise to a more complicated state equation, for which the stabilizing ranges of PID parameters may not be transformed into a polytopic problem, whereas the technique used in this chapter is suitable for a polytopic problem. One needs to find a totally different technique to solve the delay problem, which could be a future study.

4 Loop Phase Margins

The stabilizing ranges of decentralized PID parameters for MIMO systems are discussed in the previous chapter as well as the loop gain margins. This chapter will continue the stability margin discussion but focus on the loop phase margins of MIMO systems. Unlike the loop gain margins that corresponds to the stabilizing range of multi-loop P control, loop phase margins are related to multi-loop phase characteristics and not effected by P control. This gives additional difficulty in their determination. Nevertheless, two methods, respectively in time and frequency domain, are presented under the framework of LMI or constrained optimization.

4.1 Introduction

Phase margin measures how much the additional phase change can be added to the system before it becomes unstable, which reflects how far the system is away from instability when perturbations are allowed to change the directions only. Since introduced by Horowitz [106], phase margin has been well defined and fully understood for SISO systems, where it can be easily determined by Nyquist plot or Bode diagram based on Nyquist stability theorem. It is also broadly accepted and applied in control engineering due to its simple calculation and clear physical meaning. However, such a success in SISO systems can hardly be extended to MIMO systems straightforwardly because of the coupling among loops for the latter as well as complexity of matrix perturbations of unity size with different directions [97]. Although Gershgorin bands based on the generalized Nyquist stability theorem can be used to define the phase margin for MIMO systems parallel to its counterpart for SISO cases [77], such a definition may be too conservative and bring some limitation of their applications. Note that phase change in the feedback path has no effects on the gain of a system, it actually can be viewed as a unitary mapping from system output to input. From this point of view, Bar-on and Jonckheere [107] defined the phase margin for the multivariable system as the minimal tolerant phase perturbation of a unitary matrix in the feedback path, beyond which there always exists one unitary matrix which can deteriorate the stability of the closed-loop system. Such a definition allows the perturbations to be in the entire set of unitary matrices, not necessarily to be diagonal. While this is a nice formulation, permissible

Q.-G. Wang et al.: PID Control for Multivariable Processes, LNCIS 373, pp. 71–95, 2008.
springerlink.com
© Springer-Verlag Berlin Heidelberg 2008

perturbations in this class are simply too rich to imagine intuitively and connect to phase changes of individual loops, which practical control engineers have been used to.

A more direct and useful definition of phase margin for MIMO systems is for the individual loop, within which stability of the closed-loop system is guaranteed. This corresponds to a multivariable control system where each loop has some phase perturbation but no gain change. Even in this case, the problem is not so simple as one can not calculate phase margin from each loop separately due to loop interactions. One-loop's phase margin depends on all other-loop's ones so that the graphical method for SISO phase margin evaluation is not possible to extend to the MIMO case, which is elaborated in detail in Sect. 4.4.1. Since literatures on phase margins of multivariable systems are very few, no other definition or method has been reported to our best knowledge. Only loop gain margins for MIMO systems are recently discussed by Wang et al. [108].

It is well known that the phase lag can be linked to a time delay. This motivates us to obtain the loop phase margin in two steps. Step 1 is to consider a MIMO system under a decentralized delay feedback and obtain the stabilizing ranges of all time delays. Step 2 is to convert the stabilizing ranges of time delays into the stabilizing ranges of phases, which is taken as the loop phase margins. For Step 1, we present a delay-dependent stability criterion for systems with multiple delays by using the free-weighting-matrix method proposed in [109, 110, 111, 112, 113], and take into account the stability interaction among the delays. An algorithm is established to compute the ranges of delays guaranteeing the stability of closed-loop systems. For step 2, we determine a fixed frequency based on a proposition in [107]. Finally, loop phase margins are obtained by multiplying the stabilizing ranges of time delays with the fixed frequency. A numerical example is given to illustrate the effectiveness of the proposed approach.

More remarks on Step 1 are drawn as follows for clarification of our contribution in this area. Stability criteria for time-delay systems can be classified into two categories: delay-dependent and delay-independent criteria. Since delay-dependent criteria make use of information on the size of delays, they are less conservative than delay-independent ones. During the last decade, considerable attention has been devoted to the problem of delay-dependent stability analysis and controller design for time-delay systems (e.g. [114, 115, 116, 117, 118, 119, 120, 109, 110, 111, 112, 113, 121, 122]). In fact, to investigate the controller design for systems with input delays, the delay-dependent criteria are more efficient. Fixed model transformations are the main methods to deal with delay-dependent stability problems. Among them, the descriptor model transformation method combined with Park's or Moon et al.'s inequalities [114, 115] is very efficient [116, 117, 120]. Recently, in order to reduce the conservatism, a free-weighting matrix method is proposed in [109, 110, 111, 93, 113] to study the delay-dependent stability for systems delay, in which the bounding techniques on some cross product terms are not involved. On the other hand, although some delay-dependent stability criteria are presented in [116, 117] for systems with multiple delays, they do not take the relationship among the delays into account. He et al. proposed a delay-dependent criterion for systems with multiple time delays by considering the relationships among the time delays in [113] using the free-weighting matrix method. However, the time delays addressed in [113] should be in a queue according to their sizes when the number of time delays is more than two. This may not be easily employed to calculate the loop phase margins. Instead, an improved delay-dependent stability criterion which does not care if the sizes of the time delays are in a queue is presented in this chapter by using the free-weighting matrix method, which is then employed to calculate the loop phase margins conveniently.

Besides the above time domain method, a frequency domain approach to computing loop phase margins of multivariable systems is also proposed. Based on the work of Bar-on and Jonckheere [107], the stability analysis based on the generalized Nyquist theorem is converted to a constrained optimization problem with the help of mapping between two unitary vectors on complex parameter space, which is then solved numerically by Lagrange multiplier method and Newton-Raphson algorithm. The major improvement to Bar-on and Jonckheere's method is that new constraints are added in the optimization problem to guarantee the diagonal structure of phase perturbations. Accordingly, loop phase margin are well defined and easily determined for multivariable systems.

4.2 Problem Formulation

To demonstrate mutual dependence of loop phase perturbations which will preserve closed-loop stability in a coupled multivariable control system, consider the TITO system given by the following transfer function matrix:

$$G(s) = \begin{bmatrix} \dfrac{2.5}{s+1} & \dfrac{1}{s+1} \\[2mm] \dfrac{3}{s+1} & \dfrac{4}{s+1} \end{bmatrix}. \tag{4.1}$$

The class of all permissible perturbations is the decentralized unitary matrix perturbation in form of $K = \mathrm{diag}\{k_1, k_2\} := \mathrm{diag}\{e^{j\phi_1}, e^{j\phi_2}\}$. One only needs to consider $\phi_i \in [-\pi, \pi), i = 1, 2$, because $e^{j\phi_i}$ is the periodic function with the period of 2π. This diagonal phase perturbation matrix, K, is inserted to the unity negative feedback configuration, as depicted in Fig. 4.1.

Note first that the graphical method for SISO phase margin evaluation is not possible to extend to the MIMO case. The characteristic loci [98] of $G(j\omega)$, namely $\lambda_1(\omega)$ and $\lambda_2(\omega)$, are shown in Fig. 4.2, where A and B are intersection points of $\lambda_1(\omega)$ and $\lambda_2(\omega)$ with the unit circle, respectively; C is the critical point $(-1, j0)$; and O is the origin. Since the open-loop is stable and the characteristic loci do not encircle the critical point, the closed-loop system is stable based on the generalized Nyquist stability criterion. From Fig. 4.2, we have $\angle AOC = 1.7667$ and $\angle BOC = 2.3951$. But unlike the SISO case, these angles can not be taken as the phase margins for the loops. For example, $\phi_1 = 1.5$ and $\phi_2 = 2$ meet $\phi_1 < \angle AOC$ and $\phi_2 < \angle BOC$. However, the characteristic loci of $G(j\omega)K$ encircle the critical point $(-1, j0)$, which implies that the decentralized perturbation $K = \mathrm{diag}\{e^{j\phi_1}, e^{j\phi_2}\}$ makes the closed-loop system unstable. This is because the product of two matrices does not comply the commutative property of multiplication. In fact, by a similarity transformation, $G = T\Lambda T^{-1}$, where Λ is diagonal and T is unitary, one sees that $GK = T\Lambda T^{-1}K \neq T\Lambda KT^{-1}$.

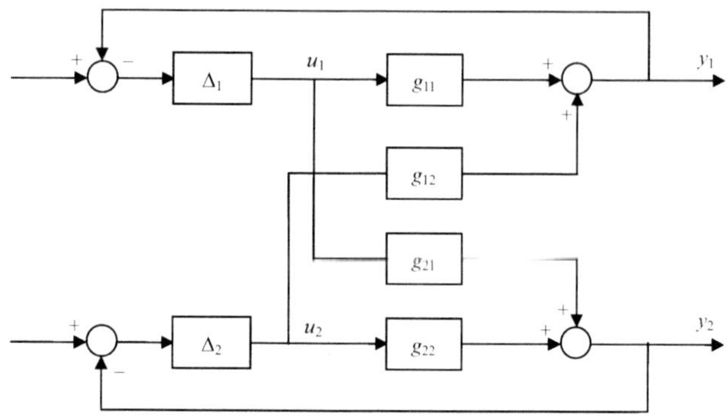

Fig. 4.1. Block diagram of TITO system

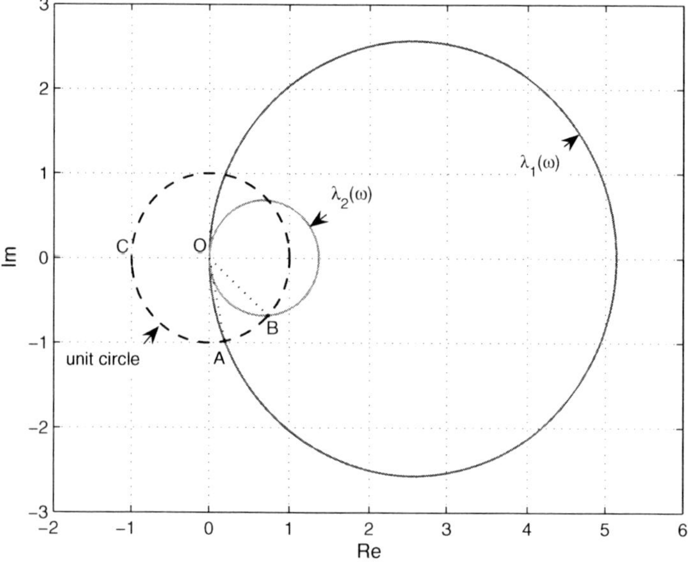

Fig. 4.2. Characteristic loci of $G(j\omega)$

To find the stabilizing region for (ϕ_1,ϕ_2), we locate its boundary, that is, consider the case when some locus, $\lambda_i(\omega)$, $i = 1$ or 2, passes through the critical point $(-1, j0)$, i.e.

$$\det(I+GK) = \begin{vmatrix} 1 + \dfrac{2.5}{j\omega+1}e^{j\phi_1} & \dfrac{1}{j\omega+1}e^{j\phi_2} \\ \dfrac{3}{j\omega+1}e^{j\phi_1} & 1 + \dfrac{4}{j\omega+1}e^{j\phi_2} \end{vmatrix} = 0,$$

or

$$\left[-\frac{7}{(j\omega+1)^2}e^{j\phi_1} - \frac{4}{j\omega+1} \right] e^{j\phi_2} = 1 + \frac{2.5}{j\omega+1}e^{j\phi_1}. \tag{4.2}$$

Taking modular on both sides of (4.2) yields

$$\left\| \frac{7}{(j\omega+1)^2}e^{j\phi_1} + \frac{4}{j\omega+1} \right\| = \left\| 1 + \frac{2.5}{j\omega+1}e^{j\phi_1} \right\|,$$

or

$$\omega^4 + 5\omega^3\sin\phi_1 + \omega^2(5\cos\phi_1 - 7.75) - 51\omega\sin\phi_1 - 51\cos\phi_1 - 57.75 = 0. \quad (4.3)$$

Let the solution to (4.3) be $\omega = \Omega(\phi_1)$. Then, substituting it back to (4.2) yields

$$\begin{aligned}
\phi_2 &= \arg\left[1 + \frac{2.5}{j\Omega(\phi_1)+1}e^{j\phi_1}\right] - \arg\left[-\frac{7}{(j\Omega(\phi_1)+1)^2}e^{j\phi_1} - \frac{4}{j\Omega(\phi_1)+1}\right] \\
&:= f(\phi_1), \quad (4.4)
\end{aligned}$$

which shows that ϕ_1 and ϕ_2 are mutually dependent of each other. Note that (4.2)–(4.4) have no analytical solution and their forms for a general system is even more complex and hard to solve numerically. Some effective and efficient method is required and the goal of this chapter is to develop it. Here we solve (4.2)–(4.4) by try and error and use the solution to demonstrate the key feature of loop phase margins. Generally, there are four solutions to (4.3). After eliminating all the complex roots, $\omega = \Omega(\phi_1)$ is usually a multiple-valued function, and so does $\phi_2 = f(\phi_1)$, which was shown as the solid curves in Fig. 4.3. $ABCD$ is the region encompassed by these curves. The region is the stabilizing region for (ϕ_1, ϕ_2) with border $ABCD$ and is denoted by Φ. Since the closed-loop system is stable for $\phi_1 = \phi_2 = 0$, the origin is stabilizing and indeed we have $(0,0) \in \Phi$.

The following lemma shows the property of the stabilizing boundary and can be extended to the general MIMO case.

Lemma 4.1. *The stabilizing boundary ABCD is symmetric with respect to the origin* $(0,0)$.

Proof. Suppose that (ϕ_1, ϕ_2) is the point on $ABCD$, then there exists some ω_c such that

$$\det[I + G(j\omega_c)K] = \det[I + G(j\omega_c)\mathrm{diag}\{e^{j\phi_1}, e^{j\phi_2}\}] = 0.$$

Taking conjugate on both sides of the above equation yields

$$\begin{aligned}
\det{}^*[I + G(j\omega_c)K] &= \det[I + G(j\omega_c)K]^* = \det[I + G^*(j\omega_c)K^*] \\
&= \det[I + G(-j\omega_c)\mathrm{diag}\{e^{-j\phi_1}, e^{-j\phi_2}\}] = 0,
\end{aligned}$$

which implies that for the pair $(-\phi_1, -\phi_2)$, there exists $-\omega_c$ such that the closed-loop system is marginally stable. Hence, $(-\phi_1, -\phi_2)$ is also the point on the stabilizing border $ABCD$. $\qquad\square$

To see how the stabilizing range of one loop's phase depends on the value of the other loop's phase, take $\phi_1 = 0.5$, the stabilizing range for ϕ_2 is $\phi_2 \in (-1.9262, 2.0231)$ from Fig. 4.3. If $\phi_1 = 1.5$, the stabilizing range for ϕ_2 becomes $\phi_2 \in (-1.8421, 1.8848)$ from Fig. 4.3 again. Note that loop 1 inevitably has some uncertainty on its parameters.

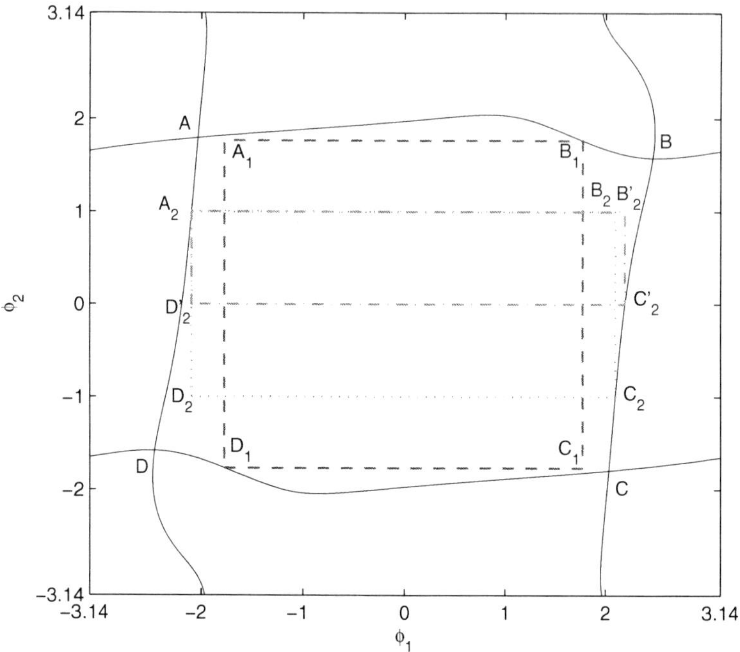

Fig. 4.3. Stabilization region of (ϕ_1, ϕ_2)

Therefore, the value of ϕ_1 cannot be known precisely. When ϕ_1 or loop 1 has some change, the previous stabilizing range for ϕ_2 may not be stabilizing any more. Such results are not very useful in the context of multivariable phase margins and their applications as they are too sensitive to other loops' phase. Instead, a more realistic and useful consideration is to prescribe a range for ϕ_1 when determining the stabilizing range for ϕ_2. In general, if ϕ_1 varies in some range which is viewed as a parameter uncertainty, the stabilizing range for ϕ_2 can be uniquely determined. For instance, if $\phi_2 \in (0,1)$, then the system remains stable for $\phi_1 \in (-2.0983, 2.1959)$. Graphically, such a stabilizing region for ϕ_1 and ϕ_2 is the rectangle with length ϕ_1 from -2.0983 to 2.1959 and width ϕ_2 from 0 to 1, shown as $A_2 B'_2 C'_2 D'_2$. When the range of ϕ_2 changes, so does the stabilizing range of ϕ_1. For instance, $\{(\phi_1, \phi_2)|\phi_2 \in (-1,1), \phi_1 \in (-2.0983, 2.0983)\}$ gives another stabilizing rectangle for ϕ_1 and ϕ_2, shown as $A_2 B_2 C_2 D_2$ in Fig. 4.3. Among all these rectangles, there exists a square, shown as $A_1 B_1 C_1 D_1$ in Fig. 4.3, where $\phi_1 \in (-1.7667, 1.7667)$ and $\phi_2 \in (-1.7667, 1.7667)$. This implies that the stabilizing range of ϕ_1 and ϕ_2 are just the same and can be defined as the common phase margin of the system. In view of the above observations, we are motivated to find such stabilizing ranges for each loop and formulate the problem as follows.

Problem 4.1. For an $m \times m$ square open-loop, $G(s)$, under the decentralized phase perturbation, $K = \text{diag}\{e^{j\phi_1}, \ldots, e^{j\phi_m}\}$, find the ranges, $(\underline{\phi}_i, \overline{\phi}_i)$, $-\pi \le \underline{\phi}_i < \overline{\phi}_i < \pi$, $i = 1, \ldots, m$, such that the closed-loop system is stable when $\phi_i \in (\underline{\phi}_i, \overline{\phi}_i)$ for all i, but marginally stable when $\phi_i = \underline{\phi}_i$ or $\phi_i = \overline{\phi}_i$ for some i.

The loop phase margins for multivariable systems can now readily be defined as follows and obtained as the solutions to Problem 4.1.

Definition 4.1. *The solution to Problem 1, $\phi_i \in (\underline{\phi_i}, \overline{\phi_i})$, is called the phase margin of the i-th loop of $G(s)$ under other loops' phases of $\phi_j \in (\underline{\phi_j}, \overline{\phi_j})$, $j \neq i$, $i = 1, \cdots, m$. If $\underline{\phi_i} = \underline{\phi_j} = \underline{\phi}$ and $\overline{\phi_i} = \overline{\phi_j} = \overline{\phi}$, then $(\underline{\phi}, \overline{\phi})$ is called the common phase margin of $G(s)$.*

It is well known that a time delay links to a phase lag with no gain change. This motivates us to obtain the loop phase margins as follows. Firstly, we consider a MIMO system under a decentralised delay feedback (the rest of this section) and obtain the stabilizing ranges of time delays (the next Section) based on LMI techniques with a delay-dependent stability criterion. Then, we convert the stabilizing ranges of time delays into the stabilizing ranges of phases by multiplying some suitable frequency (Sect. 4.3.2).

Consider the following system

$$\begin{cases} \dot{x}(t) &= Ax(t) + Bu(t), \\ y(t) &= Cx(t), \end{cases} \tag{4.5}$$

where $x \in \mathbb{R}^n$ is the state, $y \in \mathbb{R}^m$ is the output, B and C are real constant matrices with appropriate dimensions. The system is under following form of delay feedback controller: $U(s) = K(s)E(s)$, where $e(t) = r(t) - y(t)$, $r(t)$ is the set point and

$$K(s) = \begin{bmatrix} e^{-L_1 s} & 0 & \cdots & 0 \\ 0 & e^{-L_2 s} & \cdots & 0 \\ \vdots & \vdots & \ddots & \vdots \\ 0 & 0 & \cdots & e^{-L_m s} \end{bmatrix}, \tag{4.6}$$

The controller in time domain is described by

$$u(t) = - \begin{bmatrix} y_1(t - L_1) \\ y_2(t - L_2) \\ \vdots \\ y_m(t - L_m) \end{bmatrix} = - \begin{bmatrix} e_1^T Cx(t - L_1) \\ e_2^T Cx(t - L_2) \\ \vdots \\ e_m^T Cx(t - L_m) \end{bmatrix} = - \sum_{k=1}^{m} I_k Cx(t - L_k), \tag{4.7}$$

where $e_k \in \mathbb{R}^n$, $k = 1, 2, \cdots, m$, is the i-th identity column vector and I_k, $k = 1, 2, \cdots, m$, is a $m \times m$ matrix with the element (k, k) being "1" and the other elements being "0". Then, the closed-loop system is given as

$$\dot{x}(t) = Ax(t) - \sum_{k=1}^{m} BI_k Cx(t - L_k). \tag{4.8}$$

Hence, we want to find the maximum ranges of scalars L_k, $k = 1, 2, \cdots, m$ for a system (4.5) under the controller (4.7), such that the closed-loop system (4.8) is stable when L_k, $k = 1, 2, \cdots, m$, are in these ranges.

4.3 Time Domain Method

4.3.1 Finding Allowable Diagonal Delays

The following delay-dependent criterion establishes LMI conditions on delays L_k, $k = 1, 2, \cdots, m$, for stability of closed-loop systems (4.8).

Theorem 4.1. *For given scalars* $L_k \geq 0, k = 1, 2, \cdots, m$, *closed-loop system (4.8) is asymptotically stable if there exist* $P = P^T > 0$, $Q_k = Q_k^T > 0$, $W_k = W_k^T \geq 0$,

$$
Z_{ij} = Z_{ij}^T \geq 0, \ X_k = \begin{bmatrix} X_{00}^{(k)} & X_{01}^{(k)} & \cdots & X_{0m}^{(k)} \\ [X_{01}^{(k)}]^T & X_{11}^{(k)} & \cdots & X_{1m}^{(k)} \\ \vdots & \vdots & \ddots & \vdots \\ [X_{0m}^{(k)}]^T & [X_{1m}^{(k)}]^T & \cdots & X_{mm}^{(k)} \end{bmatrix} \geq 0, \ M_k = \begin{bmatrix} M_{k0} \\ M_{k1} \\ \vdots \\ M_{km} \end{bmatrix}, \ Y_{ij} =
$$

$$
\begin{bmatrix} Y_{00}^{(ij)} & Y_{01}^{(ij)} & \cdots & Y_{0m}^{(ij)} \\ [Y_{01}^{(ij)}]^T & Y_{11}^{(ij)} & \cdots & Y_{1m}^{(ij)} \\ \vdots & \vdots & \ddots & \vdots \\ [Y_{0m}^{(ij)}]^T & [Y_{1m}^{(ij)}]^T & \cdots & Y_{mm}^{(ij)} \end{bmatrix} \geq 0, N_{ij} = \begin{bmatrix} N_0^{(ij)} \\ N_1^{(ij)} \\ \vdots \\ N_m^{(ij)} \end{bmatrix}, \ k = 1, 2, \cdots, m; \ i = 1, 2, \cdots, m -
$$

1; $j = i+1, i+2, \cdots, m$, *such that the following LMIs hold:*

$$
\Phi = \bar{P}^I \bar{A} + \bar{A}^T \bar{P} + \bar{Q} + \bar{A}^T \Pi \bar{A} + \sum_{k=1}^{m} \left\{ M_k G_k + G_k^T M_k^T + L_k X_k \right\}
$$

$$
+ \sum_{i=1}^{m-1} \sum_{j=i+1}^{m} \left\{ N_{ij} H_{ij} + H_{ij}^T N_{ij}^T + |L_j - L_i| Y_{ij} \right\} < 0, \tag{4.9}
$$

$$
\Psi_k = \begin{bmatrix} X_k & M_k \\ M_k^T & W_k \end{bmatrix} \geq 0, \ k = 1, \cdots, m, \tag{4.10}
$$

$$
\Xi_{ij} = \begin{bmatrix} Y_{ij} & N_{ij} \\ N_{ij}^T & Z_{ij} \end{bmatrix} \geq 0, \ i = 1, \cdots, m-1; \ j = i+1, \cdots, m, \tag{4.11}
$$

where

$$
\bar{P} = \begin{bmatrix} P & 0 & 0 & \cdots & 0 \end{bmatrix},
$$

$$
\bar{A} = \begin{bmatrix} A & -BI_1 C & -BI_2 C & \cdots & -BI_m C \end{bmatrix},
$$

$$
\bar{Q} = \text{diag} \left\{ \sum_{k=1}^{m} Q_k, -Q_1, -Q_2, \cdots, -Q_m \right\},
$$

$$
\Pi = \sum_{i=1}^{m-1} \sum_{j=i+1}^{m} |L_j - L_i| Z_{ij} + \sum_{k=1}^{m} L_k W_k,
$$

$$
\begin{aligned}
G_k &= [I \underbrace{0 \cdots 0}_{k-1} -I \underbrace{0 \cdots 0}_{m-k}], \\
H_{ij} &= [\underbrace{0 \cdots 0}_{i} I \underbrace{0 \cdots 0}_{j-i-1} -I \underbrace{0 \cdots 0}_{m-j}].
\end{aligned}
$$

Proof. Choose the candidate Lyapunov-Krasovskii functional to be

$$
\begin{aligned}
V(x_t) \ :=\ & x^T(t)Px(t) + \sum_{k=1}^{m} \int_{t-L_k}^{t} x^T(s)Q_k x(s)\,ds \\
& + \sum_{k=1}^{m} \int_{-L_k}^{0} \int_{t+\theta}^{t} \dot{x}^T(s)W_k \dot{x}(s)\,ds\,d\theta, \\
& + \sum_{i=1}^{m-1} \sum_{j=i+1}^{m} \mathrm{sgn}(L_j - L_i) \int_{-L_j}^{-L_i} \int_{t+\theta}^{t} \dot{x}^T(s)Z_{ij}\dot{x}(s)\,ds\,d\theta,
\end{aligned}
\tag{4.12}
$$

where $P = P^T > 0$, $Q_k = Q_k^T > 0$, $W_k = W_k^T > 0$, $Z_{ij} = Z_{ij}^T \geq 0$, $k = 1,2,\cdots,m$; $i = 1,2,\cdots,m-1$; $j = i+1, i+2,\cdots, m$, are to be determined. One calculates the derivative of $V(x_t)$ along the solutions of system (4.8) as

$$
\begin{aligned}
\dot{V}(x_t) \ =\ & 2x^T(t)P\dot{x}(t) + \sum_{k=1}^{m} \left\{ x^T(t)Q_k x(t) - x^T(t-L_k)Q_k x(t-L_k) \right\} \\
& + \sum_{k=1}^{m} \left\{ L_k \dot{x}^T(t)W_k \dot{x}(t) - \int_{t-L_k}^{t} \dot{x}^T(s)W_k \dot{x}(s)\,ds \right\} \\
& + \sum_{i=1}^{m-1} \sum_{j=i+1}^{m} \mathrm{sgn}(L_j - L_i) \\
& \quad\quad \times \left\{ (L_j - L_i)\dot{x}^T(t)Z_{ij}\dot{x}(t) - \int_{t-L_j}^{t-L_i} \dot{x}^T(s)Z_{ij}\dot{x}(s)\,ds \right\} \\
=\ & \zeta_1^T(t) \left\{ \bar{P}^T \bar{A} + \bar{A}^T \bar{P} + \bar{Q} + \bar{A}^T \Pi \bar{A} \right\} \zeta_1(t) \\
& - \sum_{k=1}^{m} \int_{t-L_k}^{t} \dot{x}^T(s)W_k \dot{x}(s)\,ds \\
& - \sum_{i=1}^{m-1} \sum_{j=i+1}^{m} \left\{ \mathrm{sgn}(L_j - L_i) \int_{t-L_j}^{t-L_i} \dot{x}^T(s)Z_{ij}\dot{x}(s)\,ds \right\},
\end{aligned}
\tag{4.13}
$$

where

$$
\zeta_1(t) = \left[x^T(t) \ \ x^T(t-L_1) \ \ x^T(t-L_2) \ \ \cdots \ \ x^T(t-L_m) \right]^T.
$$

According to the Leibniz-Newton formula, for $k = 1,2,\cdots,m$; $i = 1,2,\cdots,m-1$; $j = i+1, i+2,\cdots,m$, and any appropriate dimensioned matrices N_{ij}, the following equations hold:

$$
\begin{aligned}
0 \ =\ & 2\zeta_1^T(t)M_k \left[x(t) - x(t-L_k) - \int_{t-L_k}^{t} \dot{x}(s)\,ds \right] \\
=\ & \zeta_1^T(t) \left\{ M_k G_k + G_k^T M_k^T \right\} \zeta_1(t) - 2\zeta_1^T(t)M_k \int_{t-L_k}^{t} \dot{x}(s)\,ds,
\end{aligned}
\tag{4.14}
$$

$$
\begin{aligned}
0 &= 2\zeta_1^T(t)N_{ij}\left[x(t-L_i)-x(t-L_j)-\int_{t-L_j}^{t-L_i}\dot{x}(s)\mathrm{d}s\right]\\
&= 2\zeta_1^T(t)N_{ij}\\
&\quad\times\left[x(t-L_i)-x(t-L_j)-\mathrm{sgn}(L_j-L_i)\mathrm{sgn}(L_j-L_i)\int_{t-L_j}^{t-L_i}\dot{x}(s)\mathrm{d}s\right]\\
&= \zeta_1^T(t)\left\{N_{ij}H_{ij}+H_{ij}^T N_{ij}^T\right\}\zeta_1(t)\\
&\quad -2\mathrm{sgn}(L_j-L_i)\mathrm{sgn}(L_j-L_i)\zeta_1^T(t)N_{ij}\int_{t-L_j}^{t-L_i}\dot{x}(s)\mathrm{d}s. \tag{4.15}
\end{aligned}
$$

On the other hand, for any matrices, $X_k = X_k^T \geq 0$, $Y_{ij} = Y_{ij}^T \geq 0$, $k = 1,2,\cdots,m$, $i = 1,2,\cdots,m-1$; $j = i+1,i+2,\cdots,m$, there hold:

$$
\begin{aligned}
0 &= L_k\zeta_1^T(t)\left[X_k - X_k\right]\zeta_1(t), \tag{4.16}\\
0 &= |L_j-L_i|\zeta_1^T(t)\left[Y_{ij}-Y_{ij}\right]\zeta_1(t)\\
&= |L_j-L_i|\zeta_1^T(t)Y_{ij}\zeta_1(t)-\mathrm{sgn}(L_j-L_i)(L_j-L_i)\zeta_1^T(t)Y_{ij}\zeta_1(t). \tag{4.17}
\end{aligned}
$$

Summing (4.14) and (4.16) for $k = 1,2,\cdots,m$, and (4.15) and (4.17) for $i = 1,2,\cdots$, $m-1$; $j = i+1,i+2,\cdots,m$, and adding the right side of them into $\dot{V}(x_t)$ yield

$$
\begin{aligned}
\dot{V}(x_t) &= \zeta_1^T(t)\Phi\zeta_1(t)-\sum_{k=1}^{m}\int_{t-L_j}^{t-L_k}\zeta_2^T(t,s)\Psi_k\zeta_2(t,s)\mathrm{d}s\\
&\quad -\sum_{i=1}^{m-1}\sum_{j=i+1}^{m}\int_{t-L_j}^{t-L_i}\mathrm{sgn}(L_j-L_i)\zeta_2^T(t,s)\hat{\Xi}_{ij}\zeta_2(t,s)\mathrm{d}s,
\end{aligned} \tag{4.18}
$$

where

$$
\begin{aligned}
\zeta_2(t,s) &= \left[\zeta_1^T(t)\quad \dot{x}^T(s)\right]^T,\\
\hat{\Xi}_{ij} &= \begin{bmatrix} Y_{ij} & \mathrm{sgn}(L_j-L_i)N_{ij}\\ \mathrm{sgn}(L_j-L_i)N_{ij}^T & Z_{ij} \end{bmatrix},
\end{aligned}
$$

and Φ and Ψ_k, $k = 1,2,\cdots,m$, are defined in (4.9) and (4.10), respectively. The closed-loop system (4.8) is asymptotically stable if LMIs (4.9), (4.10) and $\hat{\Xi}_{ij} \geq 0$, $i = 1,2,\cdots$, $m-1$; $j = i+1,i+2,\cdots,m$ hold, which imply (4.9), (4.10) and (4.11) by using Schur complements, respectively. $\square$

In the following, in order to determine the range of delays L_k, $k = 1,2,\cdots,m$, which guarantee the stability of closed-loop system (4.8), we define:

$$
L_k = \hat{L}_k + \Delta L_k, \quad k = 1,2,\cdots,m, \tag{4.19}
$$

where $|\Delta L_k| \leq d_k$, $k = 1,2,\cdots,m$, and $d_k \geq 0$, $k = 1,2,\cdots,m$, are given scalars. Then, we have the following corollary:

Corollary 4.1. *Suppose that scalars $L_k \geq 0$, $k = 1, 2, \cdots, m$, are given in (4.19). The closed-loop system (4.8) is asymptotically stable if there exist $P = P^T > 0$,*

$$Q_k = Q_k^T > 0,\ W_k = W_k^T \geq 0,\ Z_{ij} = Z_{ij}^T \geq 0,\ X_k = \begin{bmatrix} X_{00}^{(k)} & X_{01}^{(k)} & \cdots & X_{0m}^{(k)} \\ [X_{01}^{(k)}]^T & X_{11}^{(k)} & \cdots & X_{1m}^{(k)} \\ \vdots & \vdots & \ddots & \vdots \\ [X_{0m}^{(k)}]^T & [X_{1m}^{(k)}]^T & \cdots & X_{mm}^{(k)} \end{bmatrix} \geq 0,$$

$$Y_{ij} = \begin{bmatrix} Y_{00}^{(ij)} & Y_{01}^{(ij)} & \cdots & Y_{0m}^{(ij)} \\ [Y_{01}^{(ij)}]^T & Y_{11}^{(ij)} & \cdots & Y_{1m}^{(ij)} \\ \vdots & \vdots & \ddots & \vdots \\ [Y_{0m}^{(ij)}]^T & [Y_{1m}^{(ij)}]^T & \cdots & Y_{mm}^{(ij)} \end{bmatrix} \geq 0,\ M_k = \begin{bmatrix} M_{k0} \\ M_{k1} \\ \vdots \\ M_{km} \end{bmatrix},\ N_{ij} = \begin{bmatrix} N_0^{(ij)} \\ N_1^{(ij)} \\ \vdots \\ N_m^{(ij)} \end{bmatrix},\ k =$$

$1, 2, \cdots, m$; $i = 1, 2, \cdots, m - 1$; $j = i + 1, i + 2, \cdots, m$, such that the LMIs, (4.20), (4.10) and (4.11), hold:

$$\hat{\Phi} = \bar{P}^T \bar{A} + \bar{A}^T \bar{P} + \bar{Q} + \bar{A}^T \hat{\Pi} \bar{A} + \sum_{k=1}^{m} \left\{ M_k G_k + G_k^T M_k^T + (\hat{L}_k + d_k) X_k \right\}$$

$$+ \sum_{i=1}^{m-1} \sum_{j=i+1}^{m} \left\{ N_{ij} H_{ij} + H_{ij}^T N_{ij}^T + \left(|\hat{L}_j - \hat{L}_i| + d_j + d_i \right) Y_{ij} \right\} < 0, \qquad (4.20)$$

where $\hat{\Pi} = \displaystyle\sum_{i=1}^{m-1} \sum_{j=i+1}^{m} \left(|\hat{L}_j - \hat{L}_i| + d_j + d_i \right) Z_{ij} + \sum_{k=1}^{m} (\hat{L}_k + d_k) W_k$, and the other parameters are defined in Theorem 4.1.

Proof. Since

$$L_k = \hat{L}_k + \Delta L_k \ \leq\ \hat{L}_k + d_k,$$

$$|L_j - L_i| = |\hat{L}_j + \Delta L_j - (\hat{L}_i + \Delta L_i)| \ \leq\ |\hat{L}_j - \hat{L}_i| + d_j + d_i,$$

for $k = 1, 2, \cdots, m$; $i = 1, 2, \cdots, m - 1$; $j = i + 1, i + 2, \cdots, m$, the result follows. $\square$

The ranges of delays, L_k, $k = 1, 2, \cdots, m$, will be determined as follows. Firstly, choose the initial L_k, $k = 1, 2, \cdots, m$, such that LMIs (4.9), (4.10) and (4.11) are feasible and set $\hat{L}_k = L_k$, $k = 1, 2, \cdots, m$. Secondly, for $\hat{L}_k$, $k = 1, 2, \cdots, m$, chosen above find the maximum values of d_k, $k = 1, 2, \cdots, m$, such that LMIs (4.20), (4.10) and (4.11) are feasible when $|\Delta L_k| \leq d_k$, $k = 1, 2, \cdots, m$.

However, if the initial $\hat{L}_k$, $k = 1, 2, \cdots, m$, are chosen in the edge of the range of L_k, $k = 1, 2, \cdots, m$, the derived range of delays guaranteeing the stability of closed-loop system (4.8) is conservative. In this case, the initial $\hat{L}_k$, $k = 1, 2, \cdots, m$, should be adjusted in the center of the range of L_k, $k = 1, 2, \cdots, m$. For $i = 1, 2, \cdots, m$, we can fix $\hat{L}_k$ and d_k, $k = 1, 2, \cdots, m$, $k \neq i$ and adjust $\hat{L}_i$ and d_i. If $d_i \leq \hat{L}_i$, it means that the lower bound of the range of L_i can be enlarged, otherwise, its upper bound can be enlarged. All the above development is summarized in the following algorithm:

Algorithm 4.1. Given a state space representation in (4.5) with a decentralized controller, $K(s)$ in (4.6),

Step 1. Choose the initial L_k, $k = 1, 2, \cdots, m$, such that LMIs (4.9), (4.10) and (4.11) are feasible and set $\hat{L}_k = L_k$, $k = 1, 2, \cdots, m$.

Step 2. For $\hat{L}_k$, $k = 1, 2, \cdots, m$, chosen in Step 1, find a maximum value of $d \geq 0$, such that LMIs (4.20), (4.10) and (4.11) are feasible when $d_k = d$, $k = 1, 2, \cdots, m$. Let $d_k = d$, $k = 1, 2, \cdots, m$ and $i = 1$.

Step 3. For fixed d_k, $k = 1, 2, \cdots, m$, $k \neq i$ and given $\hat{L}_k$, $k = 1, 2, \cdots, m$, find a maximum $d_i \geq d$, such that LMIs (4.20), (4.10) and (4.11) are feasible.

Step 4. If $d_i \leq \hat{L}_i$, let $\overline{L}_i = \hat{L}_i + d_i$, then go to Procedure A; Else, let $\underline{L}_i = 0$, then go to Procedure B.

Step 5. Let $\hat{L}_i = (\overline{L}_i + \underline{L}_i)/2$ and $d_i = (\overline{L}_i - \underline{L}_i)/2$. If $i < m$, let $i = i + 1$, go to Step 3.

Step 6. The ranges of $L_k \in [\underline{L}_k, \overline{L}_k]$, $k = 1, 2, \cdots, m$, are those for guaranteeing the stability of closed-loop system (4.8).

Procedure A

Step 1. Let $r_{\text{low}} = \hat{L}_i - d_i$, $r_{\text{upp}} = \hat{L}_i + d_i$, $min = 0$, $max = r_{\text{low}}$, $\hat{L}_i = r_{\text{upp}}/2$, and $d_i = r_{\text{upp}}/2$.

Step 2. If LMIs (4.20), (4.10) and (4.11) are feasible, let $r_{\text{low}} = 0$, then go to Step 6.

Step 3. Else, let $mid = (min + max)/2$, $r_{\text{low}} = mid$, $\hat{L}_i = (r_{\text{upp}} + r_{\text{low}})/2$, and $d_i = (r_{\text{upp}} - r_{\text{low}})/2$.

Step 4. If LMIs (4.20), (4.10) and (4.11) are feasible, let $max = mid$; Else, let $min = mid$.

Step 5. If $|max - min| < \varepsilon$, a prescribed tolerance, let $r_{\text{low}} = max$; Else, go to Step 3.

Step 6. Let $\underline{L}_i = r_{\text{low}}$, then return to Step 5 of Algorithm 1.

Procedure B

Step 1. Let $r_{\text{low}} = 0$, $r_{\text{upp}} = \hat{L}_i + d_i$, $min = r_{\text{upp}}$, $max = \delta$, a given upper bound, $\hat{L}_i = max/2$, and $d_i = max/2$.

Step 2. If LMIs (4.20), (4.10) and (4.11) are feasible, let $r_{\text{upp}} = max$, then go to Step 6.

Step 3. Let $mid = (min + max)/2$, $r_{\text{upp}} = mid$, $\hat{L}_i = (r_{\text{upp}} + r_{\text{low}})/2$, and $d_i = (r_{\text{upp}} - r_{\text{low}})/2$.

Step 4. If LMIs (4.20), (4.10) and (4.11) are feasible, let $min = mid$; Else, let $max = mid$.

Step 5. If $|max - min| < \varepsilon$, a prescribed tolerance, let $r_{\text{upp}} = min$; Else, go to Step 3.

Step 6. Let $\overline{L}_i = r_{\text{upp}}$, then return to Step 5 of Algorithm 1.

4.3.2 Evaluating Phase Margins

Once the stabilizing ranges, $L_k \in (\underline{L}_k, \overline{L}_k)$, of time delays are determined, we need to find a critical frequency, ω_c, to convert the stabilizing ranges of time delays into the loop phase margins by multiplying them by ω_c. When the closed-loop system is marginally stable, there holds

$$\det(I + G(j\omega_c)\text{diag}\{e^{-j\omega_c L_k}\}) = 0, \qquad (4.21)$$

which implies that ω_c and L_k jointly contribute the phase lag and the stabilizing range for one depends on another. The functional relationship between ω_c and L_k is complicated and no analytical solutions are available. By Lemma 1, the stabilizing borders of loop phases are symmetric with respect to the origin, the values of ω_c are also symmetric with respect to the origin. Hence, one only needs to consider the positive value of ω_c to simplify our calculation. Let Ω be the set of all $\omega_c > 0$ which meet (4.21) and $\underline{\omega_c}$ be the minimum of the set. Obviously, the closed-loop system remains stable for all $0 < \phi_k < \underline{\omega_c} L_k$ because in such ranges, none of the system characteristic loci can pass through the critical point $(-1, j0)$. Since there is no easy way to find this set and its minimum, we try to under-estimate it based on a proposition in [107]. Let $\bar{\sigma}(G(j\omega))$ and $\underline{\sigma}(G(j\omega))$ be the largest and the smallest singular values of a given system, $G(j\omega)$, respectively.

Proposition 4.1 ([107]). *There exists a unitary Δ in the feedback path which destabilizes the system, $G(s)$, if and only if there exists an ω such that $\bar{\sigma}(G(j\omega)) \geq 1$ and $0 \leq \underline{\sigma}(G(j\omega)) \leq 1$.*

Let the set $\hat{\Omega} = \{\omega | 0 \leq \underline{\sigma}(G(j\omega)) \leq 1 \leq \bar{\sigma}(G(j\omega))\}$ and $\underline{\omega_g} = \min\{\omega | \omega \in \hat{\Omega}\}$, then $\Omega \subseteq \hat{\Omega}$ because Δ in Proposition 4.1 does not limit to be diagonal, which implies that Ω is over-estimated by $\hat{\Omega}$, i.e., $\Omega \subseteq \hat{\Omega}$ and $\underline{\omega_g} \leq \underline{\omega_c}$. The closed-loop system remains stable for all $\phi_k \in (\underline{\omega_g} L_k, \underline{\omega_g} \overline{L_k}) := (\underline{\phi_k}, \overline{\phi_k})$, which serves as our estimates of loop phase margins.

To obtain the set $\hat{\Omega}$, we may begin with finding the ω such that $\sigma(G(j\omega)) = 1$, where $\sigma(G(j\omega))$ is the singular value of $G(j\omega)$, which is the square root of eigenvalues of the cascade system $G^H(s)G(s)$. From (4.5), $G(s) = C(sI - A)^{-1}B$, then $G^H(s) = G^T(-s) = [C(-sI - A)^{-1}B]^T = -B^T(sI + A^T)^{-1}C^T$. The state-space representation for these systems can be written as

$$\begin{aligned} G: \dot{x}_1 &= Ax_1 + Bu, \quad y_1 = Cx_1; \\ G^H: \dot{x}_2 &= -A^T x_2 + C^T y_1, \quad y_2 = -B^T x_2. \end{aligned}$$

The state-space representation of $G^H(s)G(s)$ is

$$\begin{bmatrix} \dot{x}_1 \\ \dot{x}_2 \end{bmatrix} = \underbrace{\begin{bmatrix} A & 0 \\ C^T C & -A^T \end{bmatrix}}_{\tilde{A}} \begin{bmatrix} x_1 \\ x_2 \end{bmatrix} + \underbrace{\begin{bmatrix} B \\ 0 \end{bmatrix}}_{\tilde{B}} u, \tag{4.22}$$

$$y = \underbrace{\begin{bmatrix} 0 & -B^T \end{bmatrix}}_{\tilde{C}} \begin{bmatrix} x_1 \\ x_2 \end{bmatrix}. \tag{4.23}$$

Thus,

$$\begin{aligned} \det\left[I - G^H(s)G(s)\right] &= \det\left[I - \tilde{C}(sI - \tilde{A})^{-1}\tilde{B}\right] \\ &= \det\left[I - \tilde{B}\tilde{C}(sI - \tilde{A})^{-1}\right] \\ &= \det\left[sI - (\tilde{A} + \tilde{B}\tilde{C})\right]\det(sI - \tilde{A}). \end{aligned}$$

Suppose that $G(j\omega_i)$ has no poles on the imaginary axis, which is the case for most MIMO plants in practice, $\det(j\omega_i I - \tilde{A}) \neq 0$ for $\forall\omega$. Thus, $\det[j\omega_i I - (\tilde{A} + \tilde{B}\tilde{C})] = 0$ yields that ω_i are pure imaginary eigenvalues of $(\tilde{A} + \tilde{B}\tilde{C})$. Note that $\sigma(G(j\omega))$ is a continuous function of ω, and between the interval of two consecutive ω_i and ω_{i+1}, no other $\omega \in (\omega_i, \omega_{i+1})$ exists such that $\sigma(G(j\omega)) = 1$, otherwise, ω_i and ω_{i+1} are not consecutive any more. This implies that $\sigma(G(j\omega))$ is always greater or less than 1 for $\forall\omega \in (\omega_i, \omega_{i+1})$. Hence, by calculating $\bar{\sigma}(G(j\omega))$ and $\underline{\sigma}(G(j\omega))$ for one $\omega \in (\omega_i, \omega_{i+1})$, we know whether $(\omega_i, \omega_{i+1}) \subseteq \hat{\Omega}$. By Lemma 4.1, ω is symmetric with respect to the origin, only positive ω need to be checked, which can simplify the process of calculation.

Finally, all the above development is integrated as follows.

Algorithm 4.2. Given the stabilizing ranges of L_k, $L_k \in (\underline{L_k}, \overline{L_k})$, from Algorithm 4.1:

Step 1. Calculate the purely imaginary eigenvalues, ω_i, $i = 1, 2, \cdots$, of the matrix, $\tilde{A} + \tilde{B}\tilde{C}$, where $\tilde{A}$, $\tilde{B}$ and $\tilde{C}$ are well defined in (4.22) and (4.23);

Step 2. Choose any $\omega > 0$ and $\omega \in (\omega_i, \omega_{i+1})$ and calculate $\bar{\sigma}(G(j\omega))$ and $\underline{\sigma}(G(j\omega))$ for $i = 1, 2, \cdots$. If $\bar{\sigma}(G(j\omega)) \geq 1$ and $\underline{\sigma}(G(j\omega)) \leq 1$, then $\hat{\Omega} = \bigcup(\omega_i, \omega_{i+1})$;

Step 3. Let $\underline{\omega_g} = \min\{\omega | \omega \in \hat{\Omega}\}$, then the stabilizing range of ϕ_k is calculated as $\phi_k \in (\underline{\omega_g}\underline{L_k}, \underline{\omega_g}\overline{L_k}) := (\underline{\phi_k}, \overline{\phi_k})$.

It should be pointed out that the loop phase margins obtained with Algorithm 4.2 are indeed stability margins but may not be exact or maximum margins available. This is due to conservativeness introduced in both steps. In the first step on delay calculation, Theorem 4.1 and Corollary 4.1 give only sufficient but not necessary conditions for stability of the closed-loop system under loop delay perturbations L_k, and this sufficiency only is common in all the LMI techniques. In the second step for the critical frequency determination, we have under-estimated it with $\underline{\omega_g}$. Nevertheless, such approximations greatly simplify the problem and enable us to get a good estimation of loop phase margins with computational feasibility and efficiency, noting that there are stable and efficient algorithms for solving LMIs and singular values, which have been well developed and popularly used. Besides, there is no other systematic method available in the literature to determine loop phase margins.

4.3.3   An Example

Example 4.1. Consider system (4.5) with the following parameters [107],

$$
A = \begin{bmatrix} 0 & 1 & 0 & 0 \\ -3 & 0.75 & 1 & 0.25 \\ 0 & 0 & 0 & 1 \\ 4 & 1 & -4 & -1 \end{bmatrix}, \quad B = \begin{bmatrix} 0 & 0 \\ 0 & 1 \\ 0 & 0 \\ 0.25 & 0 \end{bmatrix}, \quad C = \begin{bmatrix} 1 & 0 & 0 & 0 \\ 0 & 0 & 1 & 0 \end{bmatrix}.
$$

Its transfer function matrix is

$$G(s) = C(sI - A)^{-1}B$$

$$= \frac{1}{s^4 + 1.75s^3 + 7.5s^2 + 4s + 8} \begin{bmatrix} 0.0625s + 0.25 & s^2 + s + 4 \\ 0.25s^2 + 0.1875s + 0.75 & s + 4 \end{bmatrix}.$$

It is clear that closed-loop system is stable. If the common delay feedback controller (4.6) is employed to control the system (4.5), i.e. $L_1 = L_2 = L$, by using the method in [111], the closed-loop system (4.8) remains stable for $L \in [0, 0.1981]$. For $L_1 \neq L_2$, the stabilizing ranges of time delays are involved with their initial values, L_1^0 and L_2^0. By using Algorithm 2 for two different initial L_1 and L_2, we have the following results.

The positive pure imaginary eigenvalues of $\tilde{A} + \tilde{B}\tilde{C}$ is 0.643i and 1.613i. At $\omega = 0$, $\sigma_1 = 0.713$ and $\sigma_2 = 0.0439$, this shows that $\hat{\Omega} = (0.643, 1.613)$. Let $\omega_c = \min\{\omega | \omega \in \hat{\Omega}\} = 0.643$, then for two different initial conditions, the phase margin of the multi-loop system is calculated as follows:

- $L_1^0 = 0$, $L_2^0 = 0$, $L_1 \in [0, 0.1979]$, and $L_2 \in [0, 0.1967]$, $\phi_1 \in [0, 0.643] \times 0.1979 = [0, 0.1272]$, $\phi_2 \in [0, 0.643] \times 0.1967 = [0, 0.1265]$;
- $L_1^0 = 0.1$, $L_2^0 = 0$, $L_1 \in [0, 0.2920]$, and $L_2 \in [0, 0.1914]$, $\phi_1 \in [0, 0.643] \times 0.2920 = [0, 0.1878]$, $\phi_2 \in [0, 0.643] \times 0.1914 = [0, 0.1231]$.

From the above calculation, the common phase margin is $\phi \in [0, 0.1265]$.

4.4 Frequency Domain Method

4.4.1 The Proposed Approach

It follows from the definition in Sect. 4.2 that loop phase margins of a given multivariable system is the polytope in m-dimensional real vector space representing m independent loop phase perturbations. To find such a stabilizing region, we try to locate its boundary.

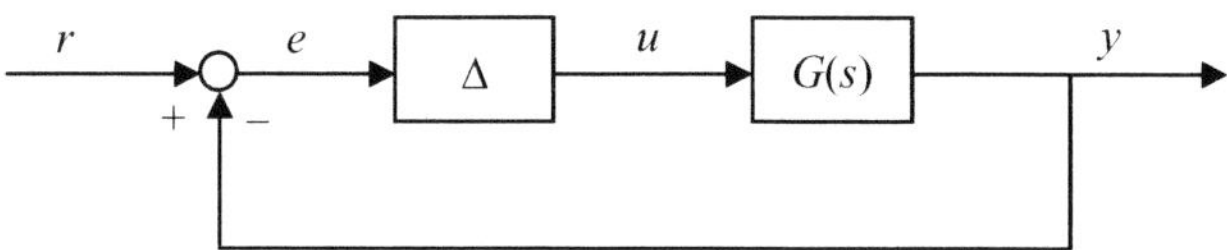

Fig. 4.4. Diagram of a MIMO control system

Consider the unity output feedback system depicted in Fig. 4.4, where $G(s)$ represents the open-loop transfer function matrix of size of $m \times m$, and $\Delta(s) = \text{diag}\{e^{j\phi_i}\}$, $i = 1, 2, \cdots, m$, is the diagonal phase perturbation matrix. Note that unlike a common robust stability analysis where the nominal case means $\Delta(s) = 0$, our nominal case means no phase perturbations, i.e., $\phi_i = 0, i = 1, 2, \cdots, m$, and thus $\Delta(s) = I_m$, the identity matrix. Except the above difference, we follow the typical robust stability analysis framework. In particular, we assume, throughout this chapter, nominal stabilization of the closed-loop system, that is, the closed-loop system is stable when $\Delta(s) = I_m$. By the assumed nominal stabilization, the system can be de-stabilized if and only if there is a phase perturbation Δ such that

$$\det(I + G(j\omega)\Delta) = 0, \tag{4.24}$$

which is equivalent to the existence of some unit vector $\mathbf{z} \in \mathbb{C}^m$ such that

$$\mathbf{z} = \Delta\mathbf{v} = -\Delta G\mathbf{z}, \tag{4.25}$$

where "$-$" denotes the negative feedback configuration. Thus, Δ is a unitary matrix which maps the unit vector $\mathbf{v}$ into $\mathbf{z}$. If all solutions to (4.25), $\mathbf{z}$ and $\mathbf{v}$, can be found, boundary points, ϕ_i, $i = 1, 2, \cdots, m$, are simply the phase angle of divisions by the corresponding elements from $\mathbf{z}$ and $\mathbf{v}$. However, solutions to (4.25) do not always exist for $\forall \omega \in (-\infty, +\infty)$ since solutions to (4.24) are frequency-dependent. Hence, the basic idea of the proposed method is composed of two parts. Firstly, with the help of unitary mapping, the frequency range, Ω, is determined to guarantee the existence of all solutions to (4.25); Secondly, in a framework of the constrained optimization, numerical solutions to (4.25) are found by the Newton-Raphson algorithm.

It follows from Proposition 4.1 that Ω can be over-estimated by $\hat{\Omega}$, where $\hat{\Omega} = \{\omega | 0 \leq \underline{\sigma}(G(j\omega)) \leq 1 \leq \overline{\sigma}(G(j\omega))\}$. For every $\omega \in \hat{\Omega}$, $\mathbf{z}$ can be found from (4.25) by solving an equivalent constrained optimization problem. Since $\hat{\Omega}$ is over-estimated for Ω, some $\omega \in \hat{\Omega}$ may cause the Newton-Raphson algorithm divergent, which implies that no diagonal phase perturbation exists to destabilize the closed-loop system at that frequency. In the following, we show how to find $\mathbf{z}$ in the framework of constrained optimization.

Let $\mathbf{z} = [z_1, z_2, \cdots, z_m]^T$ and $\mathbf{v} = [v_1, v_2, \cdots, v_m]^T$. A diagonal unitary mapping via $\mathbf{z} = \Delta\mathbf{v}$ yields $|z_k| = |v_k|$, i.e., $z_k^* z_k = v_k^* v_k$, $k = 1, 2, \cdots, m$. One can write $z_k^* z_k = \mathbf{z}^* H_k \mathbf{z}$, where $H_k = [h_{i,j}] \in \mathbb{R}^{m \times m}$ is given by

$$h_{i,j} = \begin{cases} 1, & i = j = k; \\ 0, & \text{otherswise}, \end{cases}$$

and $v_k^* v_k = \mathbf{v}^* H_k \mathbf{v} = \mathbf{z}^* G^* H_k G\mathbf{z}$ since $\mathbf{v} = -G\mathbf{z}$. Thus, $z_k^* z_k = v_k^* v_k$ yields $\mathbf{z}^* (H_k - G^* H_k G)\mathbf{z} = 0$. Unit $\mathbf{z}$ and $\mathbf{v}$ yield $\mathbf{z}^* \mathbf{z} = 1$ and $\mathbf{v}^* \mathbf{v} = \mathbf{z} G^* G\mathbf{z} = 1$. Due to the diagonal nature of Δ, $\mathbf{v}^* \mathbf{v} = \sum_{k=1}^m v_k^* v_k = \sum_{k=1}^m z_k^* z_k = \mathbf{z}^* \mathbf{z} = 1$, which implies only $m + 1$ independent constraints as follows:

$$\begin{cases} \mathbf{z}^* \mathbf{z} = 1, \\ \mathbf{z}^* (H_k - G^* H_k G)\mathbf{z} = 0, k = 1, 2, \cdots, m. \end{cases} \tag{4.26}$$

Once z_k and v_k which meet the above constraints can be obtained and, $z_k / v_k = e^{j\phi_k}$, where ϕ_k is the phase change from v_k to z_k. However, solutions to (4.26) is not unique because $\phi_k \pm 2k\pi$, $k \in \mathbb{N}$, is also a solution. Here, we limit $\phi_k \in [-\pi, \pi)$ since the nominal system ($\phi_i = 0, i = 1, 2, \cdots, m$) is stable according to our assumption. Suppose that $\overline{\phi} = \max\{|\phi_k|\}$ and $\underline{\phi} = \min\{|\phi_k|\}$, the inner product of $\mathbf{v}$ and $\mathbf{z}$ is

$$\langle \mathbf{v}, \mathbf{z} \rangle = \mathbf{v}^* \mathbf{z} = \sum_{k=1}^{m} v_k^* z_k = \sum_{k=1}^{m} e^{j\phi_k} v_k^* v_k = \sum_{k=1}^{m} |v_k|^2 \cos\phi_k + j \sum_{k=1}^{m} |v_k|^2 \sin\phi_k,$$

where

$$\sum_{k=1}^{m} |v_k|^2 \cos\phi_k = \sum_{k=1}^{m} |v_k|^2 \cos|\phi_k| \geq \cos\overline{\phi} \sum_{k=1}^{m} |v_k|^2 = \cos\overline{\phi}.$$

To ensure $\overline{\phi} = \max\{|\phi_k|\}$ really hold, $\cos\overline{\phi}$ has to be minimized, which can be achieved by minimizing its upper bound $\sum_{k=1}^{m} |v_k|^2 \cos\phi_k$. Likewise,

$$\sum_{k=1}^{m} |v_k|^2 \cos\phi_k = \sum_{k=1}^{m} |v_k|^2 \cos|\phi_k| \leq \cos\underline{\phi} \sum_{k=1}^{m} |v_k|^2 = \cos\underline{\phi},$$

and $\cos\underline{\phi}$ has to be maximized to ensure $\underline{\phi} = \min\{|\phi_k|\}$ really hold, which can be achieved by maximizing its lower bound $\sum_{k=1}^{m} |v_k|^2 \cos\phi_k$. Clearly,

$$2\sum_{k=1}^{m} |v_k|^2 \cos\phi_k = \mathbf{v}^* \mathbf{z} + \mathbf{z}^* \mathbf{v} = -[\mathbf{z}^*(G^* + G)\mathbf{z}].$$

Maximizing $\sum_{k=1}^{m} |v_k|^2 \cos\phi_k$ is equivalent to minimizing $\mathbf{z}^*(G^* + G)\mathbf{z}$ with the constraints (4.26). Thus, finding the stabilizing boundary of loop phase perturbation is then equivalently converted to the constrained minimization problem as follows:

$$\min[\mathbf{z}^*(G^* + G)\mathbf{z}] \tag{4.27}$$

$$\text{s.t.} \begin{cases} \mathbf{z}^* \mathbf{z} = 1, \\ \mathbf{z}^*(H_k - G^* H_k G)\mathbf{z} = 0, \; k = 1, 2, \cdots, m. \end{cases}$$

On the contrary, if $\sum_{k=1}^{m} |v_k|^2 \cos\phi_k$ needs to be minimized, an equivalent constrained maximization framework can be constructed in a similar way. Here, we focus on the constrained minimization problem (4.27) only and omitted its counter part since the numerical algorithm proposed to solve both of them are completely the same.

With the approach of Lagrange multiplier [123], let

$$F(\kappa) = \mathbf{z}^*(G^* + G)\mathbf{z} + \lambda_1(\mathbf{z}^* \mathbf{z} - 1) + \sum_{k=1}^{m} \lambda_{k+1} \mathbf{z}^*(H_k - G^* H_k G)\mathbf{z},$$

where $\kappa = [z_1, \cdots, z_m, \lambda_1, \lambda_2, \cdots, \lambda_{m+1}]^T$. The constrained optimization problem (4.27) can be solved by finding zeros of the following function

$$f(\kappa) = \frac{\partial F(\kappa)}{\partial \kappa} = \begin{bmatrix} (G^* + G)\mathbf{z} + \lambda_1 \mathbf{z} + \sum_{k=1}^{m} \lambda_{k+1}(H_k - G^* H_k G)\mathbf{z} \\ \mathbf{z}^* \mathbf{z} - 1 \\ \mathbf{z}^*(H_1 - G^* H_1 G)\mathbf{z} \\ \vdots \\ \mathbf{z}^*(H_m - G^* H_m G)\mathbf{z} \end{bmatrix}, \tag{4.28}$$

Numerical solutions to (4.28) are obtained by the Newton-Raphson algorithm:

$$\kappa_{n+1} = \kappa_n - J^{-1}[f(\kappa_n)]f(\kappa_n),   \tag{4.29}$$

where

$$J[f(\kappa)] = \frac{\partial f(\kappa)}{\partial \kappa} = \frac{\partial^2 F(\kappa)}{\partial \kappa^2} =$$

$$\begin{bmatrix} (G^* + G) + \lambda_1 I_m \\ + \sum_{k=1}^{m} \lambda_{k+1}(H_k - G^* H_k G) & \mathbf{z} & (H_1 - G^* H_1 G)\mathbf{z} & \cdots & (H_m - G^* H_m G)\mathbf{z} \\ 2\mathbf{z}^* & 0 & 0 & \cdots & 0 \\ 2\mathbf{z}^*(H_1 - G^* H_1 G) & 0 & 0 & \cdots & 0 \\ \vdots & \vdots & \vdots & \ddots & \vdots \\ 2\mathbf{z}^*(H_m - G^* H_m G) & 0 & 0 & \cdots & 0 \end{bmatrix}$$

is the Jacobian matrix of $f(\kappa)$. If J is singular, then a Moore-Penrose inverse is used [124]. Once the iteration routine converges to a zero of $f(\kappa)$, the eigenvalues of the Hessian matrix

$$H = \frac{\partial^2 F}{\partial \mathbf{z}^2} = \left[(G^* + G) + \lambda_1 I_m + \sum_{k-1}^{m} \lambda_{k+1}(H_k - G^* H_k G) \right]   \tag{4.30}$$

is calculated to see whether it is a local minimum or maximum. For the local minimum (or maximum), a new initial search direction is chosen as the negative of the eigenvector of H corresponding to the most positive (or negative) eigenvalue to achieve the local maximum (or minimum). Since the cost function and the constraints in (4.27) are quadratic form of $\mathbf{z}$, the local minimum (or maximum) is also the global minimum (or maximum).

It should be pointed out here that $\mathbf{z}, \mathbf{v} \in \mathbb{C}^m$ will lead to the failure of solving the optimization problem (4.27) because neither the cost function nor the constraints are holomorphic functions of $\mathbf{z}$ or ω [125]. Fortunately, the standard technique of converting (4.27) to an equivalent real constrained optimization problem is applicable by the process of decomplexification, which makes use of a canonical isomorphism between $\mathbb{C}^m$ and $\mathbb{R}^{2m}$. Let $z_k = x_k + jy_k$, $x_k, y_k \in \mathbb{R}$, $k = 1, 2, \cdots, m$; $\mathbf{z_c} = [x_1, y_1, \cdots, x_m, y_m]^T \in \mathbb{R}^{2m}$; $G_{i,j} = x_{i,j} + jy_{i,j}$, $x_{i,j}, y_{i,j} \in \mathbb{R}$, $i, j = 1, 2, \cdots, m$; and

$$G_c = \begin{bmatrix} x_{1,1} & -y_{1,1} & \cdots & x_{1,m} & -y_{1,m} \\ y_{1,1} & x_{1,1} & \cdots & y_{1,m} & x_{1,m} \\ \vdots & \vdots & \ddots & \vdots & \vdots \\ x_{m,1} & -y_{m,1} & \cdots & x_{m,m} & -y_{m,m} \\ y_{m,1} & x_{m,1} & \cdots & y_{m,m} & x_{m,m} \end{bmatrix} \in \mathbb{R}^{2m \times 2m},$$

there holds $\mathbf{z}^*(G^*+G)\mathbf{z} = \mathbf{z_c}^T(G_c^T+G_c)\mathbf{z_c}$; $\mathbf{z}^*\mathbf{z} = \mathbf{z_c}^T\mathbf{z_c}$; $\mathbf{z}^*(H_k - G^*H_kG)\mathbf{z} = \mathbf{z_c}^T(H_k^c - G_c^TH_k^cG_c)\mathbf{z_c}$, $k = 1,2,\cdots,m$, where $H_k^c = [h_{i,j}] \in \mathbb{R}^{2m \times 2m}$ with

$$h_{i,j} = \begin{cases} 1, & i = j = 2k \text{ or } 2k-1; \\ 0, & \text{otherswise.} \end{cases}$$

Thus, the constrained optimization (4.27) in $\mathbb{C}^m$ is equivalent to the optimization problem in $\mathbb{R}^{2m}$ as follows:

$$\min[\mathbf{z_c}^T(G_c^T+G)\mathbf{z_c}] \tag{4.31}$$

$$\text{s.t.} \quad \begin{cases} \mathbf{z_c}^T\mathbf{z_c} = 1, \\ \mathbf{z_c}^T(H_k^c - G_c^TH_k^cG_c)\mathbf{z_c} = 0, k = 1,2,\cdots,m. \end{cases}$$

Newton-Raphson iteration algorithm is then used to calculate the stabilizing boundary of the diagonal phase perturbation. Once the boundary is obtained, hypercubes are ready to be prescribed and the loop phase margin can be easily determined according to Definition 4.1.

The algorithm to find loop phase margins is summarized as follows:

Step 1. Determine the frequency range Ω such that the solutions to (4.24) or (4.25) exist;
Step 2. Construct the framework of the constrained optimization (4.27), which is then converted equivalently to its isomorphism in real space as (4.31);
Step 3. For every $\omega \in \Omega$, solve (4.31) with Lagrange multiplier and find $\mathbf{z}$ by Newton-Raphson iteration (4.29);
Step 4. Use the similar procedures in Step 3 to solve maximum of (4.31) with different initial values;
Step 5. The points on the stabilizing boundary of loop phase margins are given by $\phi_i = \arg\{z_i/v_i\}$, $i = 1,2,\cdots,m$, and loop phase margins are hypercubes prescribed in the stabilizing region.

4.4.2 Illustration Examples

Example 4.2. Consider again the 2×2 system presented in Sect. 4.2, where

$$G(s) = \begin{bmatrix} \dfrac{2.5}{s+1} & \dfrac{1}{s+1} \\ \dfrac{3}{s+1} & \dfrac{4}{s+1} \end{bmatrix} \quad \text{and} \quad \Delta = \begin{bmatrix} e^{j\phi_1} & 0 \\ 0 & e^{j\phi_2} \end{bmatrix}.$$

Its state-space representation gives

$$A = \begin{bmatrix} -1 & 0 \\ 0 & -1 \end{bmatrix}, B = \begin{bmatrix} 2 & 0 \\ 0 & 2 \end{bmatrix}, C = \begin{bmatrix} 1.25 & 0.5 \\ 1.5 & 2 \end{bmatrix}.$$

Then,

$$\tilde{A} + \tilde{B}\tilde{C} = \begin{bmatrix} A & 0 \\ C^T C & -A^T \end{bmatrix} + \begin{bmatrix} B \\ 0 \end{bmatrix} \begin{bmatrix} 0 & -B^T \end{bmatrix} = \begin{bmatrix} -1 & 0 & -4 & 0 \\ 0 & -1 & 0 & -4 \\ 3.8125 & 3.625 & 1 & 0 \\ 3.625 & 4.25 & 0 & 1 \end{bmatrix},$$

whose eigenvalues are $\pm 0.7737i$ and $\pm 5.4453i$. Since only positive pure imaginary eigenvalues are need to be considered, we have $\omega_1 = 0.7737$ and $\omega_2 = 5.4453$, and $\sigma(G(j\omega_i)) = 1$, $i = 1, 2$. Thus, the frequency range $[0, +\infty)$ is divided into three intervals as $[0, 0.7737]$, $[0.7737, 5.4435]$, and $[5.4435, +\infty)$. Choose a frequency in every intervals and calculate the singular values of $G(j\omega)$ respectively, we can determine $\hat{\Omega}$ by checking whether $\underline{\sigma}(G(j\omega)) \leq 1 \leq \overline{\sigma}(G(j\omega))$ holds. For example, $\omega = 0.5$ yields $\underline{\sigma}(G(j\omega)) = 1.1309 > 1$, so $[0, 0.7737] \not\subseteq \hat{\Omega}$. $\omega = 6$ yields $\overline{\sigma}(G(j\omega)) = 0.9102$, so $[5.4453, +\infty) \not\subseteq \hat{\Omega}$. Only $\omega = 1$ yields $\underline{\sigma}(G(j\omega)) = 0.8940 < 1$ and $\overline{\sigma}(G(j\omega)) = 3.9148 > 1$. So, $\hat{\Omega} = [0.7737, 5.4453]$. For every $\omega \in \hat{\Omega}$, obtain the solution $\mathbf{z}$ to the constrained optimization (4.31) with the Newton-Raphson iteration (4.29), where initial values are arbitrarily chosen. However, (4.29) is convergent only for $\omega \in [0.9254, 1.7315] \cup [3.3547, 5.0396]$, shown as the solid lines in Fig. 4.5, which implies that $\Omega = [0.9254, 1.7315] \cup [3.3547, 5.0396]$. Similarly, the maximum of (4.31) is also found for all $\omega \in \Omega$, shown as the dashed lines in Fig. 4.5. One sees that the minimum and maximum loci constitute two closed contours. As the cost function moves along these contours, the pair (ϕ_1, ϕ_2) moves along their stabilizing boundary. After $\mathbf{z}$ is known, $\mathbf{v} = -G(j\omega)\mathbf{z}$ and $\phi_i = \arg\{z_i/v_i\}$, $i = 1, 2$, $\forall \omega \in \Omega$. With the help of Lemma 4.1, the stabilizing boundary for the pair (ϕ_1, ϕ_2) is obtained and shown in Fig. 4.6, where the solid and dashed lines correspond to the minimum and maximum loci, respectively, whose symmetric parts with respect to the origin are presented by the dotted lines.

Comparing Fig. 4.6 with Fig. 4.3, one sees that the stabilizing regions are the same. As shown in Sect. 4.2, loop phase margins are not unique. A reasonable one can be determined in the following way. Refer to Fig. 4.5, when $\omega \in [0.9254, 1.7315]$, it follows from Fig. 4.6 that $\phi_1 \in [-2.4960, -1.9509]$ and $\phi_2 \in (-\pi, \pi)$ for (ϕ_1, ϕ_2) on the stabilizing boundary. From the symmetry in Lemma 4.1, the stability of the closed-loop system requires $\phi_1 \in (-1.9505, 1.9505)$ if ϕ_2 is allowed to vary arbitrarily in $(-\pi, \pi)$. Likewise, when $\omega \in [3.3547, 5.0396]$, the stabilizing boundary yields $\phi_2 \in [-2.0442, -1.5783]$ and $\phi_1 \in (-\pi, \pi)$. Closed-loop system stability requires $\phi_2 \in (-1.5783, 1.5783)$. Let $\text{PM} = \{(\phi_1, \phi_2) | \phi_1 \in (-1.9505, 1.9505) \text{ and } \phi_2 \in (-1.5783, 1.5783)\}$, it is clear that PM is a rectangle prescribed in Ω, i.e., $\text{PM} \subseteq \Omega$. According to Definition 4.1, $(-1.9505, 1.9505)$ and $(-1.5783, 1.5783)$ are phase margins for loop 1 and 2, respectively. The common phase margin can be obtained by $(-1.9505, 1.9505) \cap (-1.5783, 1.5783) = (-1.5783, 1.5783)$.

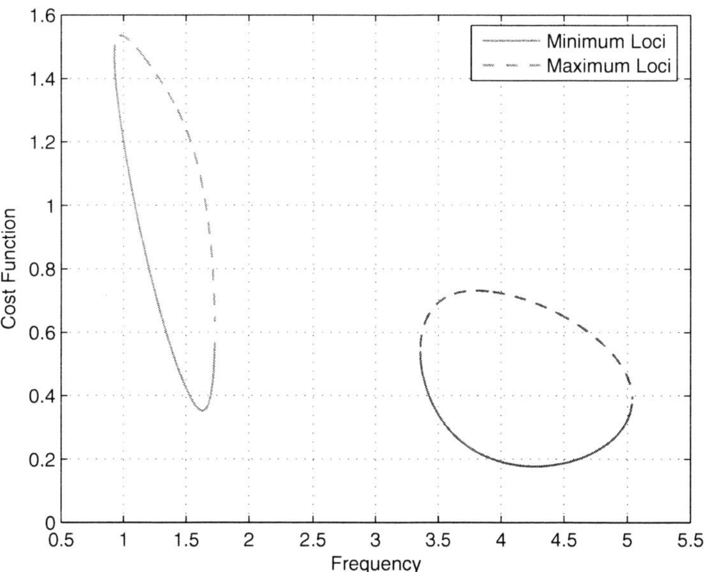

Fig. 4.5. Solving the constrained optimization for $\omega \in \Omega$

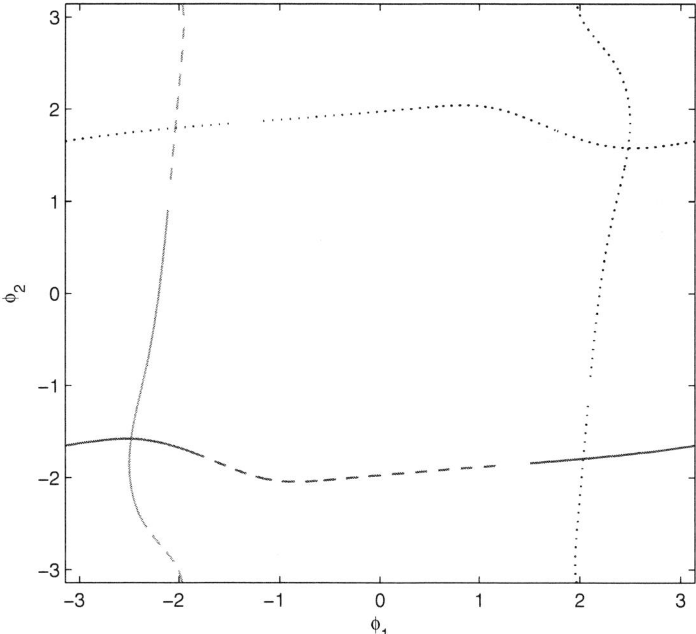

Fig. 4.6. Stabilization Region of (ϕ_1, ϕ_2)

Example 4.3 ([107]). Consider the system (A,B,C) as follows

$$
A = \begin{bmatrix} 0 & 1 & 0 & 0 \\ -3 & -0.75 & 1 & 0.25 \\ 0 & 0 & 0 & 1 \\ 4 & 1 & -4 & -1 \end{bmatrix}, \ B = \begin{bmatrix} 0 & 0 \\ 0 & 1 \\ 0 & 0 \\ 0.25 & 0 \end{bmatrix}, \ C = \begin{bmatrix} 1 & 0 & 0 & 0 \\ 0 & 0 & 1 & 0 \end{bmatrix}.
$$

Its transfer function matrix is

$$
\begin{aligned}
G(s) \ &= \ C(sI - A)^{-1}B \\[2mm]
&= \ \frac{1}{s^4 + 1.75s^3 + 7.5s^2 + 4s + 8} \begin{bmatrix} 0.0625s + 0.25 & s^2 + s + 4 \\ 0.25s^2 + 0.1875s + 0.75 & s + 4 \end{bmatrix}.
\end{aligned}
$$

The pure imaginary eigenvalues of $\tilde{A} + \tilde{B}\tilde{C}$ are $\pm 0.643\mathrm{i}$ and $\pm 1.613\mathrm{i}$. Thus, the frequency range $[0, +\infty)$ is divided into three intervals as $[0, 0.643]$, $[0.643, 1.613]$, and $[1.613, +\infty)$. By checking the holding of $\underline{\sigma}(G(j\omega)) \leq 1 \leq \bar{\sigma}(G(j\omega))$ for any given ω in these intervals, $\hat{\Omega} = (0.643, 1.613)$. For every $\omega \in \hat{\Omega}$, obtain the solution $\mathbf{z}$ to the constrained optimization (4.31) with the Newton-Raphson iteration (4.29), where initial values are arbitrarily chosen. The frequency range for the convergence of (4.29) yields $\Omega = (0.764, 0.884) \cup (1.533, 1.572)$, shown as Fig. 4.7, where the solid and dashed lines represent the minimum and maximum loci of the cost function, respectively, who constitute two closed contours, denoted by A and B. As the cost function moves along contour A (or B), the pair (ϕ_1, ϕ_2) moves along their stabilizing boundary A (or B) correspondingly, shown as Fig. 4.8, where the dotted lines are determined by the symmetry with respect to $(0, 0)$ from Lemma 4.1.

It needs to be clarified that two (or more) boundaries may be obtained in the limited range $[-\pi, \pi)$ like this example shows, which is different from the case in Example 1 that only one boundary can be found. If multiple boundaries exist, the stabilizing region of ϕ_i is the polytope encompassed by the nearest boundary (Boundary B for this example). By comparing Fig. 4.8 with Fig. 4.6, one sees that ϕ_1 is allowed to vary arbitrarily in $[-\pi, \pi)$ in this example if $|\phi_2|$ is less than some value, in another word, there is no stabilizing boundary for ϕ_1 if $|\phi_2|$ is small enough, which is another difference from the case in Example 1. This is because one of characteristic loci of $G(s)$ always lies in the unit circle and never goes through the critical point $(-1, j0)$.

To show the stabilizing region of (ϕ_1, ϕ_2) for this example more clearly, zoom Boundary B in Fig. 4.9, where $\phi_1 \in [-\pi, \pi)$ and $\phi_2 \in [-0.2423, 0.2423]$ for $\forall \omega \in (1.533, 1.572)$. With the help of Lemma 4.1, $\phi_1 \in (-\pi, \pi)$ and $\phi_2 \in (-0.2423, 0.2423)$ are one of the phase margins for loop 1 and 2, respectively. Since there is no boundary for ϕ_1, the common phase margin for this example can be determined by letting $\phi_1 = \phi_2$, which is $(-0.3108, 0.3108)$, or $(-17.808°, 17.808°)$.

Table 4.1 listed some comparison of the proposed method with the existing frequency domain and time domain methods of [107] and [126]. The method of [107] actually gives the common phase margin only, which is $(-0.2701, 0.2701)$, or $(-15.476°, 15.476°)$, and smaller than the result given by the proposed method. This is because

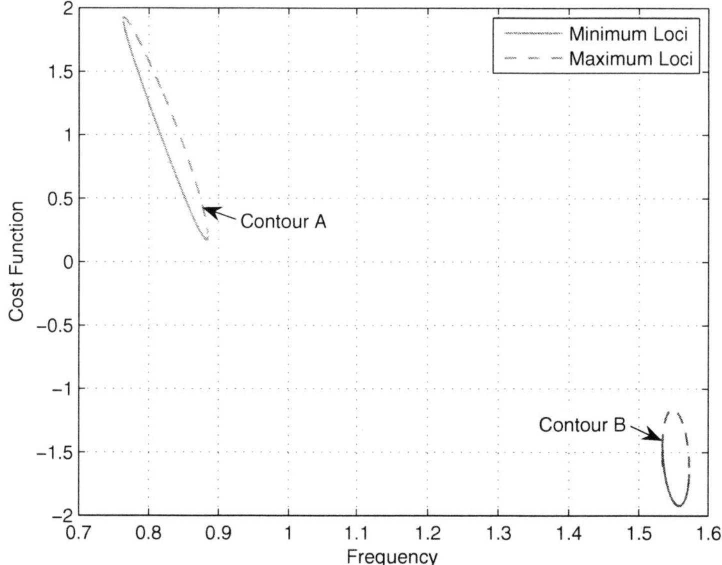

Fig. 4.7. Solving the constrained optimization for $\omega \in \Omega$

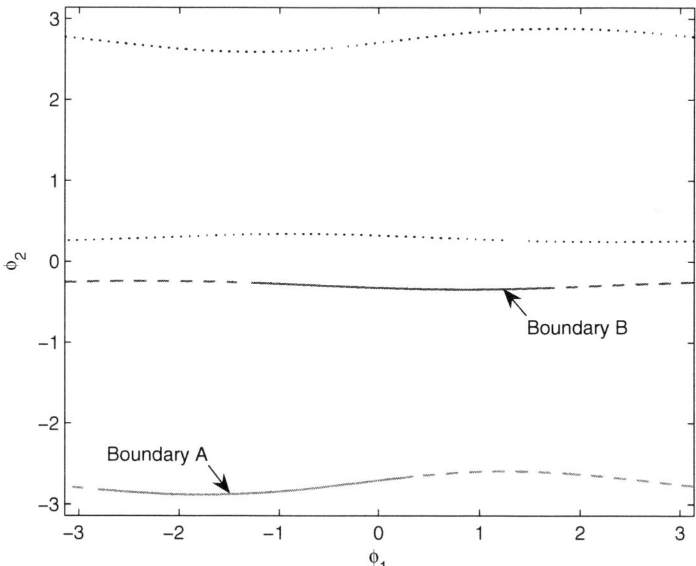

Fig. 4.8. Stabilization boundaries for (ϕ_1, ϕ_2)

only the decentralized control is considered here and the phase perturbations are not necessarily ergodic in the entire set of unitary matrices. The proposed method also gives a larger loop and common phase margins than the method of [126] does, which

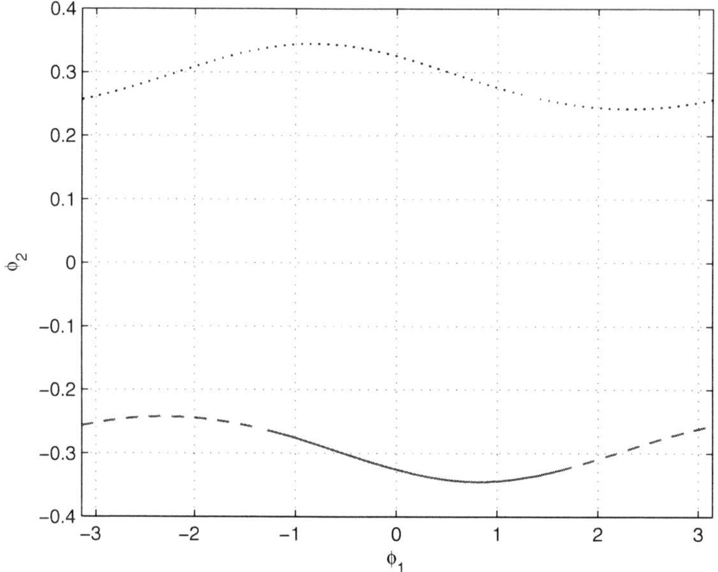

Fig. 4.9. Stabilization Region of (ϕ_1, ϕ_2)

Table 4.1. Comparison with other methods

Method	Loop Phase Margin		Common Margin
	ϕ_1	ϕ_2	
Frequency domain	$(-\pi, \pi)$	$(-0.2423, 0.2423)$	$(-0.3108, 0.3108)$
Bar-on and Jonckheere [107]	—	—	$(-0.2701, 0.2701)$
Time domain	$[0, 0.1878]$	$[0, 0.1231]$	$[0, 0.1265]$

shows that the proposed method evidently improves the LMI results by reducing the possible conservativeness.

4.5 Conclusion

In this chapter, the loop phase margins of multivariable control systems are defined as the allowable individual loop phase perturbations within which stability of the closed-loop system is guaranteed. Two methods in time and frequency domain respectively are proposed to obtain the loop phase margins. Time domain method is composed of two steps. Firstly, delay-dependent stability criteria for systems with multiple delays are proposed to establish an algorithm to calculate the ranges of delays guaranteeing the stability of closed-loop system. Then, a fixed frequency is determined to convert the stabilizing ranges of time delays into the respective loop phase margins. Frequency domain method is presented to accurately computing these phase margins, which is converted using the Nyquist stability analysis to the problem of some simple constrained optimization with the help of unitary mapping between two complex vector space. Numerical solutions are then found with the Lagrange multiplier and Newton-Raphson iteration algorithm. Comparing with time domain method, frequency domain method can provide exact margins and thus improves the LMI results by reducing the possible conservativeness. Finally, numerical examples are given for illustration of the effectiveness of both methods.

5 Multi-loop PID Control Based on dRI Analysis

Chapters 2–4 mainly focus on multivariable system analysis. From this chapter onwards, several design methods for multi-loop/multivariable PID control are presented. Based on the loop paring criterion proposed in Chapter 2, this chapter goes further to seek a simple but effective design method for decentralized PID controllers.

5.1 Introduction

Despite the rapid development of advanced process control techniques, decentralized PID control is still the most commonly adopted method in the process control industries for controlling multivariable processes. The main reasons for such popularity are that PID controllers are easily understandable by control engineers and the decentralized PID controllers require fewer parameters to tune than the multivariable controllers. Another advantage of the decentralized PID controllers is that loop failure tolerance of the resulting closed-loop control system can be guaranteed at the design stage. Even though the design and tuning of single loop PID controllers have been extensively researched [127,128,2,129], they cannot be directly applied to the design of decentralized control systems because of the existence of interactions among control loops.

Many methods had been proposed to extend SISO PID tuning rules to decentralized control by compensating foe the effects of loop interactions. A common way is first to an design individual controller for each control loop by ignoring all interactions and then to detune each loop by a detuning factor. Luyben proposed the biggest log modulus tuning (BLT) method for multiloop PI controllers [20,130]. In the BLT method, the well know Ziegler-Nichols rule is modified with the inclusion of a detuning factor, which determines the tradeoff between stability and performance of the system. Similar methods have also been addressed by Chien et al. [131]. There, the designed PID controllers for the diagonal elements are detunned according to the relative gain array values. Despite simple computations involved, the design regards interactions as elements obstructing the system stability and attempts to dispose of them rather than control them to increase the speed of the individual control loops. It is hence too conservative to exploit process structures and characteristics for the best achievable performance.

Q.-G. Wang et al.: PID Control for Multivariable Processes, LNCIS 373, pp. 97–130, 2008.
springerlink.com
© Springer-Verlag Berlin Heidelberg 2008

Another strategy is to simultaneously consider loop interactions when designing individual control loop. In the sequential design method [132,133], by taking interactions from the closed loops into account in a sequential fashion, multiple single-loop design strategies can be directly employed. The main drawback of this method is that the design must proceed in a very ad hoc manner. Design decisions are based on loops that have already closed which may have deleterious effects on the behavior of the remaining loops. The interactions are well taken care of only if the loops are of considerably different bandwidths and the closing sequence starts from the fastest loop. Lee et al. [134, 135] proposed an analytical method for multiloop PID controller design by using the frequency-dependent properties of the closed-loop interactions and the generalized internal model control (IMC)-PID tuning rule for SISO systems. The proportional and derivative terms are designed simply by neglecting the off-diagonal elements, whereas the integral term is designed by taking the off-diagonal elements into consideration. Bao et al. [136] formulated the multi-loop design as a nonlinear optimization problem with matrix inequality constraints. As has been illustrated, the formulation does not include the systems that have different input delays, which happens to be very common in MIMO processes. Using a genetic algorithm to search the optimal settings was also proposed [137]. However, the results are very much dependent on the conditions defined in the objective function, and the controllers may result in an unstable control system, such as in the case of loop failure or where loops are closed in different orders.

In recent years, the new trend in a designing decentralized control system for multivariable processes is to handle the loop interactions first and then apply SISO PID tuning rules. The independent design methods have been used by several authors, in which each controller is designed based on the paired transfer functions while satisfying some constraints due to the loop interactions [138,139]. The constraints imposed on the individual loops are given by criteria such as the μ-interaction measure [131] and the Gershgorin bands [140]. Usually, stability and failure tolerance are automatically satisfied. Since the detailed information on controller dynamics in other loops is not used, the resulting performance may be poor [132]. In the trial-and-error method [141], Lee et al. extended the iterative continuous cycling method for SISO systems to decentralized PI controller tuning. It refined the Nyquist array method to provide less conservative stability conditions, and ultimate gains for decentralized tuning are then determined. The main disadvantages are due to not only the need for successive experiments but also the weak tie between the tuning procedure and the loop performance. To overcome the difficulty of controllers interacting with each other, Wang et al. [142] used a modified Ziegler-Nichols method to determine the controller parameters that will give specified gain margins. Although it presents an interesting approach, the design of a multi-loop controller by simultaneously solving a set of equations is numerically difficult. Huang et al. [143] formulated the effective transmission in each equivalent loop as the effective open-loop process (EOP), the design of controllers can then be carried out without referring to the controller dynamics of other loops. However, for high dimensional processes, the calculation of EOPs is complex, and the controllers have to be conservative for the inevitable modeling errors encountered in the formulation.

Since the controllers interact with each other, the performance of one loop cannot be evaluated without knowing the controllers in the other loops. Even though many researchers have tried to overcome this difficulty from different angles, the success has been largely limited to the low dimensional (less than 3×3) processes. To provide a unified approach for the synthesis and design of decentralized control systems for general multivariable processes, a novel loop-pairing criterion based on the generalized interaction has recently been proposed [66]. Through defining the decomposed relative interaction array (DRIA) to evaluate the loop-by-loop interactions, the control structure configuration is uniquely determined by searching the minimal loop interaction energy. Furthermore, by applying the left-right factorization to the DRIA, the relative interaction (RI) between loops can be represented by elements summation of the decomposed relative interaction sequence. The maximum interactions under different combinations are determined by the maximum values of this sequence according to a failure index. Consequently, the necessary and sufficient conditions for decentralized closed-loop integrity of an individual loop under both single-loop and multiple-loop failures have been provided [144]. These results have laid a unified framework for the synthesis and design of decentralized control systems of general multivariable processes.

This chapter extends the work of [66, 144] to develop a simple yet effective method for designing a decentralized PID controller based on dynamic interaction analysis. By implementation of a controller to each individual diagonal control loop using the SISO PID tuning rules, the dynamic relative interaction (dRI) to an individual control loop from all other loops is estimated. With the obtained dRI, the multiplicated model factor (MMF) is calculated and approximated by a time delay function at the neighborhood of the critical frequency to construct the equivalent transfer function for individual control loop. Subsequently, an algorithm for designing appropriate controller settings for the decentralized PID controllers is provided. The proposed method is very simple and effective and easily implemented especially for higher dimensional processes. 2×2 and 4×4 processes are used to demonstrate the design procedures. Examples for a variety of 2×2, 3×3 and 4×4 systems are used to demonstrate that the overall control system performance is much better than that of other tuning methods, such as the BLT method [20, 130], the trial and error method [141], and the independent design method based on Nyquist stability analysis [140], especially for higher dimensional processes.

5.2 Preliminaries

Consider an $n \times n$ system with a decentralized feedback control structure as shown in Fig. 5.1, where, r, u and y are vectors of references, inputs and outputs respectively, $G(s) = [g_{ij}(s)]_{n \times n}$ is the system's transfer function matrix and its individual element, $g_{ij}(s)$, described a second-order plus dead-time (SOPDT) model:

$$g_{ij}(s) = \frac{k_{ij}e^{-\theta_{ij}s}}{(\tau_{ij}s + 1)(\tau'_{ij}s + 1)}, \quad \tau'_{ij} \leq \tau_{ij}. \tag{5.1}$$

and controller $C(s) = \text{diag}\{c_1(s), \cdots, c_n(s)\}$ is the decentralized PID type with its individual element given in parallel form as

$$c_i(s) = k_{Pi}(1 + \frac{1}{\tau_{Ii}s} + \tau_{Di}s), \qquad (5.2)$$

or in series form as

$$c_i(s) = k_{Pi}(1 + \frac{1}{\tau_{Ii}s})(\tau_{Di}s + 1). \qquad (5.3)$$

It is assumed that $G(s)$ has been arranged so that the pairings of the inputs and outputs in the decentralized feedback system correspond to the diagonal elements of $G(s)$.

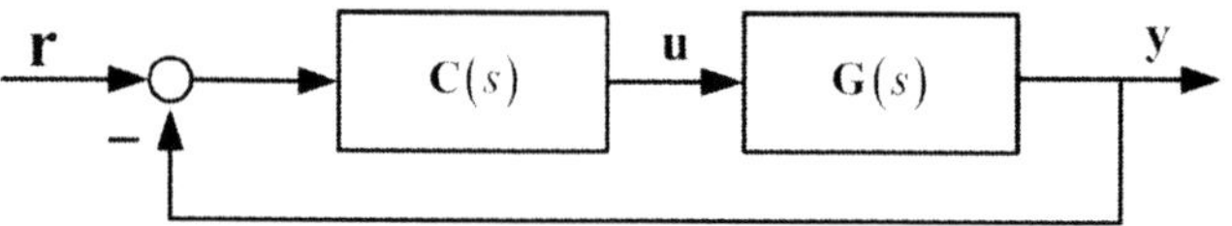

Fig. 5.1. General decentralized control system

For MIMO process, when one controller, $c_i(s)$, acting in response to the setpoint change and/or the output disturbance, it affects the overall system through the off-diagonal elements of $G(s)$, forcing other controllers to take actions, as well, these controllers reversely influence the ith loop via other off-diagonal elements, and this interacting processes among control loops continue throughout the whole transient until a steady state is reached. To examine the transmittance of interactions between an individual control loop and the others, the decentralized control system can be structurally decomposed into n individual SISO control loops with the coupling among all loops explicitly exposed and embedded in each loop. Figure 5.2 shows the structure of an arbitrary control loop y_i–u_i after the structural decomposition.

In Fig. 5.2, the interaction to an individual control loop y_i–u_i from the other $n-1$ control loops is represented by the RI $\phi_{ii,n-1}(s)$, and the equivalent transfer function of an individual control loop y_i–u_i, denoted by $\hat{g}_{ii}(s)$, can be obtained in terms of $\phi_{ii,n-1}(s)$ by

$$\hat{g}_{ii}(s) = g_{ii}(s)\rho_{ii,n-1}(s), \qquad (5.4)$$

with

$$\rho_{ii,n-1}(s) = 1 + \phi_{ii,n-1}(s), \qquad (5.5)$$

where $\rho_{ii,n-1}(s)$ is defined as multiplicate model factor (MMF) to indicate the model change of an individual control loop y_i–u_i after the other $n-1$ control loops are closed. Once the RI, $\phi_{ii,n-1}(s)$, is available, the corresponding MMF, $\rho_{ii,n-1}(s)$, and the equivalent process transfer function $\hat{g}_{ii}$ can be obtained, which can be directly used to independently design the PID controller for each individual control loop y_i–u_i.

In the following development, we use RI as a basic interaction measure to investigate the interactions among control loops and derive equivalent transfer function for each loop. Sometimes, we will omit the Laplace operator s for simplicity unless otherwise specified.

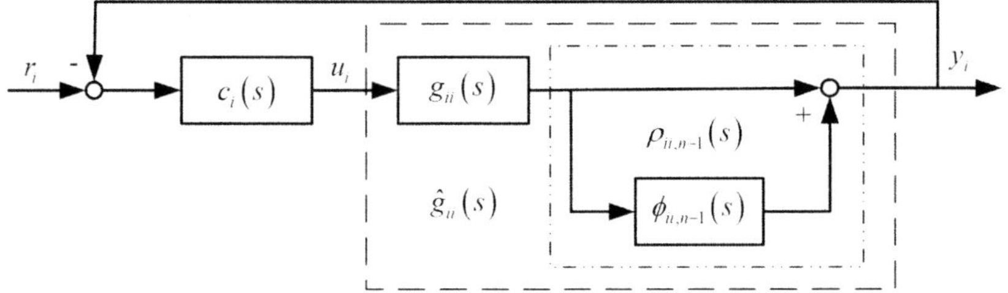

Fig. 5.2. Structure of loop y_i–u_i by structural decomposition

The RI for control loop y_i–u_j is defined as the ratio of two elements [60]: the increment of the process gain after all other control loops are closed, and the apparent gain in the same loop when all other control loops are open, that is

$$\phi_{ij,n-1} = \frac{(\partial y_i/u_j)_{y_{k\neq i}\text{constant}} - (\partial y_i/u_j)_{u_{l\neq j}\text{constant}}}{(\partial y_i/u_j)_{u_{l\neq j}\text{constant}}} \qquad k,l = 1,\cdots,n.$$

Since the RI cannot offer effective measure on the reverse effect of individual control loop and loop-by-loop interactions, He and Cai decomposed the RI as the elements summation of decomposed relative interaction array (DRIA) to give important insights into the cause-effects of loop interactions [66]. However, the obtained results are limited to the steady state, which are less useful for controller design than the dynamic representations. Hence, it is necessary to derive the dynamic interaction among control loops represented explicitly by the process models and controllers.

5.3   Dynamic Relative Interaction

Because the dynamic interactions among control loops are controller dependent [145, 146], appropriate controllers have to be designed and implemented into the control system for investigation of the dynamic interactions. For an arbitrarily decentralized PID control system of a multivariable process, we can redraw Fig. 5.1 as Fig. 5.3 for the convenience of analyzing the interactions between an arbitrary control loop y_i–u_i and the others, where $\underline{y}_i$ is a vector indicating the effects of u_i on other outputs while $\bar{y}_i$ indicates the reverse effect of y_i by all of the other closed control loops; r^i, u^i, y^i and $C^i(s)$ indicate r, u, y and $C(s)$ with their ith elements, r_i, u_i, y_i and $c_i(s)$, removed, respectively.

Because the dRIs are input independent, without loss of generality, the references of the other $n-1$ control loops are set as constants, i.e.

$$\frac{\mathrm{d}r_k}{\mathrm{d}t} = 0 \qquad \text{or} \qquad r_k(s) = 0, \qquad k = 1,\cdots,n; \; k \neq i,$$

in the analysis of the dynamic interaction between control loop y_i–u_i and the other controlled closed-loops. Then, we have

$$y^i = G^{ii}u^i + \underline{y}_i, \tag{5.6}$$

$$u^i = -C^i y^i, \tag{5.7}$$

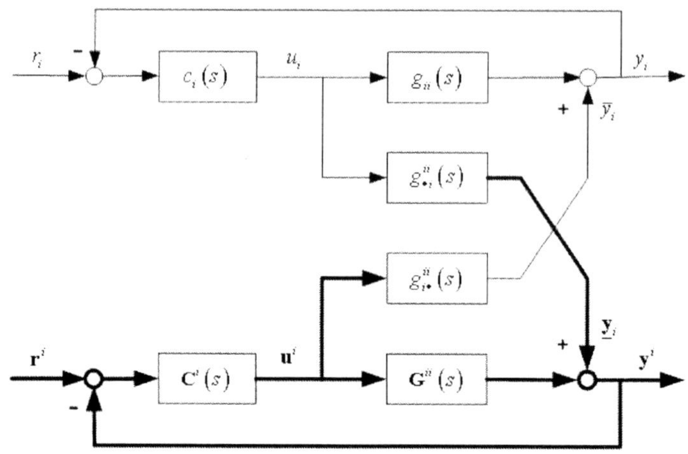

Fig. 5.3. Closed-loop system with control loop y_i–u_i presented explicitly

and

$$\underline{y}_i = g^{ii}_{*i} u_i, \tag{5.8}$$

$$\bar{y}_i = g^{ii}_{i*} u^i, \tag{5.9}$$

where G^{ii} is the transfer function matrix G with its ith row and the ith column removed, and g^{ii}_{*i} and g^{ii}_{i*} indicate the ith row and the ith column of G with the iith element, g_{ii}, removed, respectively.

Combining (5.6) and (5.7), we can write

$$u^i = -(\bar{G}^{ii})^{-1} \underline{y}_i, \tag{5.10}$$

where

$$\bar{G} = G + C^{-1}. \tag{5.11}$$

Furthermore, $\bar{y}_i$ in (5.9) can be represented by the summation of the following row vector

$$\bar{y}_i = \left\| \begin{bmatrix} g_{i1}u_1 & \cdots & g_{ik}u_k & \cdots & g_{in}u_n \end{bmatrix} \right\|_\Sigma, \qquad k = 1, \cdots, n; \; k \neq i. \tag{5.12}$$

where $\|A\|_\Sigma$ is the summation of all elements in a matrix A.

Using (5.8)–(5.12), we obtain

$$\begin{aligned}
\bar{y}_i &= \left\| -g^{ii}_{i*} (\bar{G}^{ii})^{-1} g^{ii}_{*i} \right\|_\Sigma u_i \\
&= \left\| -g^{ii}_{*i} g^{ii}_{i*} \otimes (\bar{G}^{ii})^{-T} \right\|_\Sigma u_i \\
&= \left\| -\frac{g^{ii}_{*i} g^{ii}_{i*}}{g_{ii}} \otimes (\bar{G}^{ii})^{-T} \right\|_\Sigma g_{ii} u_i.
\end{aligned}$$

Consequently, those steady state relationships provided in [42] can be extended as follows.

Define

$$\triangle G_{ii,n-1} = -\frac{1}{g_{ii}} g_{*i}^{ii} g_{i*}^{ii}$$

as the incremental process gain matrix of subsystem G^{ii} when control loop y_i–u_i is closed, then the dynamic DRIA (dDRIA) of individual control loop y_i–u_i in an $n \times n$ system can be described as

$$\Psi_{ii,n-1} = \triangle G_{ii,n-1} \otimes (\bar{G}^{ii})^{-T}, \tag{5.13}$$

and the dRI, $\phi_{ii,n-1}$, is the summation of all elements of $\Psi_{ii,n-1}$, i.e.

$$\phi_{ii,n-1} = \|\Psi_{ii,n-1}\|_\Sigma = \left\|\triangle G_{ii,n-1} \otimes (\bar{G}^{ii})^{-T}\right\|_\Sigma, \tag{5.14}$$

where $\otimes$ is the Hadamard product and $(G^{ij})^{-T}$ is the transpose of the inverse of matrix G^{ij}.

From (5.11), $\bar{G}$ can be factorized as

$$\bar{G} = \begin{bmatrix} g_{11}+1/c_1 & g_{12} & \cdots & g_{1n} \\ g_{21} & g_{22}+1/c_2 & \cdots & g_{2n} \\ \vdots & \vdots & \ddots & \vdots \\ g_{n1} & g_{n2} & \cdots & g_{nn}+1/c_n \end{bmatrix}$$

$$= G \otimes P,$$

where

$$P = \begin{bmatrix} \dfrac{1+g_{11}c_1}{g_{11}c_1} & 1 & \cdots & 1 \\ 1 & \dfrac{1+g_{22}c_2}{g_{22}c_2} & \cdots & 1 \\ \vdots & \vdots & \ddots & \vdots \\ 1 & 1 & \cdots & \dfrac{1+g_{nn}c_n}{g_{nn}c_n} \end{bmatrix}. \tag{5.15}$$

Hence, the dDRIA and dRI can be obtained respectively by,

$$\Psi_{ii,n-1} = \triangle G_{ii,n-1} \otimes (G^{ii} \otimes P^{ii})^{-T}, \tag{5.16}$$

and

$$\phi_{ii,n-1} = \left\|\triangle G_{ii,n-1} \otimes (G^{ii} \otimes P^{ii})^{-T}\right\|_\Sigma. \tag{5.17}$$

Remark 5.1. In dDRIA and dRI of (5.16) and (5.17), there exists an additional matrix P, which explicitly reveals interactions to an arbitrary loop by all the other $n-1$ closed loops under band-limited control conditions. Thus, for the given decentralized PID controllers, the dynamic interaction among control loops at an arbitrary frequency can be easily investigated through the matrix P, moreover, the calculation remains simple even for high dimensional processes.

Remark 5.2. For decentralized PI or PID control, we have

$$C^{-1}(j0) = \mathbf{0},$$

and

$$\bar{G}(j0) = G(j0)$$

which implies that the RI and the dRI are equivalent at steady state. However, because the dRI measure interactions under practical unperfect control conditions at some specified frequency points, it is more accurate in estimating the dynamic loop interactions and more effective in designing decentralized controllers.

The significance of above development are as follows:

(i) The interaction to an individual control loop from the other loops is derived in matrix form, and the relationship between RI and DRIA is extended to the whole frequency domain from the steady state;

(ii) Equations (5.13) and (5.14) indicate that the dRI is a combination of the interactions to individual control loops from the others; there may exist cause-effect cancelation among them such that selecting of loop pairings by dRI may be inaccurate [66].

(iii) Because interactions among control loops are controllers dependent, the dDRIA represents how the decentralized controllers interact with each other, while the dRI provides the overall effect to an individual control loop from the others;

(iv) The dRI can be obtained easily once process transfer function elements and controllers are available, which is not limited by the system dimension.

5.4 SIMC Based Design

5.4.1 Determination of MMF

The proposed method of designing decentralized controllers for multivariable processes involves three main steps:

(i) the design of individual controllers by ignoring the loop interactions;

(ii) the determination of the equivalent transfer function for each individual loop through use of the dRI;

(iii) the fine-tuning of each controller parameter based on the equivalent transfer function.

Because the interactions among control loops are ignored in the first step, all available SISO PID controller design techniques can be applied to the design of the initial individual controllers for the diagonal elements. Thus, according to the expected control performance, one can select the most suitable tuning rules, such as Ziegler and Nichols tuning [127], IMC tuning [128], and some other optimal design methods [147]. Obviously, different tuning rules provide different controller settings and, correspondingly, different estimations of loop interactions. Even these different estimations will affect the decentralized controllers designed in the third step; they have no influence on the design mechanism of our method. Because our main objective in this chapter is to propose a much simpler method for designing decentralized PID controller based on the dynamic interaction analysis, the SIMC-PID tuning rule, which is derived from IMC theory, is applied in the following development because of its very simple and better tradeoff between disturbance response and robustness, especially for the lag-dominant processes ($\tau_{ii} \gg \theta$ in (5.1)) [129].

For the SOPDT SISO system given by (5.1), the SIMC-PID settings in the form of (5.3) was suggested as

$$k_{Pi} = \frac{1}{k_{ii}} \frac{\tau_{ii}}{\tau_{Ci} + \theta_{ii}}, \tag{5.18}$$

$$\tau_{Ii} = \min\{\tau_{ii}, 4(\tau_{Ci} + \theta_{ii})\}, \tag{5.19}$$

$$\tau_{Di} = \tau'_{ii}, \tag{5.20}$$

where, τ_{Ci} is the desired closed-loop time constant. It is recommended to select the value of τ_{Ci} as θ_{ii} for a tradeoff between:

(i) the fast speed of the response and good disturbance rejection (favored by a small value of τ_{Ci});
(ii) the stability, robustness, and small input variation (favored by a large value of τ_{Ci}).

Let $\tilde{G} = \mathrm{diag}\{G\}$, the initial controllers $\tilde{C}$ can be designed for $\tilde{G}$ through the SIMC-PID tuning rules, such that

$$\tilde{G}\tilde{C} = \begin{bmatrix} g_{11}\tilde{c}_1 & 0 & \cdots & 0 \\ 0 & g_{22}\tilde{c}_2 & \cdots & 0 \\ \vdots & \vdots & \ddots & \vdots \\ 0 & 0 & \cdots & g_{nn}\tilde{c}_n \end{bmatrix},$$

with the open-loop transfer function for each individual control loop is

$$g_{ii}\tilde{c}_i = \frac{1}{(\tilde{\tau}_{Ci} + \theta_{ii})s} e^{-\theta_{ii}s}. \tag{5.21}$$

Using first order Taylor series expansion for (5.21), the closed-loop transfer function, open-loop gain crossover frequency ω_{ci} and matrix P, can be obtained respectively by,

$$\frac{y_i}{r_i} = \frac{1}{\tilde{\tau}_{Ci}s + 1} e^{-\theta_{ii}s},$$

$$\tilde{\omega}_{ci} = \frac{1}{\tilde{\tau}_{Ci} + \theta_{ii}}, \tag{5.22}$$

and

$$P = \begin{bmatrix} (\tilde{\tau}_{C1}s + 1)e^{\theta_{11}s} & 1 & \cdots & 1 \\ 1 & (\tilde{\tau}_{C2}s + 1)e^{\theta_{22}s} & \cdots & 1 \\ \vdots & \vdots & \ddots & \vdots \\ 1 & 1 & \cdots & (\tilde{\tau}_{Cn}s + 1)e^{\theta_{nn}s} \end{bmatrix}. \tag{5.23}$$

Because each controller is designed around the critical frequency of its transfer function, the dRI of individual control loop y_i–u_i can be estimated at the critical frequency $j\omega_{cri}$

$$\phi_{ii,n-1}(j\omega_{cri}) = \left\| \triangle G_{ii,n-1}(j\omega_{cri}) \otimes (G^{ii}(j\omega_{cri}) \otimes P^{ii}(j\omega_{cri}))^{-T} \right\|_{\Sigma}. \tag{5.24}$$

For interactive multivariable process G, it is desirable to have the same open-loop transfer function as multivariable control system $\tilde{G}\tilde{C}$ [45]. Because controller $\tilde{C}$ is designed for the diagonal elements of G without considering the couplings between control loops, the designed controller should be detunned by

$$c_i = \frac{\tilde{c}_i}{\rho_{ii,n-1}} = \frac{\tilde{c}_i}{1 + \phi_{ii,n-1}} \tag{5.25}$$

to result in approximately the same closed-loop control performance. However, it is impractical to use $\rho_{ii,n-1}$ directly to fine-tune the controller because it has different values at different frequencies. To solve this problem, one possible way is to use some appropriate transfer functions to identify those MMFs. Because at the neighborhood of the critical point the transmission interaction can be considered as a linear function, it is reasonable to represent the MMF by a low-order transfer function involving the first two items of its Taylor series, which can be further simplified for controller design by a pure time-delay transfer function as

$$\rho_{ii,n-1} = k_{\rho i,n-1} e^{-\theta_{\rho i,n-1}s}, \tag{5.26}$$

where,

$$k_{\rho i,n-1} = |\rho_{ii,n-1}(j\omega_{cri})| = |1 + \phi_{ii,n-1}(j\omega_{cri})| \tag{5.27}$$

and

$$\theta_{\rho i,n-1} = -\frac{\arg(\rho_{ii,n-1}(j\omega_{cri}))}{\omega_{cri}} = -\frac{\arg(1 + \phi_{ii,n-1}(j\omega_{cri}))}{\omega_{cri}}, \tag{5.28}$$

with ω_{cri} indicating the critical frequency of the ith control loop y_i–u_i. Obviously, the positive/negative $\theta_{\rho i,n-1}$ indicates that the interactions from other control loops bring a phase lag/lead of individual control loops.

From (5.1) and (5.4), the equivalent transfer function for loop y_i–u_i can be expressed by

$$\hat{g}_{ii}(s) = \frac{f_{ki}k_{ii}}{(\tau_{ii}s + 1)(\tau'_{ii}s + 1)} e^{-f_{\theta i}\theta_{ii}s}, \tag{5.29}$$

where,

$$f_{ki} = \max\{1, k_{\rho i,n-1}\}, \tag{5.30}$$

and

$$f_{\theta i} = \max\left\{1, 1 + \frac{\theta_{\rho i,n-1}}{\theta_{ii}}\right\}. \tag{5.31}$$

Remark 5.3. The open-loop transfer function for each individual control loop given by (5.21) is determined by the SIMC-PID tuning settings. If other tuning rules are applied, some different forms may be derived, and $g_{ii}\tilde{c}_i$ always has the form

$$g_{ii}\tilde{c}_i = \tilde{h}(s)\frac{ke^{-\theta s}}{s},$$

where, $\tilde{h}(s)$ is proper and rational with $\tilde{h}(j0) = 0$ [147].

Remark 5.4. Because the region between $\tilde{\omega}_{ci}$ ($|g_{ii}\tilde{c}_i| = 1$) and $\tilde{\omega}_{180}$ ($\arg(g_{ii}\tilde{c}_i) = -180°$) is most critical for individual control loop design, the crossover frequency of $g_{ii}\tilde{c}_i$ can be adopted as the critical frequency point for determining the dRI to the particular loop $y_i–u_i$ to obtain $\hat{g}_{ii}$ [48].

Remark 5.5. In (5.30) and (5.31), the factors f_{ki} and $f_{\theta i}$ are selected to be not smaller than 1, such that the equivalent open loop gain and the time delay of $\hat{g}_{ii}$ are no smaller than those of g_{ii}. The reason for such a selection is to make the resultant controller settings more conservative than those of $\hat{g}_{ii}$ so that loop failure tolerance property can be preserved.

5.4.2 Design of a Decentralized Controller

Using tuning rules of (5.18)–(5.20), the controller parameters for the individual equivalent transfer function of (5.29) can be determined as,

$$k_{Pi} = \frac{1}{f_{ki}k_{ii}} \frac{\tau_{ii}}{\tau_{Ci} + f_{\theta i}\theta_{ii}}, \tag{5.32}$$

$$\tau_{Ii} = \min\{\tau_{ii}, 4(\tau_{Ci} + f_{\theta i}\theta_{ii})\}, \tag{5.33}$$

$$\tau_{Di} = \tau'_{ii}. \tag{5.34}$$

Furthermore, by selecting the desired closed-loop time constant τ_{Ci} to be the same value as the time delay $f_{\theta i}\theta_{ii}$, we can obtain a set of simple rules for designing the parameters of a decentralized PID controller as,

$$k_{Pi} = \frac{1}{f_{ki}} \frac{1}{2f_{\theta i}} \frac{1}{k_{ii}} \frac{\tau_{ii}}{\theta_{ii}}, \tag{5.35}$$

$$\tau_{Ii} = \min\{\tau_{ii}, 8f_{\theta i}\theta_{ii}\}, \tag{5.36}$$

$$\tau_{Di} = \tau'_{ii}. \tag{5.37}$$

Once the closed-loop properties of the diagonal elements and the critical frequencies of all individual control loops are determined, the dRI and MMF can be obtained through calculation of the matrix P and dDRIA. Then, the parameters of the PID controllers can be calculated based on the equivalent transfer function of each control loop and the SIMC-PID tuning rules. In such a design procedure, the overall control system stability is assured if the loop pairing is structurally stable, which can be explained as follows:

(i) Each $\tilde{c}_i$ is designed without considering loop interaction; it is more aggressive to control loop $y_i–u_i$ than the final control setting c_i. Generally, we have $\bar{\sigma}(\Delta_1) \leq \bar{\sigma}(\Delta_2)$ as $\max\{1, |\rho_{ii,n-1}|\} \geq 1$, where $\bar{\sigma}(A)$ is the maximum singular value of matrix A, $\Delta_1 = \tilde{G}C(1 + \tilde{G}C)^{-1}$ and $\Delta_2 = \tilde{G}\tilde{C}(1 + \tilde{G}\tilde{C})^{-1}$.

(ii) Let $M = -(G - \tilde{G})\tilde{G}^{-1}$, and following the definition of the structured singular value (SSV) [78]:

$$\mu_\Delta(M) \equiv \frac{1}{\min\{\bar{\sigma}(\Delta)\,|\,\det(I - k_m M\Delta) = 0 \text{ for structured } \Delta\}},$$

we have $\mu_{\Delta_1}(M) \leq \mu_{\Delta_2}(M)$, which implies there exist smaller interactions among control loops when the detunned controller C is applied [45].

(iii) Because the bigger dRI $\phi_{ii,n-1}$ is used to determine the detuning factors and f_{ki} and $f_{\theta i}$ are no smaller than 1, the resultant controller c_i will be more conservative, with a smaller gain and crossover frequency compared with $\tilde{c}_i$.

(iv) Conservative control action in each loop presents smaller interaction to other loops. Because the design based on $\hat{g}_{ii}$ for loop y_i–u_i will end up with a more conservative c_i, the stability margin for each individual loop will be further increased compared with that using the true $\phi_{ii,n-1}$.

As a summary of the above results, an algorithm for the design of a decentralized PID controller for general multivariable processes is given as below.

Algorithm 5.1

Step 1.   For multivariable process G, pair the inputs and outputs based on the method given in ref [66];

Step 2.   Design an individual controller for the diagonal elements of the process transfer function matrix following (5.18)–(5.20);

Step 3.   Use (5.23) to construct matrix P;

Step 4.   Let $i = 1$, go to step 5;

Step 5.   Calculate critical frequency ω_{cri}, dRI and $\phi_{ii,n-1}(j\omega_{cri})$, by (5.22) and (5.24), respectively;

Step 6.   Determine $k_{\rho i,n-1}$ and $\theta_{\rho i,n-1}$ for MMF, $\rho_{ii,n-1}$, by (5.26)–(5.28);

Step 7.   Obtain the factors f_{ki} and $f_{\theta i}$ by (5.30) and (5.31);

Step 8.   Fine-tune the controller parameters following (5.32)–(5.34) or (5.35–5.37);

Step 9.   If $i < n$, $i = i + 1$, go to step 5. Otherwise, go to step 10;

Step 10. End.

The procedure for the independent design of a decentralized PID controller by using the proposed method is illustrated in Fig. 5.4.

5.4.3   Examples

The main advantage of the proposed method is to provide a simple yet effective way to design decentralized PID controller especially for a high-dimension process (a process larger than 2×2 process). Many 2×2, 3×3 and 4×4 processes have been tested on the closed-loop performance of the proposed method. In the following, we will present the results on a 2×2 Vinante and Luyben (VL) column process and a 4×4 Alatiqi(A1) column process, of which the VL case is used to illustrate the step-by-step design procedures while the A1 case is used to demonstrate the effectiveness of the proposed method for higher dimensional processes.

Example 5.1 (VL Column). The VL column system with its transfer function matrix was given by Luyben [20]:

$$G = \begin{bmatrix} -\dfrac{2.2e^{-s}}{7s+1} & \dfrac{1.3e^{-0.3s}}{7s+1} \\ -\dfrac{2.8e^{-1.8s}}{9.5s+1} & \dfrac{4.3e^{-0.35s}}{9.2s+1} \end{bmatrix}.$$

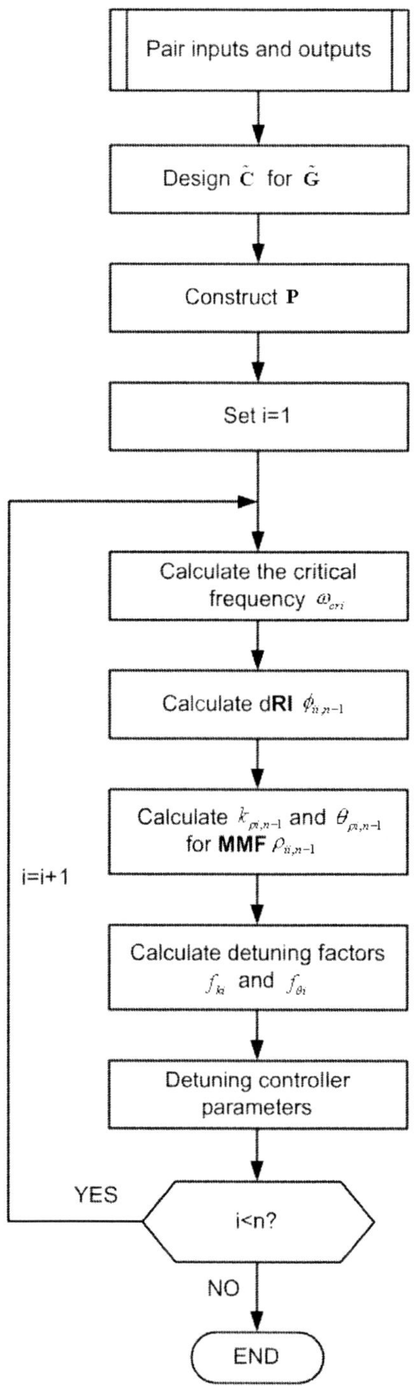

Fig. 5.4. Procedure for the independent design of a decentralized PID controller

According to (5.18)–(5.20) and selecting $\tau_{Ci} = \theta_{ii}$, the SIMC-PID controller for diagonal system $\tilde{G}$ is obtained as,

$$\tilde{C} = \begin{bmatrix} -1.5909\left(1 + \dfrac{1}{7s}\right) & 0 \\ 0 & 3.0565\left(1 + \dfrac{1}{2.8s}\right) \end{bmatrix}$$

resulting in the critical frequencies for control loops y_1–u_1 and y_2–u_2 as $\omega_{cr1} = \tilde{\omega}_{cr1} = 0.5$ rad/s and $\omega_{cr2} = \tilde{\omega}_{cr2} = 1.4286$ rad/s, respectively (equation (5.22)). Then from (5.23) and (5.24), we have

$$P = \begin{bmatrix} (s+1)e^s & 1 \\ 1 & (0.35s+1)e^{0.35s} \end{bmatrix}.$$

and

$$\begin{aligned}
\phi_{11,1}(j0.5) &= -0.2739 + j0.2451, \\
\phi_{22,1}(j1.4286) &= 0.2026 - j0.0674.
\end{aligned}$$

Subsequently,

$$\begin{aligned}
\rho_{11,1} &= 0.7663e^{-(-0.6510)s}, \\
\rho_{22,1} &= 1.2047e^{-0.0392s},
\end{aligned}$$

and the equivalent transfer functions of both control loops are constructed as

$$\begin{aligned}
\hat{g}_{11} &= -\frac{2.2e^{-s}}{7s+1}, \\
\hat{g}_{22} &= \frac{5.1802e^{-0.3892s}}{9.2s+1}.
\end{aligned}$$

It is noted that because $k_{\rho 1,1} < 1$ and $\theta_{\rho 1,1} < 0$, both f_{k1} and $f_{\theta 1}$ are set to 1 such that $\hat{g}_{11} = g_{11}$ to guarantee the loop failure tolerance of control loop y_1–u_1. While $f_{k2} = 1.2047$ and $f_{\theta 2} = 1.1120$ are used to fine-tune the PI controller setting of control loop y_2–u_2. Consequently, the decentralized PI controller parameters are determined as $k_{P1} = -1.5909$, $\tau_{I1} = 7.0000$, and $k_{P2} = 2.2817$, $\tau_{I2} = 3.1135$.

Fig. 5.5 shows the system inputs and responses for step changes in the set-points for y_1 and y_2. The simulation results indicate that both setpoint responses and magnitude of inputs of the proposed PI controllers are comparable with those of the BLT PI controller ($k_{P1} = -1.0700$, $\tau_{I1} = 7.1000$, and $k_{P2} = 1.9700$, $\tau_{I2} = 2.5800$) [20].

Example 5.2 (Alatiqi Column (A1)). The transfer function matrix for the A1 column system is given by

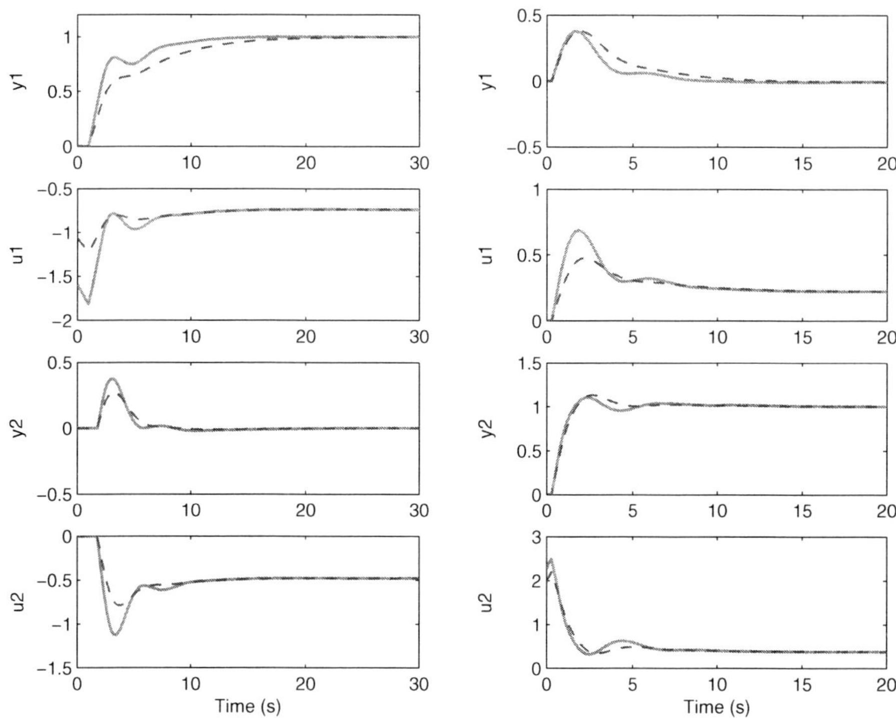

Fig. 5.5. Setpoint change response for the Vinante and Luyben column (solid line, Proposed design; dashed line, BLT design)

$$G = \begin{bmatrix} \dfrac{2.22e^{-2.5s}}{(36s+1)(25s+1)} & \dfrac{-2.94(7.9s+1)e^{-0.05s}}{(23.7s+1)^2} & \dfrac{0.017e^{-0.2s}}{(31.6s+1)(7s+1)} & \dfrac{-0.64e^{-20s}}{(29s+1)^2} \\[3mm] \dfrac{-2.33e^{-5s}}{(35s+1)^2} & \dfrac{3.46e^{-1.01s}}{32s+1} & \dfrac{-0.51e^{-7.5s}}{(32s+1)^2} & \dfrac{1.68e^{-2s}}{(28s+1)^2} \\[3mm] \dfrac{-1.06e^{-22s}}{(17s+1)^2} & \dfrac{3.511e^{-13s}}{(12s+1)^2} & \dfrac{4.41e^{-1.01s}}{16.2s+1} & \dfrac{-5.38e^{-0.5s}}{17s+1} \\[3mm] \dfrac{-5.73e^{-2.5s}}{(8s+1)(50s+1)} & \dfrac{4.32(25s+1)e^{-0.01s}}{(50s+1)(5s+1)} & \dfrac{-1.25e^{-2.8s}}{(43.6s+1)(9s+1)} & \dfrac{4.78e^{-1.15s}}{(48s+1)(5s+1)} \end{bmatrix}.$$

According to Algorithm 1, the resulting decentralized controllers and those obtained by using the BLT method of Luyben [20], the trial-and-error method [141], and an independent design method of Chen and Seborg [140] are shown in Table 5.1.

The closed-loop responses as well as system inputs for unit step changes of the set-points r_1–r_4 are shown in Fig. 5.6–5.9. The simulation results indicate that the control performance is better than those of the other three design methods.

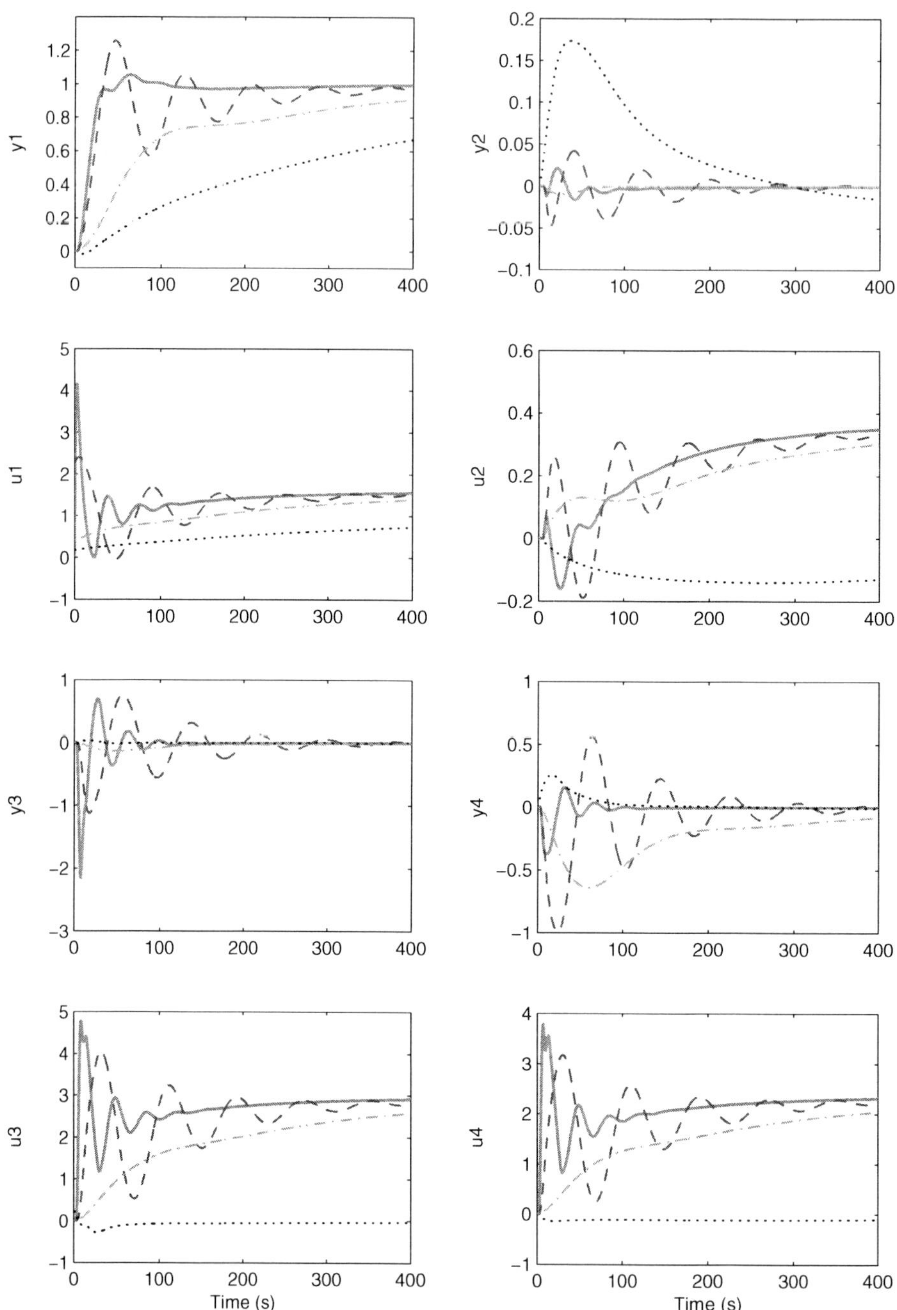

Fig. 5.6. Setpoint change response of y_1 for the A1 column example (solid line, proposed design; dashed line, BLT design; dashed-dotted line, Lee at el.; dotted line, Chen at el.)

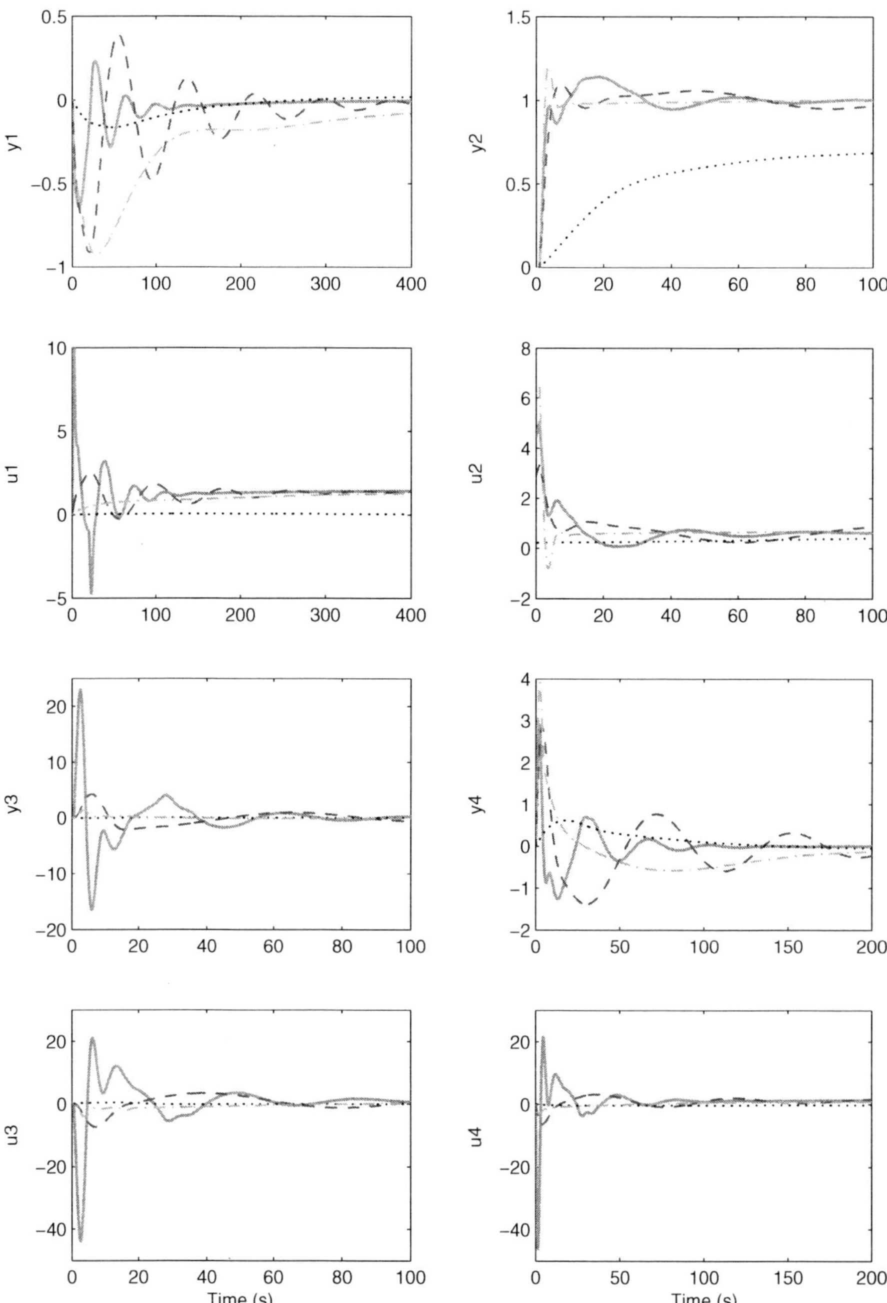

Fig. 5.7. Setpoint change response of y_2 for the A1 column example (solid line, proposed design; dashed line, BLT design; dashed-dotted line, Lee at el.; dotted line, Chen at el.)

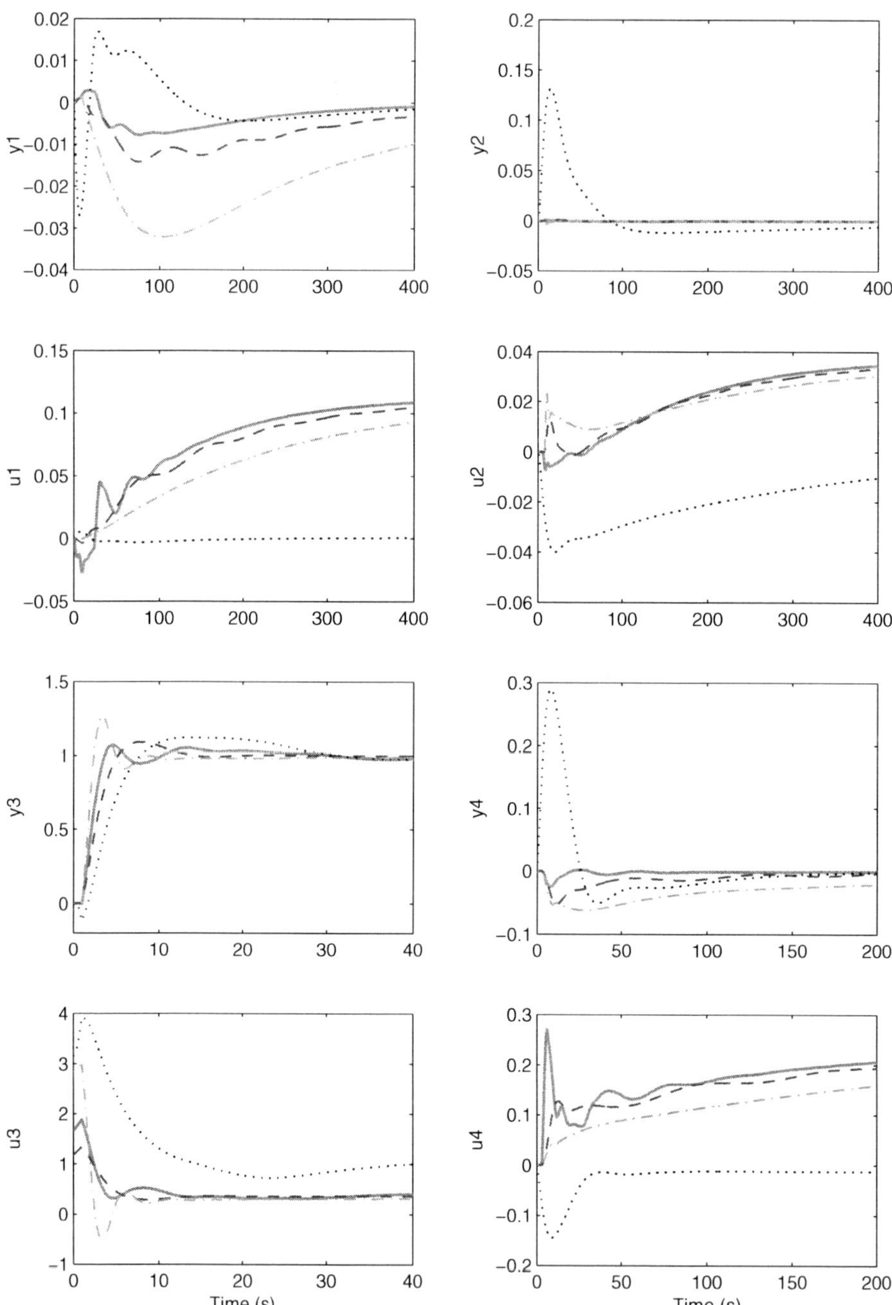

Fig. 5.8. Setpoint change response of y_3 for the A1 column example (solid line, proposed design; dashed line, BLT design; dashed-dotted line, Lee at el.; dotted line, Chen at el.)

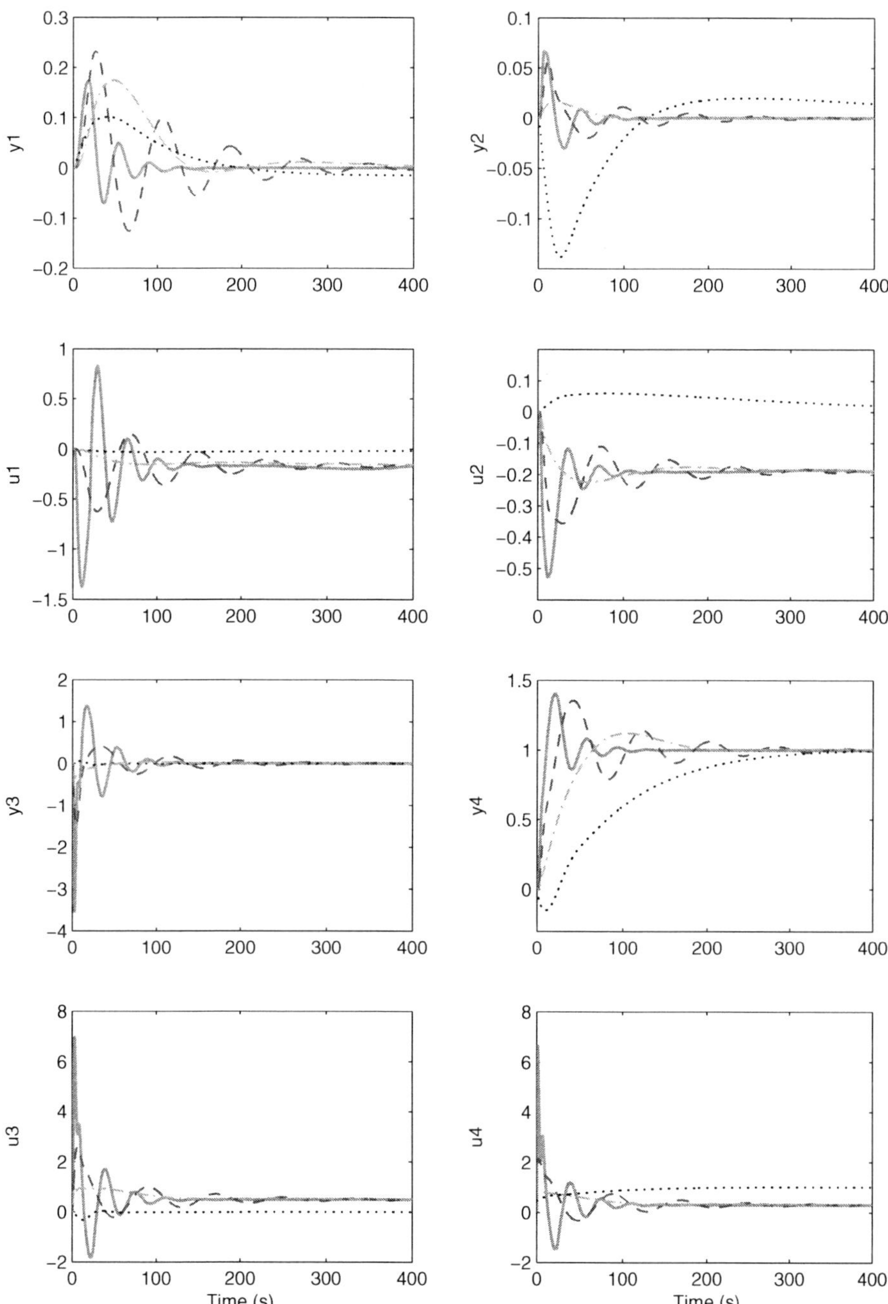

Fig. 5.9. Setpoint change response of y_4 for the A1 column example (solid line, proposed design; dashed line, BLT design; dashed-dotted line, Lee at el.; dotted line, Chen at el.)

Table 5.1. Proposed PID controllers for Alatiqi column

loop	proposed design			BLT design		ICC design		Chen's design	
	k_{Pi}	τ_{Ii}	τ_{Di}	k_{Pi}	τ_{Ii}	k_{Pi}	τ_{Ii}	k_{Pi}	τ_{Ii}
1	2.1822	29.7250	25	2.28	72.2	0.385	34.72	0.176	62.9
2	4.4807	8.0800	0	2.94	7.48	6.190	21.80	0.220	31.0
3	1.6656	8.0800	0	1.18	7.39	2.836	19.22	3.15	8.03
4	4.3660	9.2000	5	2.02	27.8	0.732	36.93	0.447	47.5

5.5 IMC Based Design

5.5.1 Estimation of Equivalent Transfer Function

The proposed method of designing decentralized controllers for multivariable processes involves three main steps:

(i)   Design individual controllers by ignoring the loop interaction;
(ii) Estimate the equivalent transfer function for each individual loop through dynamic interaction analysis;
(iii) Design decentralized controller based on the equivalent transfer function.

Let $\tilde{G} = \text{diag}\{G\}$ and ignore the interaction effect among control loops, the initial controller $\tilde{C}$ can be designed by applying the well known IMC tuning rules to each element in $\tilde{G}$ [128]. The IMC design procedure is brief studied as follows. The process model $\tilde{g}$ is factorized into an all-pass portion $\tilde{g}_+$ and minimum phase portion $\tilde{g}_-$, that is

$$\tilde{g} = \tilde{g}_+ \tilde{g}_- .$$

The all-pass portion $\tilde{g}_+$ includes all the open right-half-plane zeros and delays of $\tilde{g}$ and has the form

$$\tilde{g}_+ = e^{-\theta s} \prod_i (-\beta_i s + 1), \qquad \text{Re}\{\beta_i\} > 0,$$

where $\theta > 0$ is the tim delay and β_i^{-1} is the right-half-plane zero in the process model.

Then the IMC controller and the complementary sensitivity function are derived respectively as

$$g_c = \tilde{g}_-^{-1} f,$$

and

$$T = \tilde{g}_+ f,$$

where f is the IMC filter and has the form

$$f = \frac{1}{(\tau_C s + 1)^r},$$

where the filter order r is selected large enough to make g_c proper, and the adjustable filter parameter τ_C provides the tradeoff between performance and robustness. The key advantage of the IMC design procedure is that all controller parameters are related in a unique, straightforward manner to the model parameters. There is only one adjustable parameter τ_C which has intuitive appeal because it determines the speed of response of the system. Furthermore, τ_C is approximately proportional to the closed-loop bandwidth which must always be smaller than the bandwidth over which the process model is valid.

Table 5.2. Parameters of IMC-PID controller for typical low order systems[*]

	g	g_{CL}	ω_c	k_P	τ_I	τ_D [†]
A[‡]	$ke^{-\theta s}$			$-$	k_I	$-$
B	$\dfrac{ke^{-\theta s}}{s}$			$\dfrac{1}{k(\tau_C + \theta)}$	$-$	$-$
C	$\dfrac{ke^{-\theta s}}{\tau s + 1}$	$\dfrac{e^{-\theta s}}{\tau_C s + 1}$	$\dfrac{1}{\tau_C + \theta}$	$\dfrac{\tau}{k(\tau_C + \theta)}$	τ	$-$
D	$\dfrac{ke^{-\theta s}}{s(\tau s + 1)}$			$\dfrac{1}{k(\tau_C + \theta)}$	$-$	τ
E	$\dfrac{ke^{-\theta s}}{(\tau s + 1)(\tau' s + 1)}$			$\dfrac{\tau + \tau'}{k(\tau_C + \theta)}$	$\tau + \tau'$	$\dfrac{\tau\tau'}{\tau + \tau'}$
F	$\dfrac{ke^{-\theta s}}{\tau^2 s + 2\xi\tau s + 1}$			$\dfrac{2\xi\tau}{k(\tau_C + \theta)}$	$2\xi\tau$	$\dfrac{\tau}{2\xi}$

[*] The given settings are IAE and ISE optimal for step setpoint changes when $\tau_C = 0$ and $\tau_C = \theta$ respectively. It is recommended to select $\tau_C > \theta$ for practical design.
[†] To achieve much better performance, the derivation can be added by following traditional rule [148].
[‡] For pure time delay system, the pure integral controller $c(s) = (k_I/s)$ is applied and $k_I \equiv (k_P/\tau_I) = [1/k(\tau_C + \theta)]$ [129].

Even though more precise higher order process models can be obtained by either physical model construction (following the mass and energy balance principles) or the classical parameter identification methods, from a practical point of view, the lower order process model is more convenient for controller design. Six commonly used low order process models and parameters of their IMC-PID controllers are listed in Table 5.2, where g_{CL} and ω_c are the close-loop transfer function and the crossover frequency of gc, respectively.

From (5.15) and according to Table 5.2, we have

$$
P = \begin{bmatrix} \dfrac{1+g_{11}\tilde{c}_1}{g_{11}\tilde{c}_1} & 1 & \cdots & 1 \\ 1 & \dfrac{1+g_{22}\tilde{c}_2}{g_{22}\tilde{c}_2} & \cdots & 1 \\ \vdots & \vdots & \ddots & \vdots \\ 1 & 1 & \cdots & \dfrac{1+g_{nn}\tilde{c}_n}{g_{nn}\tilde{c}_n} \end{bmatrix}
$$

$$
= \begin{bmatrix} g_{11,CL}^{-1} & 1 & \cdots & 1 \\ 1 & g_{22,CL}^{-1} & \cdots & 1 \\ \vdots & \vdots & \ddots & \vdots \\ 1 & 1 & \cdots & g_{nn,CL}^{-1} \end{bmatrix}
$$

$$
= \begin{bmatrix} (\tilde{\tau}_{C1}s+1)e^{\theta_{11}s} & 1 & \cdots & 1 \\ 1 & (\tilde{\tau}_{C2}s+1)e^{\theta_{22}s} & \cdots & 1 \\ \vdots & \vdots & \ddots & \vdots \\ 1 & 1 & \cdots & (\tilde{\tau}_{Cn}s+1)e^{\theta_{nn}s} \end{bmatrix} . \tag{5.38}
$$

Since each controller is designed around the critical frequency of its transfer function, the dRI of individual control loop y_i–u_i can be estimated at the critical frequency $j\omega_{cri}$

$$
\phi_{ii,n-1}(j\omega_{cri}) = \left\| \triangle G_{ii,n-1}(j\omega_{cri}) \otimes \left(G^{ii}(j\omega_{cri}) \otimes P^{ii}(j\omega_{cri}) \right)^{-T} \right\|_{\Sigma}. \tag{5.39}
$$

For interactive multivariable process G, it is desirable to have the same open-loop transfer function as multivariable control system $\tilde{G}\tilde{C}$ [45]. As controller $\tilde{C}$ is designed for the diagonal elements of G without considering the couplings between control loops, the designed controller should be detuned by

$$
c_i = \frac{\tilde{c}_i}{\rho_{ii,n-1}} = \frac{\tilde{c}_i}{1+\phi_{ii,n-1}}
$$

to result approximately the same closed-loop control performance. However, it is impractical to use $\rho_{ii,n-1}$ directly to fine-tune the controller, because it has different values at different frequencies. To solve this problem, one possible way is to use some appropriate transfer functions to identify those MMFs. Since at the neighborhood of the critical point, the transmission interaction can be considered as a linear function, it is reasonable to represent the MMF by a low order transfer function involved the first two items of its Taylor series, which can be further simplified for controller design by a pure time delay transfer function as

$$
\rho_{ii,n-1} = k_{\rho i,n-1}e^{-\theta_{\rho i,n-1}s}, \tag{5.40}
$$

where,

$$k_{\rho i,n-1} = |\rho_{ii,n-1}(j\omega_{cri})| = |1 + \phi_{ii,n-1}(j\omega_{cri})| \tag{5.41}$$

and

$$\theta_{\rho i,n-1} = -\frac{\arg(\rho_{ii,n-1}(j\omega_{cri}))}{\omega_{cri}} = -\frac{\arg(1 + \phi_{ii,n-1}(j\omega_{cri}))}{\omega_{cri}}, \tag{5.42}$$

with ω_{cri} indicates the critical frequency of the ith control loop y_i–u_i. Then for individual control loop y_i–u_i with an arbitrary process model listed in Table 5.2, its equivalent transfer function can be represented as showed by the third column in Table 5.3, where

$$f_{ki} = \max\{1, k_{\rho i,n-1}\}, \tag{5.43}$$

and

$$f_{\theta i} = \max\left\{1, 1 + \frac{\theta_{\rho i,n-1}}{\theta_{ii}}\right\}. \tag{5.44}$$

Remark 5.6. Since the region between $\tilde{\omega}_{ci}$ ($|g_{ii}\tilde{c}_i| = 1$) and $\tilde{\omega}_{180}$ ($\arg(g_{ii}\tilde{c}_i) = -180^o$)is most critical for individual control loop design, the crossover frequency of ($g_{ii}\tilde{c}_i$) can be adopted as the critical frequency point for determining the dRI to the particular loop y_i–u_i to obtain $\hat{g}_{ii}$ [48].

Remark 5.7. In (5.43) and (5.44), the factors f_{ki} and $f_{\theta i}$ are selected to be not smaller than 1, such that the equivalent open loop gain and the time delay of $\hat{g}_{ii}$ are no smaller than that of g_{ii}. The reason for such selection is to make the resultant controller settings more conservative than that of $\hat{g}_{ii}$, such that loop failure tolerance property can be preserved.

5.5.2 Design of a Decentralized Controller

Once the closed-loop properties of the diagonal elements and the critical frequencies of all individual control loops are determined, the dRI and MMF can be obtained through calculating the matrix P and dDRIA. Then, the parameters of decentralized controllers can be calculated based on the equivalent transfer function of each control loop and the IMC-PID tuning rules as shown in Table 5.3.

In such design procedure, the overall control system stability is assured if the loop pairing is structurally stable which can be explained as follows:

(i) Each $\tilde{c}_i$ designed without considering loop interaction is more aggressive to control loop y_i–u_i than that of the final control setting c_i. Generally, we have $\bar{\sigma}(\Delta_1) \leq \bar{\sigma}(\Delta_2)$ as $\max\{1, |\rho_{ii,n-1}|\} \geq 1$, where $\bar{\sigma}(A)$ is the maximum singular value of matrix A, $\Delta_1 = \tilde{G}C(1 + \tilde{G}C)^{-1}$ and $\Delta_2 = \tilde{G}\tilde{C}(1 + \tilde{G}\tilde{C})^{-1}$.

(ii) Let $M = -(G - \tilde{G})\tilde{G}^{-1}$, and following the definition of the Structured Singular Value (SSV) [78]:

$$\mu_\Delta(M) \equiv \frac{1}{\min\{\bar{\sigma}(\Delta)|\det(I - k_m M\Delta) = 0 \text{ for structured } \Delta\}},$$

we have $\mu_{\Delta_1}(M) \leq \mu_{\Delta_2}(M)$, which implies there exist smaller interactions among control loops when the detuned controller C is applied [45].

Table 5.3. PID controllers of the equivalent processes for typical low order systems

	g_{ii}	$\hat{g}_{ii}$	k_P	τ_I	τ_D
A[†]	$k_{ii}\mathrm{e}^{-\theta_{ii}s}$	$f_{ki}k_{ii}\mathrm{e}^{-f_{\theta i}\theta_{ii}s}$	—	k_{Ii}	—
B	$\dfrac{k_{ii}\mathrm{e}^{-\theta_{ii}s}}{s}$	$\dfrac{f_{ki}k_{ii}\mathrm{e}^{-f_{\theta i}\theta_{ii}s}}{s}$	$\dfrac{1}{f_{ki}k_{ii}(\tau_{Ci}+f_{\theta i}\theta_{ii})}$	—	—
C	$\dfrac{k_{ii}\mathrm{e}^{-\theta_{ii}s}}{\tau_{ii}s+1}$	$\dfrac{f_{ki}k_{ii}\mathrm{e}^{-f_{\theta i}\theta_{ii}s}}{\tau_{ii}s+1}$	$\dfrac{\tau_{ii}}{f_{ki}k_{ii}(\tau_{Ci}+f_{\theta i}\theta_{ii})}$	τ_{ii}	—
D	$\dfrac{k_{ii}\mathrm{e}^{-\theta_{ii}s}}{s(\tau_{ii}s+1)}$	$\dfrac{f_{ki}k_{ii}\mathrm{e}^{-f_{\theta i}\theta_{ii}s}}{s(\tau_{ii}s+1)}$	$\dfrac{1}{f_{ki}k_{ii}(\tau_{Ci}+f_{\theta i}\theta_{ii})}$	—	τ_{ii}
E	$\dfrac{k_{ii}\mathrm{e}^{-\theta_{ii}s}}{(\tau_{ii}s+1)(\tau'_{ii}s+1)}$	$\dfrac{f_{ki}k_{ii}\mathrm{e}^{-f_{\theta i}\theta_{ii}s}}{(\tau_{ii}s+1)(\tau'_{ii}s+1)}$	$\dfrac{\tau_{ii}+\tau'_{ii}}{f_{ki}k_{ii}(\tau_{Ci}+f_{\theta i}\theta_{ii})}$	$\tau_{ii}+\tau'_{ii}$	$\dfrac{\tau_{ii}\tau'_{ii}}{\tau_{ii}+\tau'_{ii}}$
F[b]	$\dfrac{k_{ii}\mathrm{e}^{-\theta_{ii}s}}{\tau_{ii}^2 s+2\xi_{ii}\tau_{ii}s+1}$	$\dfrac{f_{ki}k_{ii}\mathrm{e}^{-f_{\theta i}\theta_{ii}s}}{\tau_{ii}^2 s+2\xi_{ii}\tau_{ii}s+1}$	$\dfrac{2\xi_{ii}\tau_{ii}}{f_{ki}k_{ii}(\tau_{Ci}+f_{\theta i}\theta_{ii})}$	$2\xi_{ii}\tau_{ii}$	$\dfrac{\tau_{ii}}{2\xi_{ii}}$

[†] For pure time delay system, the pure integral controller $c(s)=(k_{Ii}/s)$ is applied and $k_{Ii}\equiv(k_{Pi}/\tau_{Ii})=[1/f_{ki}k_{ii}(\tau_C+f_{\theta i}\theta_{ii})]$.

(iii) Since the bigger dRI $\phi_{ii,n-1}$ is used to determine the detuning factors, and f_{ki} and $f_{\theta i}$ are no smaller than 1, the resultant controller c_i will be more conservative with smaller gain and crossover frequency compared with $\tilde{c}_i$.

(iv) Conservative control action in each loop presents smaller interaction to other loops. As the design based on $\hat{g}_{ii}$ for loop y_i–u_i will end up with a more conservative c_i. the stability margin for each individual loop will be further increased compared with using the true $\phi_{ii,n-1}$.

Summarize the above results, a procedure for designing decentralized PID controller for general multivariable processes is illustrated as in Fig. 5.10.

Remark 5.8. As the interactions among control loops are ignored in both the dynamic interaction estimation step and the PID controller designing step, all available SISO PID controller design techniques can be adopted. Thus, according to the expected control performance, one can select the most suitable tuning rules, such as Ziegler and Nichols tuning rule [127], IMC tuning rule [128], and some other optimal design methods [2] for each step independently. Apparently, applying various tuning rules must lead to various interaction estimations, initial controller settings, final controller settings as well as overall control performance. However, has no influence to the design mechanism of our method. In the present chapter, the IMC-PID tuning rule is adopted because of its robust, generally good responses for setpoint changes and widely accepted.

5.5.3 Simulation Examples

To evaluate effectiveness of the proposed decentralized PID controller design method, 10 multivariable processes in reference [20] are studied:

- 2×2 systems: (1) Tyreus stabilizer — TS; (2) Wood and Berry — WB; (3) Vinante and Luyben — VL; (4) Wardle and Wood — WW.
- 3×3 systems: (5) Ogunnaike and Ray — OR; (6) Tyreus case 1 — T1; (7) Tyreus case 4 — T4.
- 4×4 systems: (8) Doukas and Luyben — DL; (9) Alatiqi case 1 — A1; (10) Alatiqi case 2 — A2.

Table 5.4. Process Open-Loop Transfer Functions of 2×2 Systems

	TS (Tyreus stabilizer)	WB (Wood and Berry)	VL (Vinante and Luyben)	WW (Wardle and Wood)
g_{11}	$\dfrac{-0.1153(10s+1)e^{-0.1s}}{(4s+1)^3}$	$\dfrac{12.8e^{-s}}{16.7s+1}$	$\dfrac{-2.2e^{-s}}{7s+1}$	$\dfrac{0.126e^{-6s}}{60s+1}$
g_{12}	$\dfrac{0.2429e^{-2s}}{(33s+1)^2}$	$\dfrac{-18.9e^{-3s}}{21s+1}$	$\dfrac{1.3e^{-0.3s}}{7s+1}$	$\dfrac{-0.101e^{-12s}}{(48s+1)(45s+1)}$
g_{21}	$\dfrac{-0.0887e^{-12.6s}}{(43s+1)(22s+1)}$	$\dfrac{6.6e^{-7s}}{10.9s+1}$	$\dfrac{-2.8e^{-1.8s}}{9.5s+1}$	$\dfrac{0.094e^{-8s}}{38s+1}$
g_{22}	$\dfrac{0.2429e^{-0.17s}}{(44s+1)(20s+1)}$	$\dfrac{-19.4e^{-3s}}{14.4s+1}$	$\dfrac{4.3e^{-0.35s}}{9.2s+1}$	$\dfrac{-0.12e^{-8s}}{35s+1}$

Table 5.5. Process open-loop transfer functions of 3×3 systems

	OR (Ogunnaike and Ray)	T1 (Tyreus case 1)	T4 (Tyreus case 4)
g_{11}	$\dfrac{0.66e^{-2.6s}}{6.7s+1}$	$\dfrac{-1.986e^{-0.71s}}{66.67s+1}$	$\dfrac{-1.986e^{-0.71s}}{66.67s+1}$
g_{12}	$\dfrac{-0.61e^{-3.5s}}{8.64s+1}$	$\dfrac{5.984e^{-2.24s}}{14.29s+1}$	$\dfrac{5.24e^{-60s}}{400s+1}$
g_{13}	$\dfrac{-0.0049e^{-s}}{9.06s+1}$	$\dfrac{0.422e^{-8.72s}}{(250s+1)^2}$	$\dfrac{5.984e^{-2.24s}}{14.29s+1}$
g_{21}	$\dfrac{1.11e^{-6.5s}}{3.25s+1}$	$\dfrac{0.0204e^{-0.59s}}{(7.14s+1)^2}$	$\dfrac{0.0204e^{-0.59s}}{(7.14s+1)^2}$
g_{22}	$\dfrac{-2.36e^{-3s}}{5s+1}$	$\dfrac{2.38e^{-0.42s}}{(1.43s+1)^2}$	$\dfrac{-0.33e^{-0.68s}}{(2.38s+1)^2}$
g_{23}	$\dfrac{-0.01e^{-1.2s}}{7.09s+1}$	$0.513e^{-1s}$	$\dfrac{2.38e^{-0.42s}}{(1.43s+1)^2}$
g_{31}	$\dfrac{-34.68e^{-9.2s}}{8.15s+1}$	$\dfrac{0.374e^{-7.75s}}{22.22s+1}$	$\dfrac{0.374e^{-7.75s}}{22.22s+1}$
g_{32}	$\dfrac{46.2e^{-9.4s}}{10.9s+1}$	$\dfrac{-9.811e^{-1.59s}}{11.36s+1}$	$\dfrac{-11.3e^{-3.79s}}{(21.74s+1)^2}$
g_{33}	$\dfrac{0.87(11.61s+1)e^{-s}}{(3.89s+1)(18.8s+1)}$	$\dfrac{-2.368e^{-27.33s}}{33.3s+1}$	$\dfrac{-9.811e^{-1.59s}}{11.36s+1}$

Table 5.6. Process open-loop transfer functions of 4×4 systems

	DL (Doukas and Luyben)	A1 (Alatiqi case 1)	A2 (Alatiqi case 2)
g_{11}	$\dfrac{-9.811e^{-1.59s}}{11.36s+1}$	$\dfrac{2.22e^{-2.5s}}{(36s+1)(25s+1)}$	$\dfrac{4.09e^{-1.3s}}{(33s+1)(8.3s+1)}$
g_{12}	$\dfrac{0.374e^{-7.75s}}{22.22s+1}$	$\dfrac{-2.94(7.9s+1)e^{-0.05s}}{(23.7s+1)^2}$	$\dfrac{-6.36e^{-0.2s}}{(31.6s+1)(20s+1)}$
g_{13}	$\dfrac{-2.368e^{-27.33s}}{33.3s+1}$	$\dfrac{0.017e^{-0.2s}}{(31.6s+1)(7s+1)}$	$\dfrac{-0.25e^{-0.4s}}{21s+1}$
g_{14}	$\dfrac{-11.3e^{-3.79s}}{(21.74s+1)^2}$	$\dfrac{-0.64e^{-20s}}{(29s+1)^2}$	$\dfrac{-0.49e^{-5s}}{(22s+1)^2}$
g_{21}	$\dfrac{5.984e^{-2.24s}}{14.29s+1}$	$\dfrac{-2.33e^{-5s}}{(35s+1)^2}$	$\dfrac{-4.17e^{-4s}}{45s+1}$
g_{22}	$\dfrac{-1.986e^{-0.71s}}{66.67s+1}$	$\dfrac{3.46e^{-1.01s}}{32s+1}$	$\dfrac{6.93e^{-1.01s}}{44.6s+1}$
g_{23}	$\dfrac{0.422e^{-8.72s}}{(250s+1)^2}$	$\dfrac{-0.51e^{-7.5s}}{(32s+1)^2}$	$\dfrac{-0.05e^{-5s}}{(34.5s+1)^2}$
g_{24}	$\dfrac{5.24e^{-60s}}{400s+1}$	$\dfrac{1.68e^{-2s}}{(28s+1)^2}$	$\dfrac{1.53e^{-2.8s}}{48s+1}$
g_{31}	$\dfrac{2.38e^{-0.42s}}{(1.43s+1)^2}$	$\dfrac{-1.06e^{-22s}}{(17s+1)^2}$	$\dfrac{-1.73e^{-17s}}{(13s+1)^2}$
g_{32}	$\dfrac{0.0204e^{-0.59s}}{(7.14s+1)^2}$	$\dfrac{3.511e^{-13s}}{(12s+1)^2}$	$\dfrac{5.11e^{-11s}}{(13.3s+1)^2}$
g_{33}	$\dfrac{0.513e^{-s}}{s+1}$	$\dfrac{4.41e^{-1.01s}}{16.2s+1}$	$\dfrac{4.61e^{-1.02s}}{18.5s+1}$
g_{34}	$\dfrac{-0.33e^{-0.68s}}{(2.38s+1)^2}$	$\dfrac{-5.38e^{-0.5s}}{17s+1}$	$\dfrac{-5.48e^{-0.5s}}{15s+1}$
g_{41}	$\dfrac{-11.3e^{-3.79s}}{(21.74s+1)^2}$	$\dfrac{-5.73e^{-2.5s}}{(8s+1)(50s+1)}$	$\dfrac{-11.18e^{-2.6s}}{(43s+1)(6.5s+1)}$
g_{42}	$\dfrac{-0.176e^{-0.48s}}{(6.9s+1)^2}$	$\dfrac{4.32(25s+1)e^{-0.01s}}{(50s+1)(5s+1)}$	$\dfrac{14.04e^{-0.02s}}{(45s+1)(10s+1)}$
g_{43}	$\dfrac{15.54e^{-s}}{s+1}$	$\dfrac{-1.25e^{-2.8s}}{(43.6s+1)(9s+1)}$	$\dfrac{-0.1e^{-0.05s}}{(31.6s+1)(5s+1)}$
g_{44}	$\dfrac{4.48e^{-0.52s}}{11.11s+1}$	$\dfrac{4.78e^{-1.15s}}{(48s+1)(5s+1)}$	$\dfrac{4.49e^{-0.6s}}{(48s+1)(6.3s+1)}$

The process open-loop transfer function matrices of these systems are listed by Tables 5.4, 5.5 and 5.6, respectively. As some process models, such as g_{11} in TS case, are in higher order (higher than second-order), the standard order reduction method is used to make them have the forms as those presented in Table 5.2. The controller parameters are listed in Table 5.7 together with those obtained by using other three design methods: the BLT method of Luyben [20], the trial-and-error method [141], and an independent design method based on Nyquist stability analysis [140]. It should be pointed out that the Gershorin circle and Gershorin band are utilized to determine the stability region in the last method, a static decoupler is required if the processes is not open-loop column diagonal dominance.

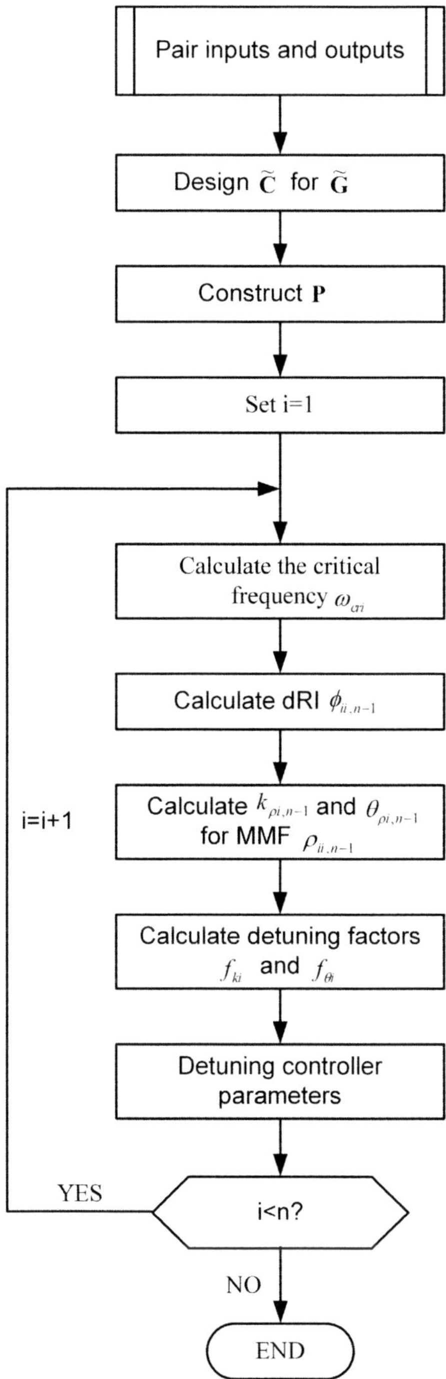

Fig. 5.10. The procedure for designing decentralized PID controller

Table 5.7. Parameters of the decentralized PID controllers for 10 classical systems[*]

	BLT design		Lee design		Chen design		Proposed design		
	k_P	τ_I	k_P	τ_I	k_P	τ_I	k_P	τ_I	τ_D
TS	-16.6	20.6					-149.0	3.460	2.300
	70.6	80.1					769.5	64.00	13.75
WB	0.375	8.29	0.850	7.21	0.436	11.0	0.932	16.70	–
	-0.075	23.6	-0.0885	8.86	-0.0945	15.5	-0.124	14.40	–
VL	-1.07	7.1	-1.31	2.26	1.21	4.64	-2.121	7.000	–
	1.97	2.58	3.97	2.42	3.74	1.10	3.951	9.200	–
WW	27.4	41.4	53.8	31.1			52.52	60.00	–
	-13.3	52.9	-20.3	29.7			-24.52	35.00	–
OR	1.51	16.4					1.676	6.700	–
	-0.295	18					-0.353	5.000	–
	2.63	6.61					4.385	8.620	–
T1	-17.8	4.5					-20.85	66.67	–
	0.749	5.61					1.168	2.860	0.715
	-0.261	139					-0.088	33.30	–
T4	-11.26	7.09					-23.64	66.67	–
	-3.52	14.5					0.622	2.860	0.715
	-0.182	15.1					-0.515	46.48	11.57
DL	-0.118	23.5					-0.410	43.48	10.87
	-7.26	11					-23.64	66.67	–
	0.429	12.1					0.589	2.860	0.715
	0.743	7.94					0.028	1.000	–
A1	2.28	72.2	0.385	34.72	0.176	62.9	3.698	61.00	14.75
	2.94	7.48	6.190	21.80	0.220	31.0	4.481	32.00	–
	1.18	7.39	2.836	19.22	3.150	8.03	1.666	16.20	–
	2.02	27.8	0.732	36.93	0.447	47.5	4.821	53.00	4.528
A2	0.923	61.7					3.884	41.30	6.632
	1.16	13.2					2.549	44.60	–
	0.727	13.2					1.311	18.50	–
	2.17	40					4.233	54.30	5.569

[*] In the proposed design, the control configurations of both Tyreus case 4 and Doukas and Luyben systems are re-selected as y_1–u_1/y_2–u_3/y_3–u_2 and y_1–u_4/y_2–u_2/y_3–u_1/y_4–u_3, respectively, by using the pairing method proposed in [66].

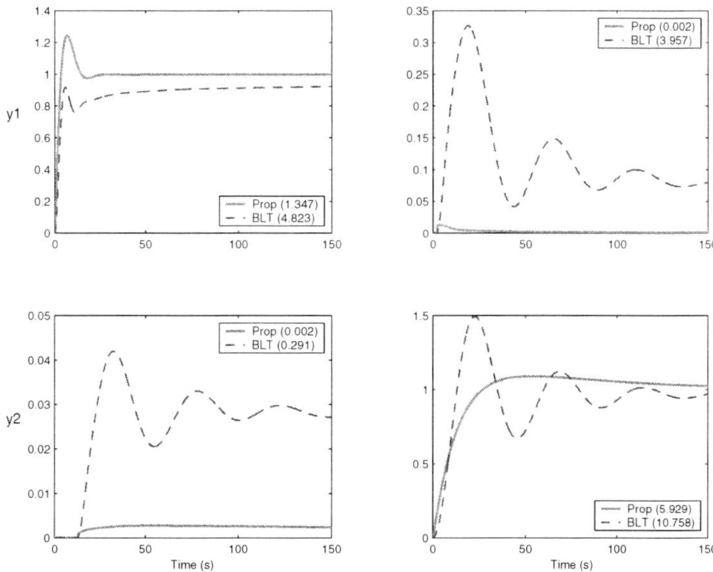

Fig. 5.11. Step response and ISE values of decentralized control for Tyreus stabilizer (solid line: Proposed design, dashed line: BLT design, dashed-dotted line: Lee at el., dotted line: Chen at el.)

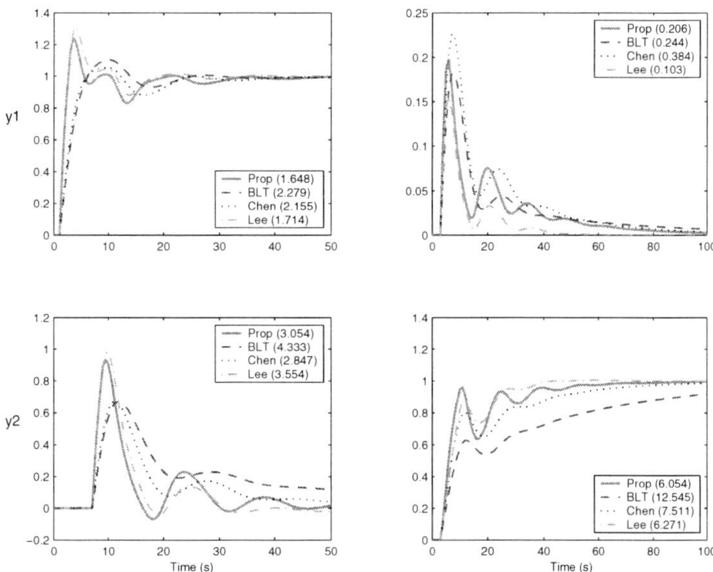

Fig. 5.12. Step response and ISE values of decentralized control for Wood and Berry (lower) systems (solid line: Proposed design, dashed line: BLT design, dashed-dotted line: Lee at el., dotted line: Chen at el.)

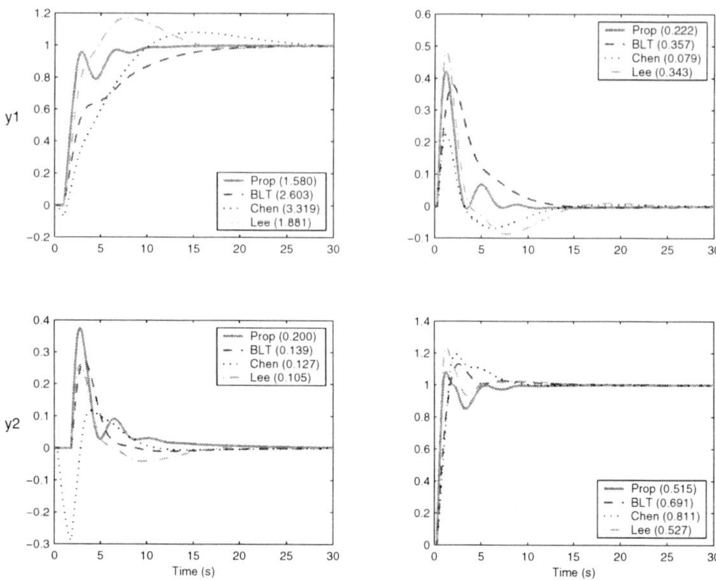

Fig. 5.13. Step response and ISE values of decentralized control for Vinate and Luyben system (solid line: Proposed design, dashed line: BLT design, dashed-dotted line: Lee at el., dotted line: Chen at el.)

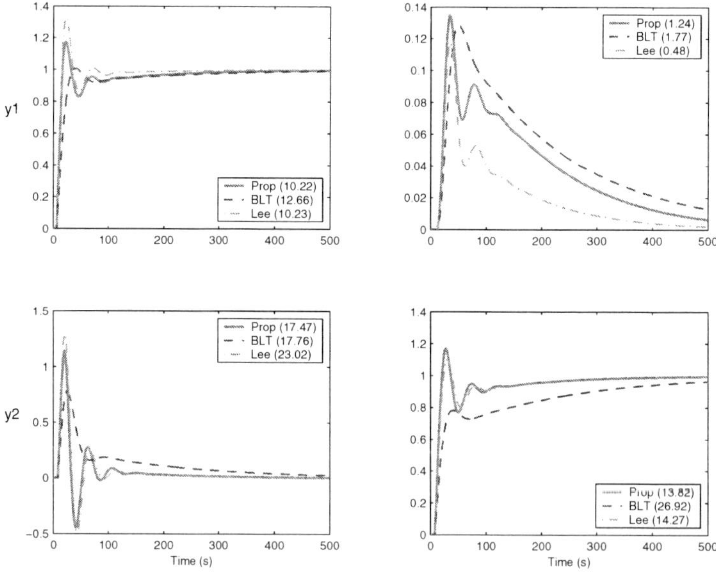

Fig. 5.14. Step response and ISE values of decentralized control for Wardle and Wood system (solid line: Proposed design, dashed line: BLT design, dashed-dotted line: Lee at el., dotted line: Chen at el.)

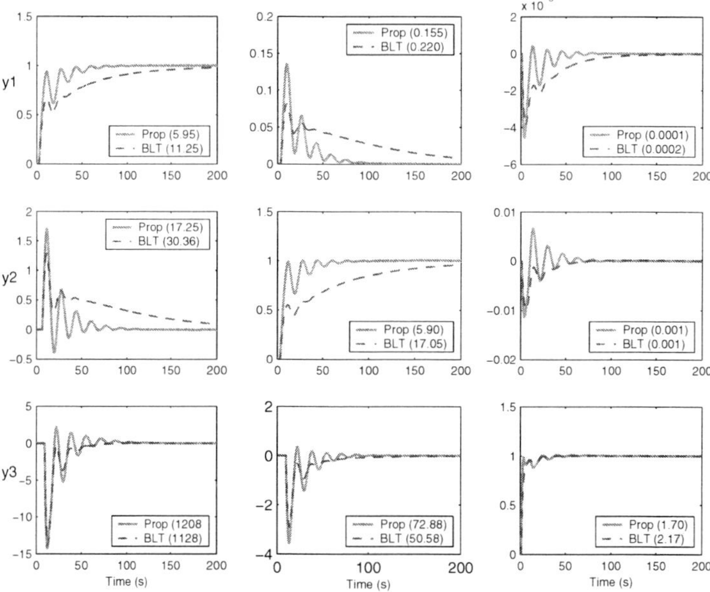

Fig. 5.15. Step response and ISE values of decentralized control for Ogunnaile and Ray system (solid line: Proposed design, dashed line: BLT design)

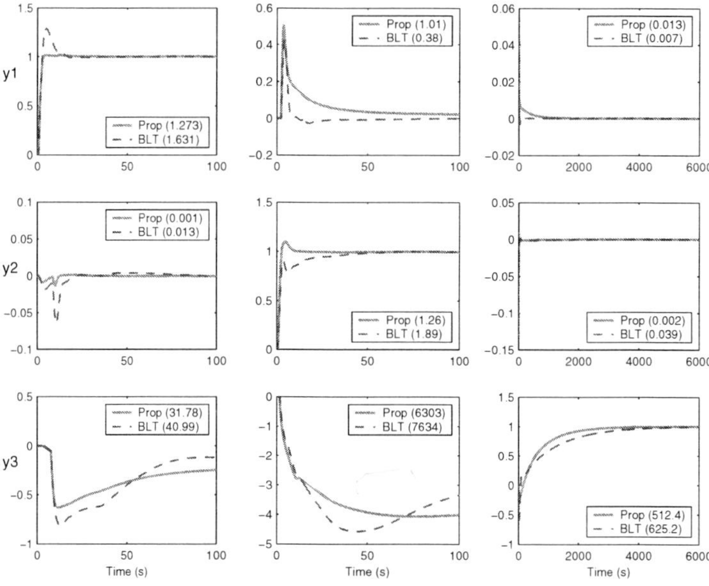

Fig. 5.16. Step response and ISE values of decentralized control for Tyreus case 1 systems (solid line: Proposed design, dashed line: BLT design)

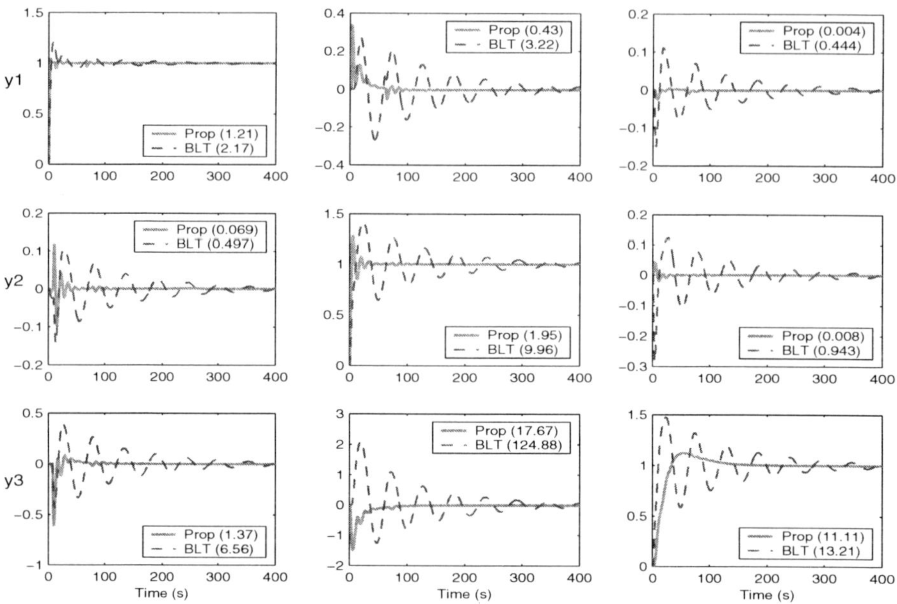

Fig. 5.17. Step response and ISE values of decentralized control for Tyreus case 4 system (solid line: Proposed design, dashed line: BLT design)

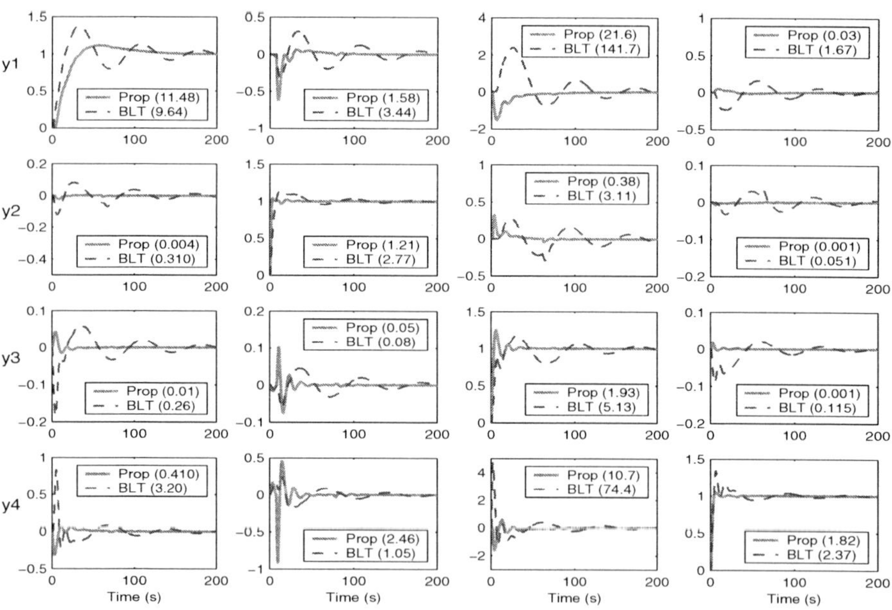

Fig. 5.18. Step response and ISE values of decentralized control for Doukas and Luyben system (solid line: Proposed design, dashed line: BLT design)

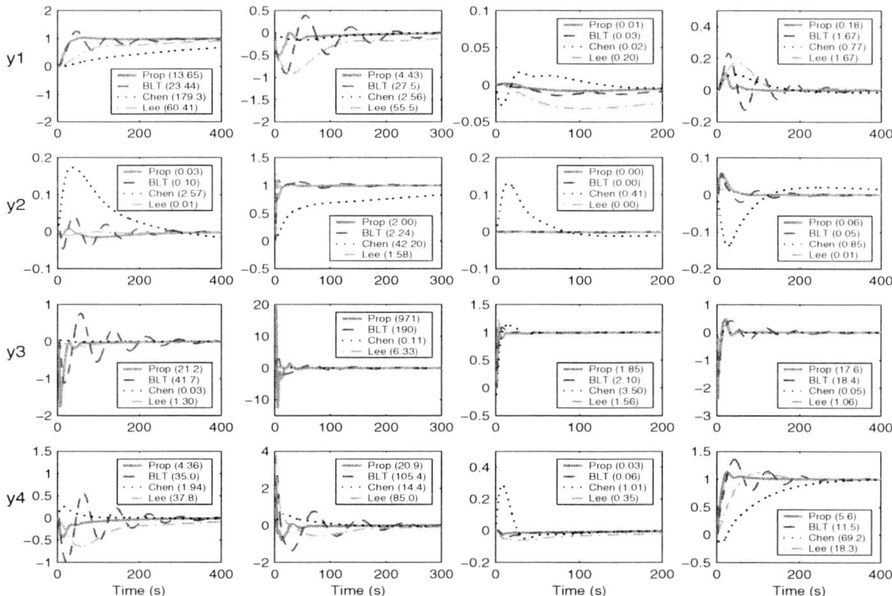

Fig. 5.19. Step response and ISE values of decentralized control for Alatiqi case 1 system (solid line: Proposed design, dashed line: BLT design, dashed-dotted line: Lee at el., dotted line: Chen at el.)

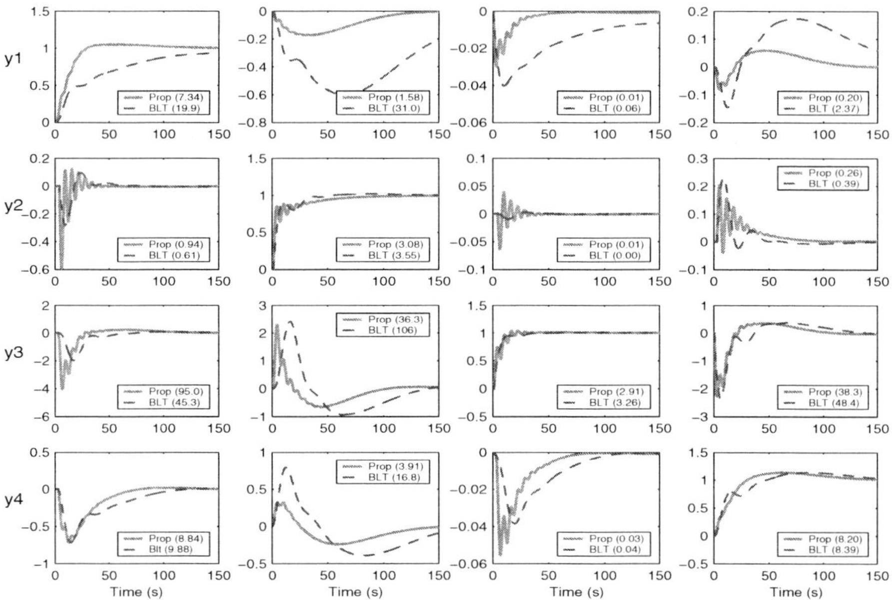

Fig. 5.20. Step response and ISE values of decentralized control for Alatiqi case 2 system (solid line: Proposed design, dashed line: BLT design)

To evaluate the output control performance, we consider a unit step setpoint change ($r_i = 1$) of all control loops one-by-one, and the integral square error (ISE) of $e_i = y_i - r_i$ is used to evaluate the control performance,

$$J_i = \int_0^\infty e_i^2 \mathrm{d}t.$$

The simulation results and ISE values are given in Fig. 5.11–5.20. The results show that, for some of the 2×2 processes, the proposed design provides better performance than both BLT method and Chen et al. method, and is quite competitive with Lee et al. method, but for higher dimensional processes, the proposed design provides less conservative controller settings as well as better control performance.

5.6 Conclusion

A simple yet effective design method for decentralized PID controller design method was proposed based on dynamic interaction analysis and internal model control principle. On the basis of structure decomposition, the dynamic relative interaction was defined and represented by the process model and controller explicitly. An initial decentralized controller was designed first by using the diagonal elements and then implemented to estimate the dRI to individual control loop from all others. By using the dRI, the MMF was derived and simplified to a pure time delay function at the neighborhood of each control loop critical frequency to obtain the equivalent transfer function. Consequently by applying the SISO SIMC- and IMC-PID tuning rules for the equivalent transfer function, appropriate controller parameters for individual control loop were determined. The proposed technique is very simple and effective, and has been applied to a variety of 2×2, 3×3 and 4×4 systems. Simulation results showed that the overall control system performance is much better than that of other tuning methods, such as the BLT method, the trial and error method, and the independent design method based on Nyquist stability analysis, especially for higher-dimensional processes,

It is noted that we use SISO SIMC- and IMC-PID controller tuning rules for controller tuning in the chapter purely for demonstration purpose; others such as Ziegler-Nichols tuning rules can also be directly applied in the design procedure without modification. Since the intension of this chapter is to present a simple and effective design method of decentralized PID controller for general multivariable processes, the decentralized closed-loop integrity was not considered here. As an important potential advantage for decentralized control structure [30, 54, 144], this issue can be found in Sect.2.7.

6 Multivariable PID Control Based on Dominant Pole Placement

In the previous chapter, an IMC design method is presented for decentralized (multi-loop) PID control. Next, Chapters 6–7 will address design methods for centralized (multivariable) PID control, where this chapter mainly considers the dominant pole placement design.

There is no easy way to guarantee the dominance of the desired poles if time delays are present in the loops because a continuous-time feedback system with time delay has infinite spectrum and it is not possible to assign such infinite spectrum with a finite-dimensional controller. In such a case, only the partial pole placement may be feasible and hopefully some of the assigned poles are dominant. In this chapter, an analytical PID design method is firstly proposed for continuous-time delay systems to achieve approximate pole placement with dominance. Its idea is to bypass continuous infinite spectrum problem by converting a delay process to a rational discrete model and getting back continuous PID controller from its discrete form designed for the model with pole placement. Then, simple and easy methods are also proposed which can guarantee the dominance of the assigned two poles for PID control systems.

6.1 Introduction

Pole placement in the state space and polynomial settings is very popular. For SISO plants, the equivalent output feedback control should be at least of the plant order minus one to achieve arbitrary pole placement. Arbitrary pole placement is otherwise difficult to achieve if one has to use a low-order output feedback controller for a high-order or time-delay plant. One typical example is that in process control, PID controller is used to regulate a plant with delay. To overcome this difficulty, the dominant pole design has been proposed. It is to choose and position a pair conjugate poles which represent the requirements on the closed-loop response, such as overshoot and settling time. However, continuous-time delay control systems are infinite-dimensional [149]. They have infinite spectrum and it is not possible to assign such infinite spectrum with a finite-dimensional controller [150]. Instead, one naturally wishes to assign a pair of

Q.-G. Wang et al.: PID Control for Multivariable Processes, LNCIS 373, pp. 131–166, 2008.
springerlink.com
© Springer-Verlag Berlin Heidelberg 2008

poles which dominate all other poles. This idea was first introduced by Persson and
Åström [151] and further explained in [2]. In [152], this idea is developed for the
tuning of lead-lag controllers. Both methods are based on a simplified model of processes and thus cannot guarantee the chosen poles to be indeed dominant in reality. In
the case of high-order systems or systems with time delay, these conventional dominant
pole designs, if not well handled, could result in sluggish response or even instability
of the closed-loop. To the best knowledge of the authors, no method is available in the
literature to guarantee dominance of the assigned poles in general.

It is thus desirable to find out ways to ensure the dominance of chosen poles and also
the closed-loop stability. This chapter aims to solve this problem and two design methods are proposed: one is for approximate pole placement; and the other is for guaranteed
pole placement.

For the approximate pole placement, the continuous delay process is firstly converted
to a low-order rational discrete model. Then, a discrete PID controller is designed to ensure dominant pole placement in discrete domain. This is a finite-dimensional problem
and the solution for pole placement is readily available. The designed discrete PID controller is finally converted back to the continuous one. The poles in continuous domain
are generally not precisely the same as originally set. It is argued that exact pole placement is not necessary as practical design specifications are commonly set as ranges
instead of precise values, and approximate one should be sufficient as long as they do
not deviate too much from the ideal ones. The dominance and error of the assigned poles
are measured and checked for the design. It is shown by simulation that the proposed
method works well with great dominance and negligible error of approximately assigned desired poles for a large range of normalized dead time up to at least 4. It should
also be pointed out that discretization of a continuous process and discrete PID calculations are purely employed as a design intermediate and can be viewed as a fictitious
process to get a workable continuous PID controller. No sampling is applied anyway.
Performance of our design should be judged from that of the so-obtained continuous
PID controller, rather from discretization errors involved.

Continuous controller design is always carried out in continuous domain, and this
causes an infinite spectrum assignment problem for a delay process under a PID control,
a hard and open problem, while the proposed method of transform into and out of a
discrete model is first of its kind and brings the infinite spectrum assignment problem
to an approximate finite spectrum assignment problem for which by a special selection
of sampling time, a simple solution is obtained. No method is available in the literature
to guarantee dominance of the assigned poles for PID control of a continuous delay
process while the proposed method can do so.

For the guaranteed pole placement, the common idea is: the chosen pair of poles
give rise to two real equations which are solved for integral and derivative terms via the
proportional gain and the locations of all other closed-loop poles can then be studied
with respect to this single variable gain by means of root locus or Nyquist techniques.
Thus, two approaches for guaranteed dominant pole placement with PID controller are
naturally developed.

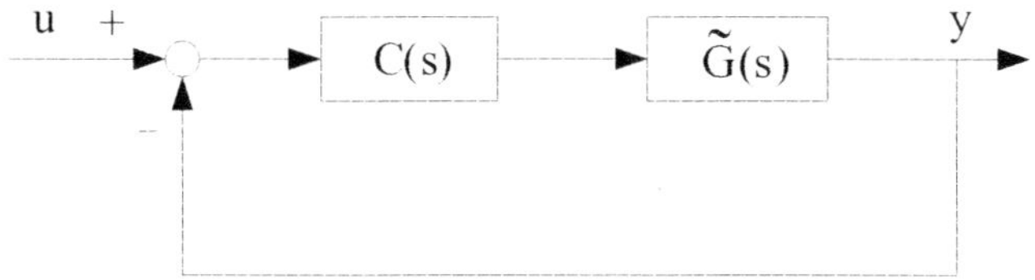

Fig. 6.1. PID control systems

6.2 Approximate Dominant Pole Placement

6.2.1 Problem Statement

A block diagram of a PID control system is shown in Fig. 6.1, where $\tilde{G}(s)$ is a continuous-time delay process and $C(s)$ is the PID controller. Suppose that control system design specifications are represented by the overshoot and settling time on its closed-loop step response. The overshoot is usually achieved by setting a suitable damping ratio, ξ. A reasonable value of the damping ratio is typically in the range of 0.4 to 1. The settling time, T_s, cannot be taken arbitrarily but largely limited by the process characteristics and available magnitude of the manipulated variable. If T_s is too large, the response is very slow, which is bad performance and should be avoided. On the other hand, if T_s is too small, this may cause very large control signal and less robust control system. From view of dominant pole placement, pole dominance is also difficult to realize [2, 25] if T_s is very small. In this chapter, through extensive simulation, we adopt the following empirical formula to choose T_s for the process with a monotonic step response:

$$T_s = T\left(4.5 + 7.5\frac{L}{T}\right)\left(\frac{0.35}{\xi} + 0.5\right), \tag{6.1}$$

where T and L are the equivalent time constant and dead time of the process. The natural frequency, ω_o, is calculated with $\omega_o = 4/(\xi T_s)$. Then, the specifications can be transferred to the corresponding desired 2nd-order dynamics:

$$s^2 + 2\xi\,\omega_o s + \omega_o^2 = 0.$$

Its two roots are denoted by $p_{s,1}$ with a positive imaginary part and $p_{s,2}$, which are the desired closed-loop poles to be achieved and be dominant by our controller design.

The actual closed-loop system has its characteristic equation:

$$1 + \tilde{G}(s)C(s) = 0.$$

Let its roots or closed-loop poles be $\tilde{p}_{s,i}, i = 1, 2, \cdots$. They are ordered such that $\tilde{p}_{s,i}$ meets $\text{Re}(\tilde{p}_{s,i}) \geq \text{Re}(\tilde{p}_{s,i+1})$ and if $\text{Re}(\tilde{p}_{s,i}) = \text{Re}(\tilde{p}_{s,i+1})$, $\text{Im}(\tilde{p}_{s,i}) > \text{Im}(\tilde{p}_{s,i+1})$, where $\text{Re}(\tilde{p}_{s,i})$ and $\text{Im}(\tilde{p}_{s,i})$ are the real and imaginary parts of $\tilde{p}_{s,i}$, respectively. Note that the actual poles, $\tilde{p}_{s,i}, i = 1, 2$, may not be the same as the desired ones: $p_{s,1}$ and $p_{s,2}$, and $\tilde{p}_{s,i}, i = 1, 2$, may not be dominant enough with respect to other poles. Thus, we introduce two measures to reflect them: the relative pole assignment error,

$$E_P = \max\left\{\left|\frac{\tilde{p}_{s,1} - p_{s,1}}{\tilde{p}_{s,1}}\right|, \left|\frac{\tilde{p}_{s,2} - p_{s,2}}{\tilde{p}_{s,2}}\right|\right\}, \tag{6.2}$$

and the relative dominance,

$$E_D = \frac{\text{Re}(\tilde{p}_{s,3})}{\text{Re}(\tilde{p}_{s,2})}. \tag{6.3}$$

Our problem of approximate pole placement with dominance is to determine a continuous controller $C(s)$ so as to produce reasonably small relative pole assignment error and large relative dominance, say, $E_P \leq 20\%$ and $E_D \geq 3$, which are used as defaults.

The difficulty of the above problem lies in existence of an infinite number of closed-loop poles for a continuous delay process under PID control. It is impossible to assign all the closed-loop poles. However, a continuous-time delay process may be converted to a low-dimensional discrete system with some special sampling time selection. In this chapter, discrete design is used as a bridge to approximate pole placement in continuous PID control systems but no sampling is done in the real control system of Fig. 6.1.

6.2.2   The Proposed Method

Let a continuous-time delay process $\tilde{G}(s)$ have a monotonic step response and be represented by a first-order time delay model:

$$G(s) = \frac{K}{Ts+1}e^{-Ls}. \tag{6.4}$$

In this chapter, we choose the sampling time h as $h = L$ to make the discretized process, $G(z)$, have the lowest order. The process has a pole at $-1/T$. This pole is mapped via $z - e^{hs}$ (adopted in pole-zero matching method in [153]), to the pole of its discrete equivalent at $\tilde{T} = e^{-L/T}$. so that $K/(Ts+1)$ is converted to $\tilde{K}/(z-\tilde{T})$, where $\tilde{K}$ is selected to match the static gain, $K/(Ts+1)|_{s=0} = \tilde{K}/(z-\tilde{T})|_{z=1}$, and thus $\tilde{K} = K(1-e^{-L/T})$. Note that The discrete equivalent of e^{-Ls} is $1/z$ under $h = L$. Overall, the process in form of (6.4) is converted to

$$G(z) = \frac{\tilde{K}}{z(z-\tilde{T})}. \tag{6.5}$$

The continuous PID controller in form of

$$C(s) = K_p\left(1 + \frac{1}{T_is} + T_ds\right), \tag{6.6}$$

is also converted to the discrete-time model,

$$C(z) = \frac{k_1z^2 + k_2z + k_3}{z-1}, \tag{6.7}$$

where k_1, k_2 and k_3 are the functions of K_p, T_i and T_d. The characteristic polynomial of the discrete closed-loop system is

$$\begin{aligned} A_{\text{cl}}(z) &= z(z-\tilde{T})(z-1)(1+G(z)C(z)) \\ &= z^3 + (k_1\tilde{K} - 1 - \tilde{T})z^2 + (\tilde{T} + k_2\tilde{K})z + k_3\tilde{K}. \end{aligned} \tag{6.8}$$

On the other hand, the given $p_{s,1}$ and $p_{s,2}$ have the desirable discrete characteristic polynomial as follows

$$A_{de}(z) = (z - p_{z,1})(z - p_{z,2})(z - p_{z,3}) = z^3 + p_1 z^2 + p_2 z + p_3, \tag{6.9}$$

where $p_{z,1} = e^{L p_{s,1}}$, $p_{z,2} = e^{L p_{s,2}}$, and $p_{z,3}$ is a user-defined parameter and set at $e^{10 L \mathrm{Re}(p_{s,1})}$ in this chapter. Equalizing $A_{cl}(z)$ with $A_{de}(z)$ yields

$$k_1 = \frac{p_1 + 1 + \tilde{T}}{\tilde{K}},$$

$$k_2 = \frac{p_2 - \tilde{T}}{\tilde{K}},$$

$$k_3 = \frac{p_3}{\tilde{K}}.$$

Once k_1, k_2 and k_3 are known, the two zeros of $C(z)$ can be calculated as $z_{1,2} = (-k_2 \pm \sqrt{k_2^2 - 4k_1 k_3})/(2k_1)$. Using the pole-zero matching method gives the continuous controller as

$$C(s) = \frac{K_c \left(s - \dfrac{\log(z_1)}{L} \right) \left(s - \dfrac{\log(z_2)}{L} \right)}{s},$$

with K_c selected to match the gain of $C(s)$ at $s = 0.1m/L$, where m is the smallest integer and meets $e^{0.1m} \neq 1, z_1$ and z_2. Finally, $C(s)$ can be then rearranged into the form in (6.6) with its settings given as follows,

$$K_p = -\frac{K_c [\log(z_1) + \log(z_2)]}{L},$$

$$T_i = -\frac{L[\log(z_1) + \log(z_2)]}{\log(z_1) \log(z_2)},$$

$$T_d = -\frac{L}{\log(z_1) + \log(z_2)}.$$

To apply the above method to a non-first-order process $\tilde{G}(j\omega)$ with monotonic step response, we have to obtain its first-order approximate model $G(s)$ in form of (6.4). The simplest technique is to match the model frequency response with the process one at two frequency points, $\omega = 0$ and $\omega = \omega_p$, the phase cross-over frequency. The formulas are well known [154]:

$$K = \tilde{G}(0), \tag{6.10}$$

$$T = \sqrt{\frac{K^2 - |\tilde{G}(j\omega_p)|^2}{|\tilde{G}(j\omega_p)|^2 \omega_p^2}}, \tag{6.11}$$

$$L = \frac{\pi + \tan^{-1}(-\omega_p T)}{\omega_p}. \tag{6.12}$$

Gain and phase margins are basic measure of the system's robustness. In this chapter, we apply these specifications to judge robustness of the design results. Tuning ξ will give suitable robust stability of the closed-loop system against the parameter uncertainties.

6.2.3 Simulation Study

Example 6.1. Consider an exact first order process with

$$\tilde{G}(s) = \frac{1}{s+1}e^{-Ls},$$

and study our design with several typical values of L.

Let $L = 0.5$ first. Suppose that the desired damping ratio is $\xi = 0.7$. T_s is calculated from (6.1) as 8.25. We have $p_{s,1} = -0.4848 + 0.4946i$, $p_{s,2} = -0.4848 - 0.4946i$. The third pole is then $p_{s,3} = 10\mathrm{Re}(p_{s,1}) = -4.848$. The proposed method with these specifications leads to the discrete PID:

$$C(z) = \frac{-0.009416z^2 + 0.366z - 0.1386}{z - 1},$$

and via the pole-zero matching method, the continuous PID:

$$C(s) = \frac{-0.0321s^2 + 0.1726s + 0.4505}{s},$$

which is rearranged in form of (6.6) as

$$C(s) = 0.1726\left(1 + \frac{1}{0.3832s} - 0.1859s\right).$$

The closed-loop poles are calculated from the roots of $1 + \tilde{G}(s)C(s) = 0$ with a 40th order Pade approximate to the time delay as $\tilde{p}_{s,1} = -0.5135 + 0.4837i$, $\tilde{p}_{s,2} = -0.5135 - 0.4837i$, $\tilde{p}_{s,3} = -5.6623$, $\tilde{p}_{s,4} = -6.4016 + 13.1493i$, $\tilde{p}_{s,5} = -6.4016 - 13.1493i$, $\cdots$. It follows that $E_P = 4.43\%$ and $E_D = 11.03$. The gain margin and phase margin are 6.64 and 63.92°, respectively. The step response and the manipulated variable are shown in Fig. 6.2. The settling time of the the control system is 8.5 and the overshoot is 3.71% with the corresponding damping ratio of 0.72. The step responses of the discrete system, $G(z)C(z)/[1 + G(z)C(z)]$, and the prototype continuous system, $2.326/(s^3 + 5.818s^2 + 5.181s + 2.326)$ with its poles at the desired $-0.4848 \pm 0.4946i$ and one extra at -4.848, are given in Fig. 6.3 for comparisons, from which one sees that the the designed continuous system is quite close to them.

Consider $L = 2$. Suppose that the desired damping ratio is $\xi = 0.7$. T_s is calculated from (6.1) as 19.5. We have $p_{s,1} = -0.2051 + 0.2093i$ and $p_{s,2} = -0.2051 + 0.2093i$. The third pole is at -2.051. The proposed method with these specifications leads to the discrete PID:

$$C(z) = \frac{-0.1083z^2 + 0.3758z - 0.008416}{z - 1},$$

and via the pole-zero matching method, the continuous PID:

$$C(s) = \frac{-0.1179s^2 - 0.1506s + 0.1384}{s},$$

which is rearranged in form of (6.6) as follows

$$C(s) = -0.1506\left(1 - \frac{1}{1.0883s} + 0.7829s\right).$$

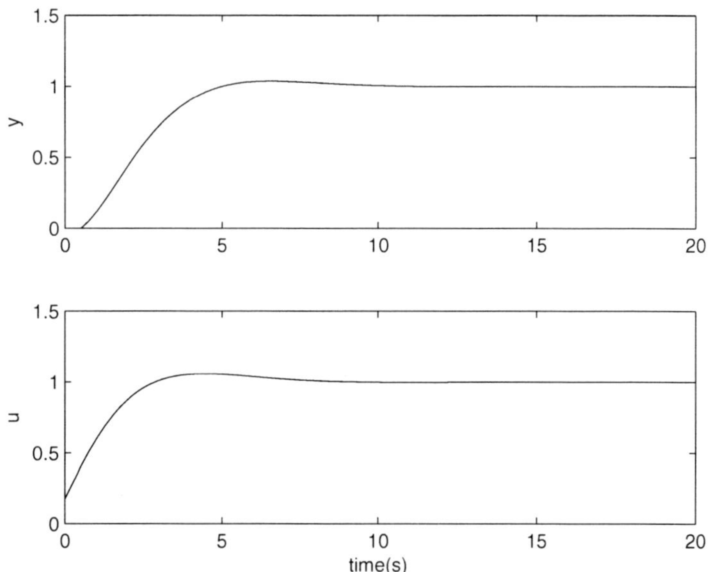

Fig. 6.2. Step response and manipulated variable of Example 6.1 with $L = 0.5$

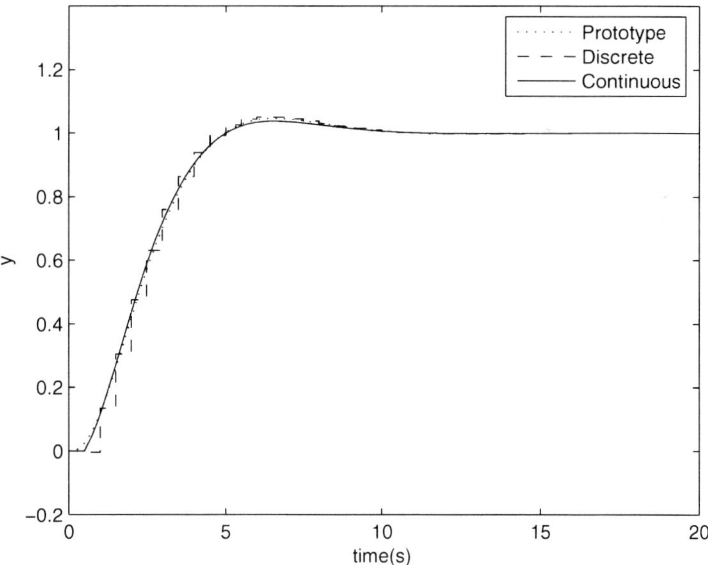

Fig. 6.3. Step response of Example 6.1 with $L = 0.5$

The closed-loop poles are $\tilde{p}_{s,1} = -0.1913 + 0.2284i$, $\tilde{p}_{s,2} = -0.1913 - 0.2284i$, $\tilde{p}_{s,3} = -1.0131 + 3.0847i$, $\tilde{p}_{s,4} = -1.0131 - 3.0847i$, $\cdots$. It follows that $E_P = 8.04\%$ and $E_D = 5.30$. The gain margin and phase margin are 2.59 and 57.25°, respectively. The

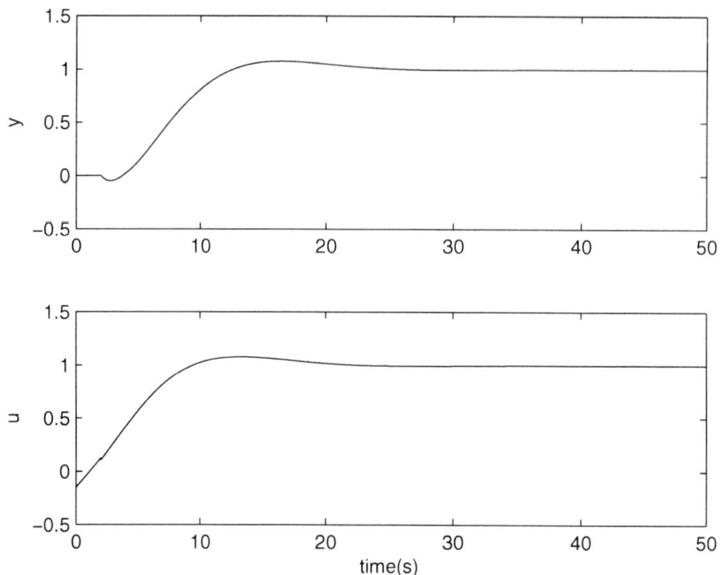

Fig. 6.4. Step response and manipulated variable of Example 6.1 with $L = 2$

step response and the manipulated variable are shown in Fig. 6.4. The settling time is 22.95 and the overshoot is 7.49% with the corresponding damping ratio of 0.64. The step responses of the discrete system, $G(z)C(z)/[1 + G(z)C(z)]$, and the prototype continuous system, $0.1761/(s^3 + 2.462s^2 + 0.9274s + 0.1761)$, with its poles at the desired $-0.2051 \pm 0.2093i$ and -2.051, are given in Fig. 6.5 for comparison.

Consider $L = 4$. Suppose that the desired damping ratio is $\xi = 0.7$. T_s is calculated from (6.1) as 34.5. We have $p_{s,1} = -0.1159 + 0.1183i$ and $p_{s,2} = -0.1159 - 0.1183i$. The third pole is at -1.159. The proposed method with these specifications leads to the discrete PID:

$$C(z) = \frac{-0.1131z^2 + 0.3953z - 0.0039}{z - 1},$$

and via the pole-zero matching method, the continuous PID:

$$C(s) = \frac{-0.207s^2 - 0.1743s + 0.07457}{s},$$

which is rearranged in form of (6.6) as follows

$$C(s) = -0.1743\left(1 - \frac{1}{2.3366s} + 1.1880s\right).$$

The closed-loop poles are $\tilde{p}_{s,1} = -0.1184 + 0.1289i$, $\tilde{p}_{s,2} = -0.1184 - 0.1289i$, $\tilde{p}_{s,3} = -0.3704 + 1.5947i$, $\tilde{p}_{s,4} = -0.3704 - 1.5947i$, $\cdots$. It follows that $E_P = 6.56\%$ and $E_D = 3.12$. The gain margin and phase margin are 2.48 and $58.02°$, respectively. The step response and the manipulated variable are shown in Fig. 6.6. The settling time

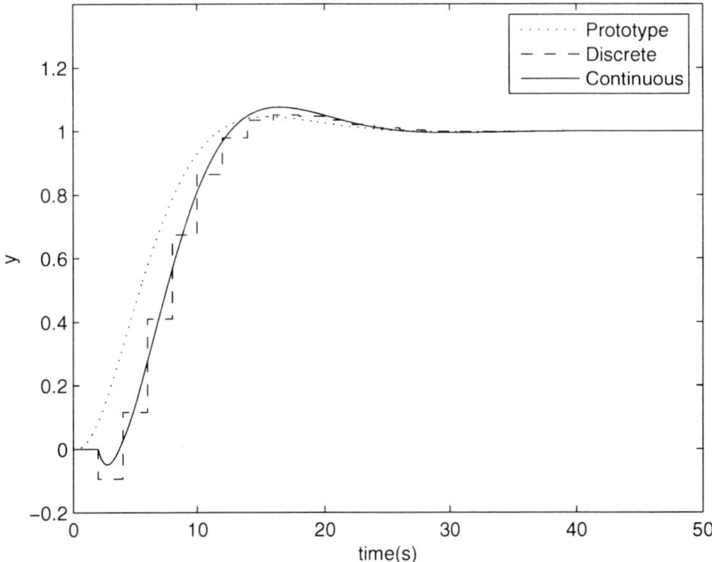

Fig. 6.5. Step response of Example 6.1 with $L = 2$

is 39.73 and the overshoot is 5.94% with the corresponding damping ratio of 0.67. The step responses of the discrete system, $G(z)C(z)/[1 + G(z)C(z)]$, and the prototype continuous system, $0.03181/(s^3 + 1.391s^2 + 0.2963s + 0.03181)$, with its poles at the desired $-0.1159 \pm 0.1183i$ and -1.159, are given in Fig. 6.7 for comparison.

Example 6.2. Consider a high-order process,

$$\tilde{G}(s) = \frac{2s + 1}{(s + 1)^2(4s + 1)}e^{-s}.$$

By Formulas (6.10), (6.11) and (6.12), we obtain its first-order approximate as

$$G(s) = \frac{1}{3.743s + 1}e^{-1.49s}.$$

Suppose that the desired damping ratio is $\xi = 0.7$. T_s is calculated from (6.1) as 28. We have $p_{s,1} = -0.1427 + 0.1456i$ and $p_{s,2} = -0.1427 - 0.1456i$. The third pole is at -1.427. The proposed method with these specifications leads to the discrete PID:

$$C(z) = \frac{-0.07994z^2 + 0.5168z - 0.2366}{z - 1},$$

and via the pole-zero matching method, the continuous PID:

$$C(s) = \frac{-0.2485s^2 + 0.1808s + 0.14}{s},$$

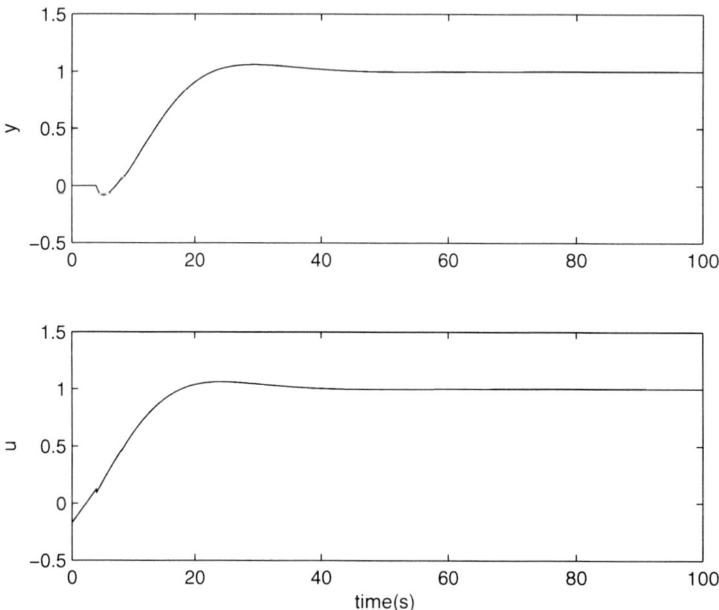

Fig. 6.6. Step response and manipulated variable of Example 6.1 with $L = 4$

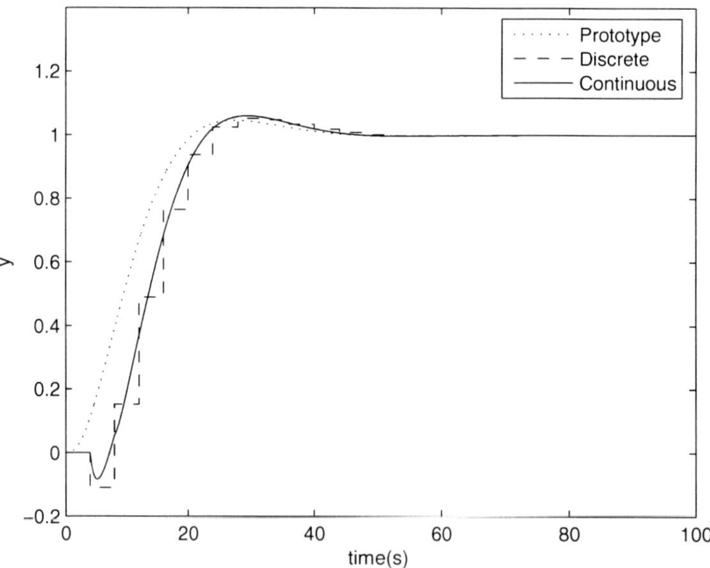

Fig. 6.7. Step response of Example 6.1 with $L = 4$

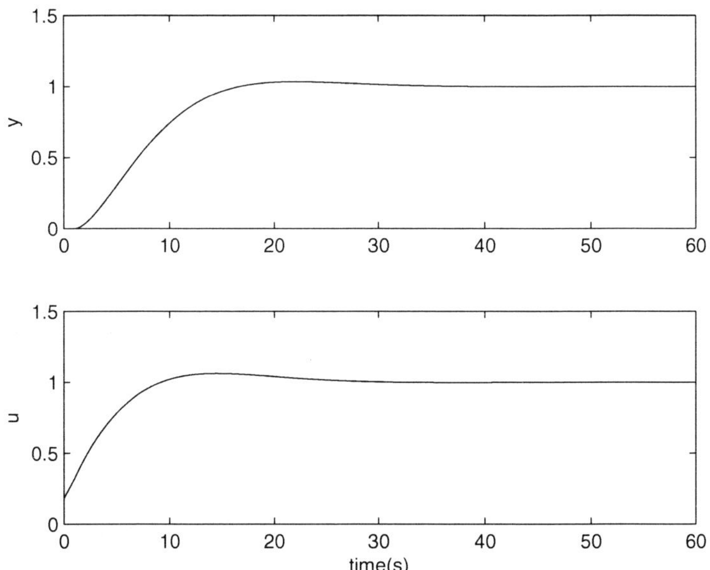

Fig. 6.8. Step response and manipulated variable of Example 6.2

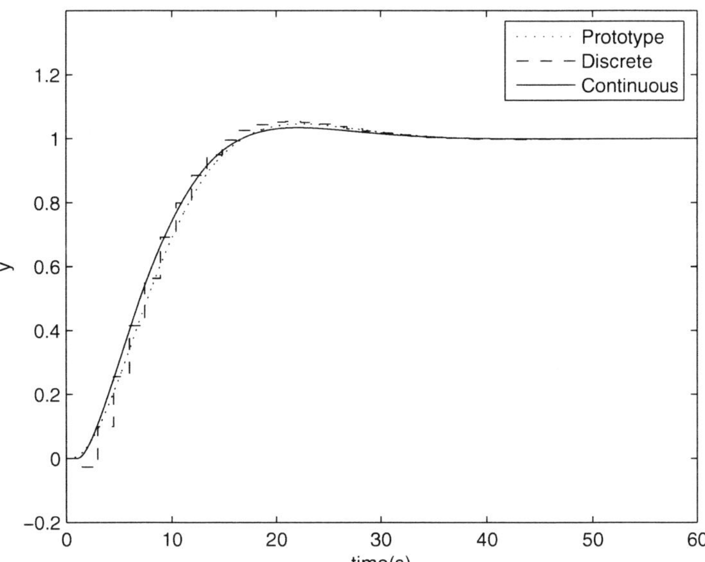

Fig. 6.9. Step response of Example 6.2

which is rearranged in form of (6.6) as

$$C(s) = 0.1808 \left(1 + \frac{1}{1.2910s} - 1.3747s \right).$$

The closed-loop poles are $\tilde{p}_{s,1} = -0.1530 + 0.1369i$, $\tilde{p}_{s,2} = -0.1530 - 0.1369i$, $\tilde{p}_{s,3} = -0.7307 + 0.3366i$, $\tilde{p}_{s,4} = -0.7307 - 0.3366i$, $\cdots$. It follows that $E_P = 6.62\%$ and $E_D = 4.77$. The gain margin and phase margin are 5.47 and 63.81°, respectively. The step response and the manipulated variable is shown in Fig. 6.8. The settling time of the control system is 28.28 and the overshoot is 3.41% with the corresponding damping ratio of 0.73. The step responses of the discrete system, $G(z)C(z)/[1 + G(z)C(z)]$, and the prototype continuous system, $0.05931/(s^3 + 1.713s^2 + 0.4489s + 0.05931)$ with its poles at $-0.1427 \pm 0.1456i$ and -1.427, are also given in Fig. 6.9 for comparison.

In practice, the measurement noise and unmodeled dynamics, such as disturbances, are generally present. For the same example, the measurement noise is simulated by adding a white noise to the output and a disturbance with the magnitude of -0.3 is added to the output at $t = 30$. The response, $y(t)$, the measured output, $y_n(t)$, and the manipulated variable, $u(t)$, are shown in Fig. 6.10. The effectiveness of our method is shown.

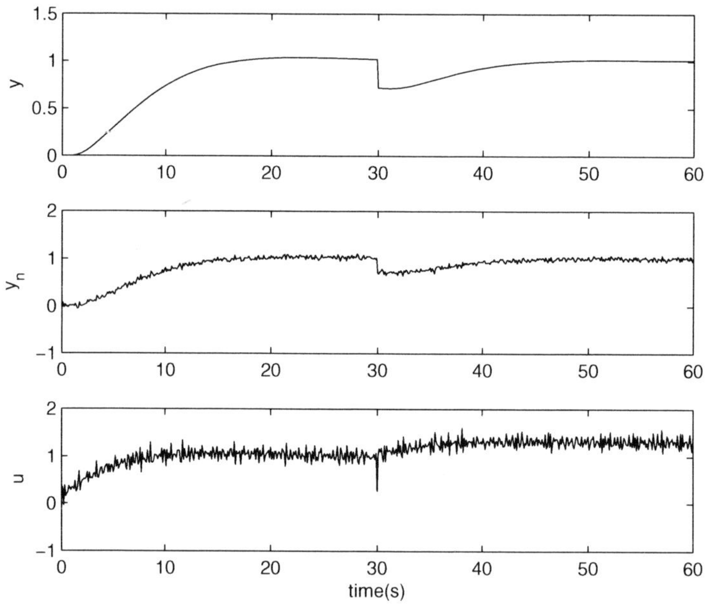

Fig. 6.10. Step response, measured response and manipulated variable of Example 6.2

6.2.4 Real Time Testing

In this section, the proposed PID tuning method is also applied to a temperature chamber system, which is made by National Instruments Corp. and shown in Fig. 6.11. The experiment setup consists of a thermal chamber and a personal computer with data

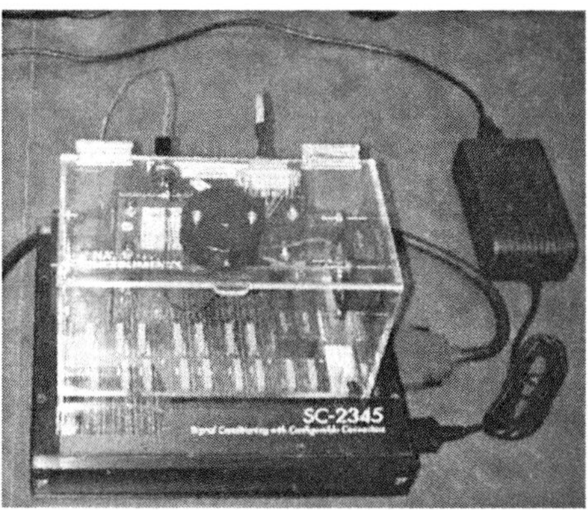

Fig. 6.11. Temperature chamber set

acquisition cards and LabView software. The system input, u, is the adjustable power supply to $20W$ Halogen bulb. The system output, y, is the temperature of the temperature chamber. The model of the process is

$$G(s) = \frac{29.49\mathrm{e}^{-0.106s}}{0.6853s + 1}.$$

The proposed method with $\xi = 0.8$ leads to the PID controller as

$$C(s) = 0.0047\left(1 + \frac{1}{0.1535s} - 1.3408s\right).$$

This ideal PID is not physically realizable and is thus replaced by

$$C(s) = 0.0047\left(1 + \frac{1}{0.1535s} - \frac{1.3408s}{(1.3408/N)s + 1}\right),$$

where $N = 4$, in the real time testing. Before the test is applied, the control system is at a steady state. At $t = 0$, the reference input is changed from 29 to 27. The process input and output are given in Fig. 6.12. The step response of the prototype continuous system, $20.8/(s^3 + 13.2s^2 + 26.09s + 20.8)$ are also given in Fig. 6.12 for comparison. The designed system has satisfying performance.

6.2.5 Positive PID Settings

It is noted from the simulation results in the preceding section that some of the PID parameters are not positive. In many applications, it is not permissible. To avoid this problem, we choose the controller in the form of

$$C(s) = K_p\left(1 + \frac{1}{T_i s}\right)\left(\frac{s + \beta}{s + \alpha}\right), \tag{6.13}$$

which corresponds to the practical form (no pure D) of PID controller in the cascaded structure [2, 155]. We choose $T_i = T$ to cancel the pole of $G(s)$. The open-loop

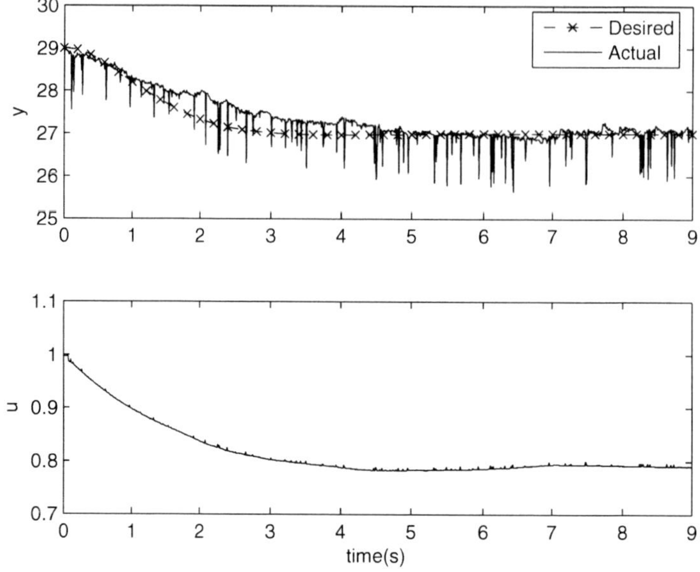

Fig. 6.12. Step response and manipulated variable of the thermal chamber

transfer function, $G(s)C(s)$, is converted by the pole-zero matching method to its discrete equivalent,

$$G(z)C(z) = \frac{\hat{K}}{z} \frac{k_1 z + k_2}{(z-1)(z+k_3)}, \tag{6.14}$$

where $\hat{K} = K/T$, and k_1, k_2, k_3 are the functions of K_p, β and α. The discrete closed-loop characteristic polynomial is

$$A_{\mathrm{cl}}(z) = z^3 + (k_3 - 1)z^2 + (\hat{K}k_1 - k_3)z + \hat{K}k_2.$$

By making $A_{\mathrm{cl}}(z) = A_{\mathrm{de}}(z)$, we can solve for k_1, k_2 and k_3 as

$$k_1 = \frac{p_1 + p_2 + 1}{\hat{K}}, \tag{6.15}$$

$$k_2 = \frac{p_3}{\hat{K}}, \tag{6.16}$$

$$k_3 = p_1 + 1. \tag{6.17}$$

Once k_1, k_2 and k_3 are known, we obtain the controller parameters in continuous domain as

$$\beta = -\log\left(\frac{-k_2}{k_1}\right)/L, \tag{6.18}$$

$$\alpha = -\log(-k_3)/L, \tag{6.19}$$

$$K_p = \frac{\dfrac{k_1 e^{0.1m} + k_2}{(e^{0.1m} + t_3)(e^{0.1m} - 1)}}{\dfrac{10L(0.1m + dL)}{m(0.1m + cL)}},$$ (6.20)

where m is the smallest integer, which meets $e^{0.1m} \neq 1, -k_3, -k_2/k_1$.

Example 6.3 (Example 6.1 continued). Consider Example 6.1 again with $L = 0.5$. Suppose that the desired damping ratio is $\xi = 0.7$ and $T_s = 8.25$ as before. The controller in form of (6.13) is obtained as

$$C(s) = 0.2195 \left(1 + \frac{1}{s}\right) \left(\frac{s + 1.8901}{s + 0.9878}\right).$$

The closed-loop poles are calculated as $\tilde{p}_{s,1} = -0.5382 + 0.4020i$, $\tilde{p}_{s,2} = -0.5382 - 0.4020i$, $\tilde{p}_{s,3} = -7.32, \cdots$. For this example, $E_P = 15.43\%$ and $E_D = 13.6$. The closed-loop pole at -1 is concealed by the closed-loop zero at -1. The gain margin and phase margin are 10.31 and $68.53°$, respectively. The step response and the manipulated variable are shown in Fig. 6.13. The settling time of the resultant control system is 5.45 and the overshoot is 1.63% with the corresponding damping ratio of 0.79. The step responses of the discrete system, $G(z)C(z)/[1 + G(z)C(z)]$, and the prototype continuous system, $2.326/(s^3 + 5.818s^2 + 5.181s + 2.326)$, are given in Fig. 6.14 for comparison.

6.2.6 Oscillation Processes

Some practical processes such as temperature loops exhibit oscillatory or essentially 2nd-order behavior in its step response. The first-order modeling is not adequate for them. Instead, one has to use the following model:

$$G(s) = \frac{K}{s^2 + as + b} e^{-Ls}.$$ (6.21)

Define $p_{g,i}$, $i = 1, 2$ as the roots of $s^2 + as + b = 0$. The equivalent time constant of the oscillation process is defined as $T = -1/\mathrm{Re}(p_{g,i})$. To set a desired 2nd-order dynamic properly, the damping ratio is chosen as before, while the following new formula,

$$T_s = T\left(1 + 15\frac{L}{T}\right)\left(\frac{0.35}{\xi} + 0.5\right),$$ (6.22)

is used for determining T_s. For this kind of processes, we exploit the controller in the form of

$$C(s) = K_p\left(1 + \frac{1}{T_i s} + T_d s\right)\left(\frac{s + \beta}{s + \alpha}\right),$$ (6.23)

and choose T_i and T_d to cancel the poles of $G(s)$:

$$T_d = \frac{1}{a},$$
$$T_i = \frac{a}{b}.$$

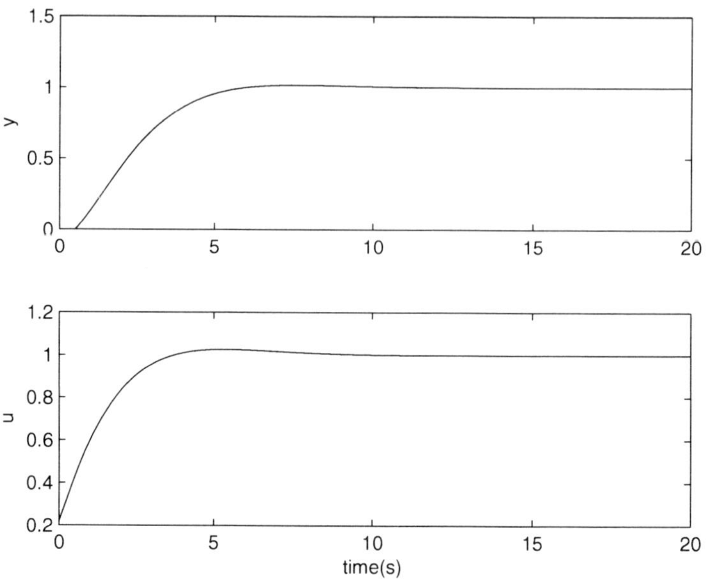

Fig. 6.13. Step response and manipulated variable of Example 6.1 with $L = 0.5$

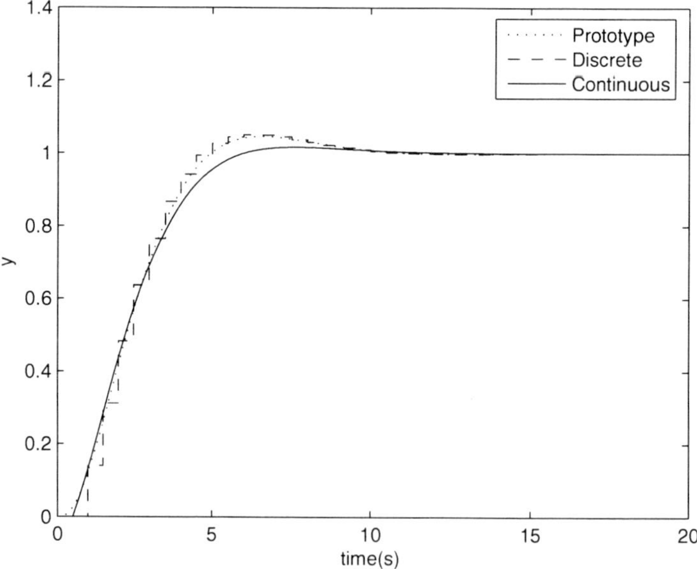

Fig. 6.14. Step response of Example 6.1 with $L = 0.5$

Then, the resulting open-loop $G(s)C(s)$ and its discrete equivalent are the same as those in Section 6.2.5 with $\hat{K} = K/a$. The procedure there applies to obtain K_p, β and α from (6.15), (6.16) and (6.17). With k_1, k_2 and k_3, we calculate K_p, β and α according to (6.18), (6.19) and (6.20).

Example 6.4. Consider a oscillation process,

$$\tilde{G}(s) = \frac{1}{s^2 + 1.2s + 1} e^{-0.7s}.$$

The equivalent time constant of the process is $T = 1.667$. Suppose that the desired damping ratio is $\xi = 0.7$. T_s is calculated from (6.22) as 13. We have $p_{s,1} = -0.3288 + 0.3354i$ and $p_{s,2} = -0.3288 - 0.3354i$. The third pole is at -3.288. The proposed method in this section with these specifications leads to the continuous controller:

$$C(s) = 0.1953 \left(1 + \frac{1}{1.2s} + 0.8333s \right) \left(\frac{s + 1.1410}{s + 0.6256} \right).$$

The closed-loop poles are calculated as $\tilde{p}_{s,1} = -0.3585 + 0.2755i$, $\tilde{p}_{s,2} = -0.3585 - 0.2755i$, $\tilde{p}_{s,3} = -5.0949$, $\tilde{p}_{s,4} = -6.1839 + 10.3973i$, $\tilde{p}_{s,5} = -6.1839 - 10.3973i$, $\cdots$. It follows $E_P = 14.24\%$ and $E_D = 14.21$. The closed-loop pole at $-0.6000 \pm 0.8000i$ are concealed by the closed-loop zeros. The gain margin and phase margin are 10.72 and $68.51°$, respectively. The step response and the manipulated variable are shown in Fig. 6.15. The settling time of the control system is 11 and the overshoot is 2.04% with the corresponding damping ratio of 0.77. The step responses of the discrete system, $G(z)C(z)/[1 + G(z)C(z)]$, and the prototype continuous system, $0.7252/(s^3 + 3.945s^2 + 2.382s + 0.7252)$, are given in Fig. 6.16 for comparison.

For comparison with first-order design method, by (6.10), (6.11) and (6.12), we obtain its first-order model as

$$G(s) = \frac{1}{1.235s + 1} e^{-1.44s}.$$

Suppose the desired damping ratio is $\xi = 0.7$. $T_s = 16.4$ is calculated from (6.1) with $T = 1.235$ and $L = 1.44$. The proposed method in Section 6.2.2 with these specification leads to the continuous PID:

$$C(s) = -0.0457 \left(1 - \frac{1}{0.2481s} + 1.6713s \right).$$

The closed-loop poles, are calculated as $\tilde{p}_{s,1} = -0.3437$, $\tilde{p}_{s,2} = -0.4066 + 0.5870i$, $\tilde{p}_{s,3} = -0.4066 - 0.5870i$, $\tilde{p}_{s,4} = -6.69 + 5.47i$, $\tilde{p}_{s,5} = -6.69 - 5.47i$, $\cdots$. The resulting dominant poles are -0.3437 and $-0.4066 \pm 0.5870i$, which are far from the desired ones.

Example 6.5. Consider a high-order oscillation process,

$$\tilde{G}(s) = \frac{1}{(0.8s + 1)(s^2 + 1.1s + 1)} e^{-2s}.$$

Applying the identification method proposed by [156], we obtain one of its estimations as

$$G(s) = \frac{0.702}{s^2 + 0.9708s + 0.7114} e^{-2.33s},$$

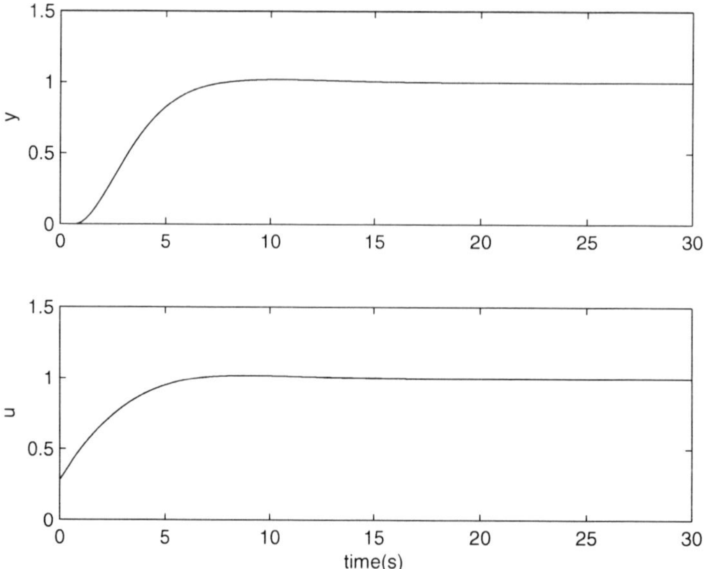

Fig. 6.15. Step response and manipulated variable of Example 6.4

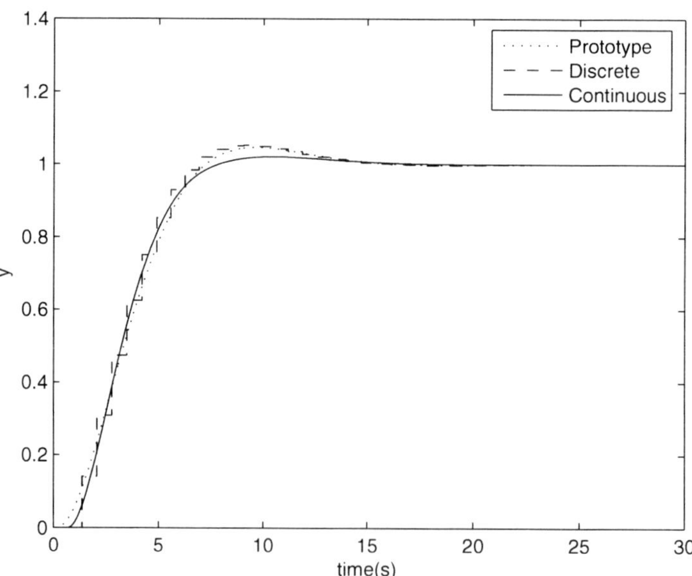

Fig. 6.16. Step response of Example 6.4

with the equivalent time constant of $T = 2.06$. Suppose that the desired damping ratio is $\xi = 0.7$. T_s is calculated from (6.22) as 37. We have $p_{s,1} = -0.1081 + 0.1103i$ and $p_{s,2} = -0.1081 - 0.1103i$. The third poles is at -1.081. The proposed method in this section with these specifications leads to the continuous controller:

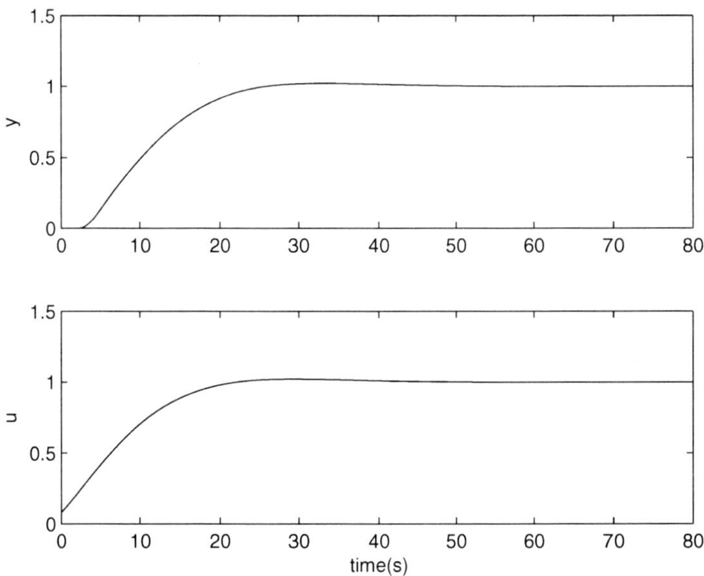

Fig. 6.17. Step response and manipulated variable of Example 6.5

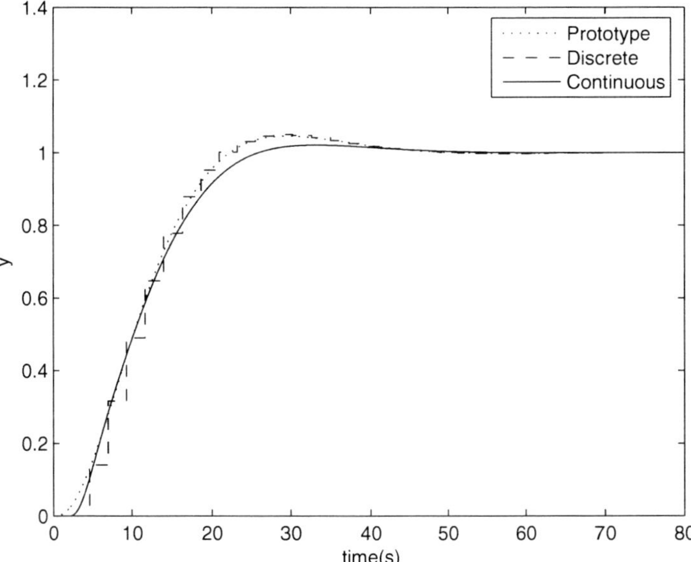

Fig. 6.18. Step response of Example 6.5

$$C(s) = 0.0637 \left(1 + \frac{1}{1.3646s} + 1.0301s \right) \left(\frac{s+0.4567}{s+0.2306} \right).$$

The closed-loop poles are calculated as $\tilde{p}_{s,1} = -0.1178 + 0.0941i$, $\tilde{p}_{s,2} = -0.1178 - 0.0941i$, $\tilde{p}_{s,3} = -0.5221 + 0.8374i$, $\tilde{p}_{s,4} = -0.5221 - 0.8374i$, $\cdots$. It follows $E_P = 12.24\%$ and $E_D = 4.43$. The gain margin and phase margin are 10.02 and 67.34°, respectively. The step response and the manipulated variable are shown in Fig. 6.17. The settling time of the resultant control system is 35.36 and the overshoot is 2.1% with the corresponding damping ratio of 0.77. The step responses of the discrete system, $G(z)C(z)/[1 + G(z)C(z)]$, and the prototype continuous system, $0.02576/(s^3 + 1.297s^2 + 0.2575s + 0.02576)$, are given in Fig. 6.18 for comparison.

6.3 Guaranteed Dominant Pole Placement

6.3.1 Problem Statement and Preliminary

Consider a plant described by its transfer function,

$$G(s) = \frac{N(s)}{D(s)} e^{-sL}, \tag{6.24}$$

where $N(s)/D(s)$ is a proper and co-prime rational function. A PID controller in the form of

$$C(s) = K_p + \frac{K_i}{s} + K_d s$$

is used to control the plant in the conventional unity output feedback configuration as depicted in Fig. 6.19. The closed-loop characteristic equation is

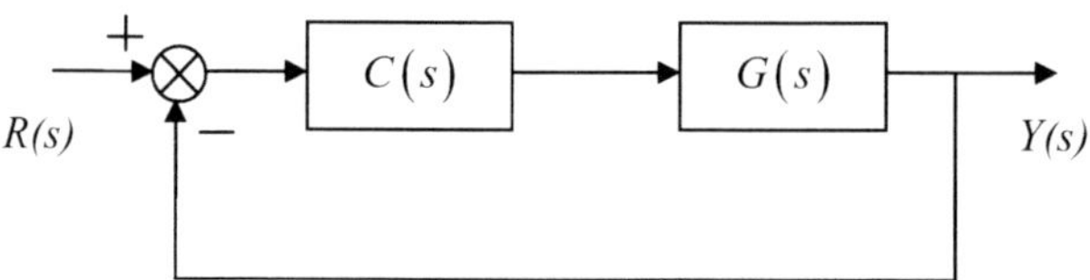

Fig. 6.19. Unity output feedback control system

$$1 + C(s)G(s) = 0. \tag{6.25}$$

The closed-loop transfer function is

$$H(s) = \frac{N(s)\left(K_d s^2 + K_p s + K_i \right)}{D(s)s + N(s)e^{-Ls}\left(K_d s^2 + K_p s + K_i \right)} e^{-Ls}.$$

Suppose that the requirements of the closed-loop control performance in frequency or time domain are converted into a pair of conjugate poles [2]: $\rho_{1,2} = -a \pm bj$. Their

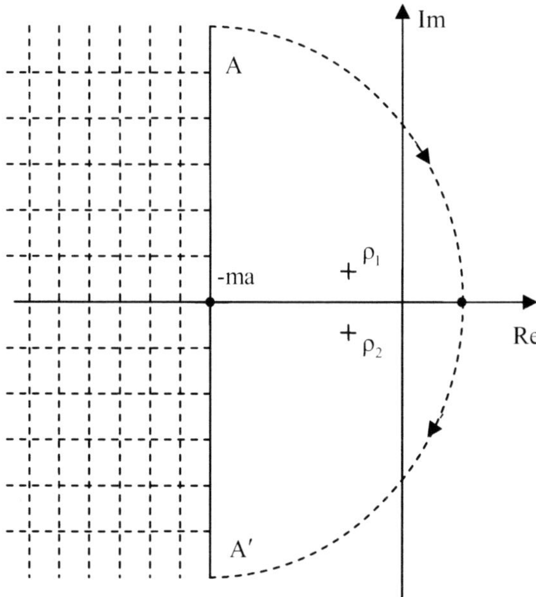

Fig. 6.20. Desired region (hatched) of other poles

dominance requires that the ratio of the real part of any of other poles to $-a$ exceeds m (m is usually 3 to 5) and there are no zeros nearby. Thus, we want all other poles to be located at the left of the line of $s = -ma$, that is, the desired region as hatched in Fig. 6.20. The problem of the guaranteed dominant pole placement is to find the PID parameters such that all the closed-loop poles lie in the desired region except the dominant poles, $\rho_{1,2}$.

Substitute $\rho_1 = -a + bj$ into (6.25):

$$K_p + \frac{K_i}{-a+bj} + K_d(-a+bj) = -\frac{1}{G(\rho_1)},$$

which is a complex equation. Solving the two equations given by its real and imaginary parts for K_i and K_d in terms of K_p yields

$$\begin{cases} K_i &= \dfrac{a^2+b^2}{2a}K_p - \left(a^2+b^2\right)X_1, \\ K_d &= \dfrac{1}{2a}K_p + X_2, \end{cases} \tag{6.26}$$

where

$$\begin{aligned} X_1 &= \frac{1}{2b}\mathrm{Im}\left[\frac{-1}{G(\rho_1)}\right] + \frac{1}{2a}\mathrm{Re}\left[\frac{-1}{G(\rho_1)}\right], \\ X_2 &= \frac{1}{2b}\mathrm{Im}\left[\frac{-1}{G(\rho_1)}\right] - \frac{1}{2a}\mathrm{Re}\left[\frac{-1}{G(\rho_1)}\right]. \end{aligned}$$

This simplifies the original problem to a one-parameter problem for which well known methods like root locus and Nyquist plot are applicable now.

6.3.2 Root Locus Method

The root-locus method is to show movement of the roots of the characteristic equation for all values of a system parameter. We plot the roots of the closed-loop characteristic equation for all the positive values of K_p and determine the range of K_p such that the roots other than the chosen dominant pair are all in the desired region.

Substituting (6.26) into (6.25) yields

$$1 + X_2 \frac{N(s)e^{-Ls}}{D(s)}s - \left(a^2 + b^2\right) X_1 \frac{N(s)e^{-Ls}}{D(s)s} \tag{6.27}$$

$$+ K_p \frac{N(s)e^{-Ls}}{D(s)} \frac{s^2 + 2as + (a^2 + b^2)}{2as} = 0. \tag{6.28}$$

Dividing both sides by the terms without K_p gives:

$$1 + K_p \overline{G}(s) = 0, \tag{6.29}$$

where

$$\overline{G}(s) = \frac{N(s)\left[s^2 + 2as + (a^2 + b^2)\right]e^{-Ls}}{2aD(s)s + 2aX_2 N(s)s^2 e^{-Ls} - 2a\left(a^2 + b^2\right)X_1 N(s)e^{-Ls}}. \tag{6.30}$$

It can be easily verified that the manipulation does not change the roots. If $G(s)$ has no time-delay term, $\overline{G}(s)$ is a proper rational transfer function since the degrees of its nominator and denominator of $\overline{G}(s)$ equal those of the closed-loop transfer function's nominator and denominator, respectively. The root locus of (6.29) can easily be drawn with Matlab as K_p varies. The interval of K_p for guaranteed dominant pole placement can be determined from the root locus. Example 6.6 shows the design procedure in detail.

Example 6.6. Consider a 4th-order process,

$$G(s) = \frac{1}{(s+1)^2(s+5)^2}.$$

If the overshoot is to be less than 5% and the rising time less than 2.5 s, the corresponding dominant poles are $\rho_{1,2} = -0.6136 \pm 0.6434\,j$. Equation (6.26) becomes

$$\begin{cases} K_i = 0.6442K_p - 0.1847, \\ K_d = 0.8149K_p - 12.4627. \end{cases}$$

And it follows from (6.30) that

$$\overline{G}(s) = \frac{s^2 + 1.227s + 0.7905}{1.227s^5 + 14.73s^4 + 56.45s^3 + 58.33s^2 + 30.68s - 0.2267}.$$

The root-locus of $\overline{G}(s)$ is exhibited in Fig. 6.21 with the solid lines while the edge of the desired region with $m = 3$ is indicated with dotted lines. Note that $\overline{G}(s)$ is of 5-th order and has five branches of root loci, of which two are fixed at the dominant poles while the other three move with the gain. From the root locus, two intersection points corresponding to root locus entering into and departing from the desired region are located and give the gain range of $K_p \in (36,51)$, which ensures all other three poles in the desired region. Besides, the positiveness of K_d and K_i requires $K_p > 15.2935$. Taking the joint solution of these two, we have $K_p \in (36,51)$. If $K_p = 50$ is chosen, the PID controller is

$$C(s) = 50 + \frac{32.0233}{s} + 28.2832s.$$

The zeros of the closed-loop system are at $s = -0.8839 \pm 0.5934j$, which are not near the dominant poles. Fig. 6.22 shows the step response of the closed-loop system.

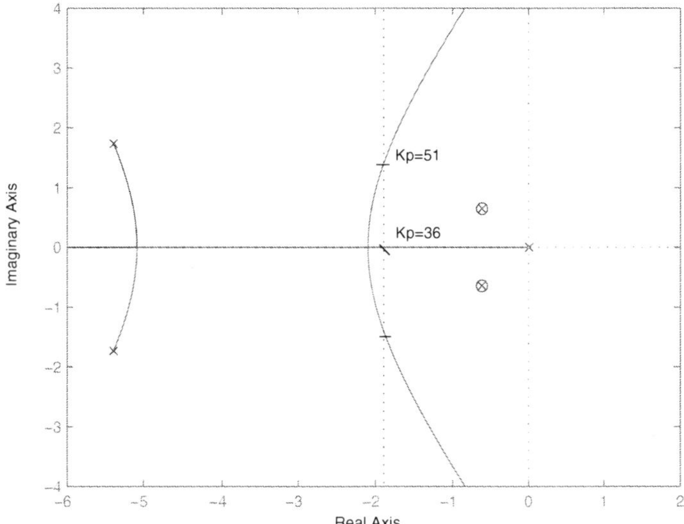

Fig. 6.21. Root-Locus for Example 6.6

6.3.3 Nyquist Plot Method

If $G(s)$ has time delay, so will be $\overline{G}(s)$. Then, drawing the root locus for it could be difficult and checking locations of infinite poles is a forbidden task. Note that the Nyquist plot works well for delay systems. The Nyquist stability criterion determines the number of unstable closed-loop poles based on the Nyquist plot and the open-loop unstable poles. We use the same idea but have to modify the conventional Nyquist contour. The Modified Nyquist contour is obtained by shifting the conventional Nyquist contour to the left by ma, as Fig. 6.20 shows. The image of G(s) when s traverses the modified Nyquist contour is called the modified Nyquist plot. The number of poles located outside the desired region plays the same role as that of unstable poles in the standard Nyquist criterion.

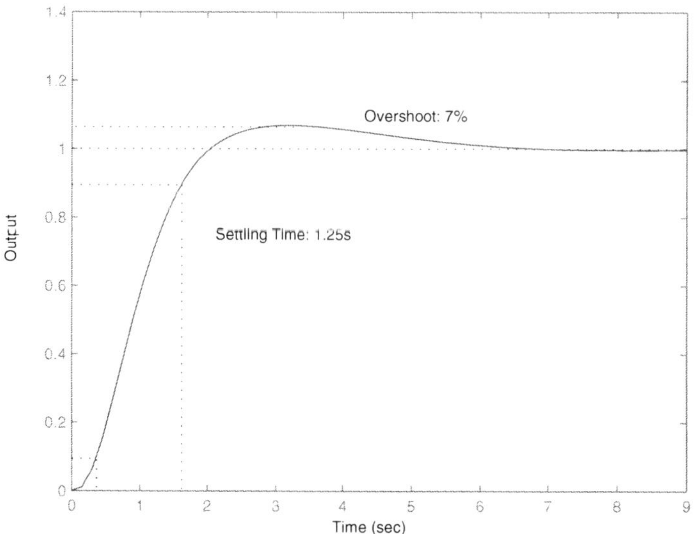

Fig. 6.22. Closed-loop Step Response for Example 6.6

Rewrite (6.29) as

$$\frac{1}{K_p} + \overline{G}(s) = 0. \tag{6.31}$$

It always has $\rho_{1,2}$ as its two roots by our construction. These two lie outside the desired region. We want no more to ensure dominant pole placement. Equivalently, we want the modified Nyquist plot of $\overline{G}(s)$ to have the number of clockwise encirclements with respect to $(-1/K_p, j0)$ equal to 2 minus the number of poles of $\overline{G}(s)$ outside the desired region. This condition will determine the interval of K_p such that roots of (6.31) other than two dominant poles are in the desired region.

To find the poles of $\overline{G}(s)$ located outside the desired region, note that they are simply the roots of its denominator. Thus, we construct another characteristic equation from the denominator of $\overline{G}(s)$ in (6.30) as follows:

$$1 + \overline{G}_o(s) = 0, \tag{6.32}$$

where

$$\overline{G}_o(s) = \frac{X_2 N(s)s^2 - \left(a^2 + b^2\right) X_1 N(s)}{D(s)s} e^{-Ls}.$$

$\overline{G}_o(s)$ has its rational part with the degrees of its nominator and denominator being equal to those of the open-loop transfer function's nominator and denominator, respectively. The number of the roots of (6.32), that is, poles of $\overline{G}(s)$ lying outside the desired region, equals the number of clockwise encirclements of the modified Nyquist plot of $\overline{G}_o(s)$ with respect to $(-1, j0)$, plus the number of poles of $\overline{G}_o(s)$ located outside the desired region. The latter is easy to find from the known denominator of $\overline{G}_o(s)$, which is, $D(s)s$.

The design procedure is summarized as follows.

Step 1. Find the poles of $\overline{G}_o(s)$ (the roots of $D(s)s$) outside the desired region and name its total number as $P^+_{\overline{G}_o}$;

Step 2. Draw the modified Nyquist plot of $\overline{G}_o(s)$, count the number of clockwise encirclements with respect to the $-1 + j0$ point as $N^+_{\overline{G}_o}$, and obtain the number of poles of $\overline{G}(s)$ outside the desired region as $P^+_{\overline{G}} = N^+_{\overline{G}_o} + P^+_{\overline{G}_o}$;

Step 3. Draw the modified Nyquist plot of $\overline{G}(s)$ and find the range of K_p during which the clockwise encirclements with respect to the $(-1/K_p, j0)$ is $2 - P^+_{\overline{G}}$.

Example 6.7 illustrates the design procedure in detail.

Example 6.7. Consider a highly oscillatory process,

$$G(s) = \frac{1}{s^2 + s + 5}e^{-0.1s}.$$

If the overshoot is to be not larger than 10% and the settling time to be less than 15 s, the dominant poles are $\rho_{1,2} = -0.2751 \pm 0.3754j$. Equation (6.26) becomes

$$\begin{cases} K_i = 0.3937K_p + 1.8773, \\ K_d = 1.8173K_p + 7.7760. \end{cases}$$

We have

$$\overline{G}_o(s) = \frac{7.776s^2 + 1.877}{s(s^2 + s + 5)}e^{-0.1s}.$$

Let $m = 3$, then we have $ma = 0.8253$ and all three poles of $\overline{G}_o(s)$ outside the desired region and $P^+_{\overline{G}_o} = 3$. Fig. 6.23 is the modified Nyquist plot of $\overline{G}_o(s)$ and there is one anti-clockwise encirclement of the point $(-1, j0)$, that is, $N^+_{\overline{G}_o} = -1$. Therefore, $\overline{G}(s)$ has two poles located in the desired region since $P^+_{\overline{G}} = N^+_{\overline{G}_o} + P^+_{\overline{G}_o} = 2$. It means the modified Nyquist plot of $\overline{G}(s)$ should have its clockwise encirclement with respect to the point $(-1/K_p, j0)$, equal to $2 - P^+_{\overline{G}} = 0$, that is zero net encirclement, for two assigned poles to dominate all others. Fig. 6.24 shows the modified Nyquist plot of $\overline{G}(s)$, from which $-1/K_p \in (-\infty, -0.2851)$ is determined to have zero clockwise encirclement. A positive K_p could always make K_d and K_i positive. Therefore, we have the joint solution as $K_p \in (0, 3.5075)$. If $K_p = 1$ is chosen, the PID controller is

$$C(s) = 1 + \frac{2.2709}{s} + 9.5933s.$$

The zeros of the closed-loop system are at $s = -0.0521 \pm 0.4837j$, which are not near the dominant poles. Fig. 6.25 shows the step response of the closed-loop system.

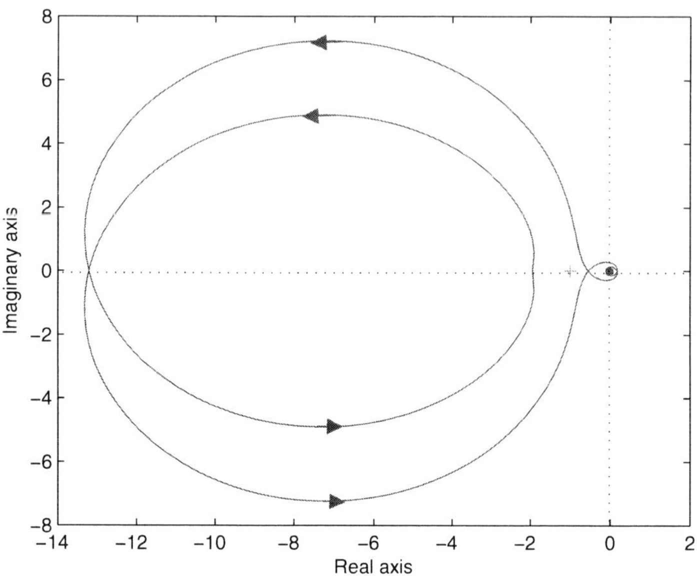

Fig. 6.23. Modified Nyquist Plot of $\overline{G}_o$ for Example 6.7

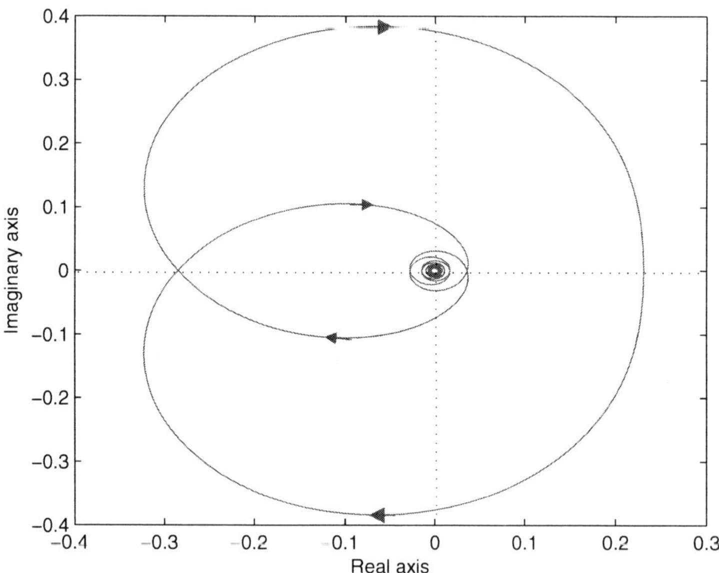

Fig. 6.24. Modified Nyquist Plot of $\overline{G}$ for Example 6.7

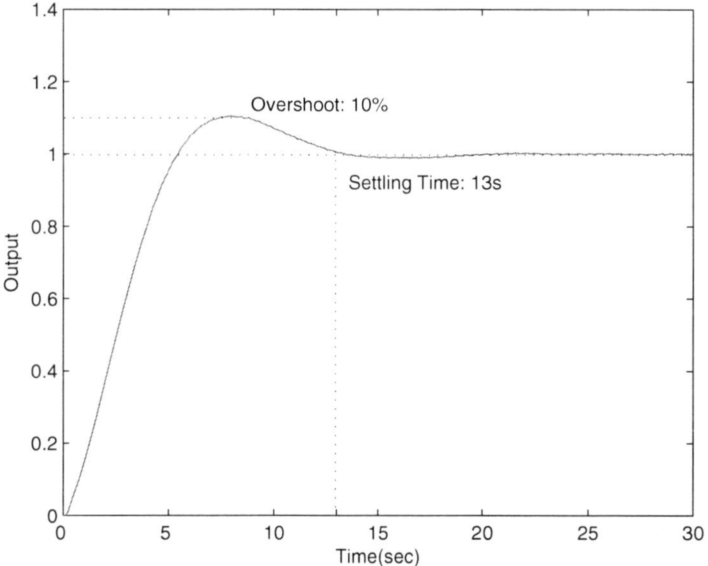

Fig. 6.25. Closed-loop Step Response for Example 6.7

6.4 Multivariable Case

In fact, many real-life industrial processes are multivariable in nature. It is of great interest and value to extend our single variable PID tuning method to multivarible PID controller design. Let $G(s) = [g_{ij}(s)]$ be the $m \times m$ multivariable process and $C(s) = [c_{ij}(s)]$ be the multivarible controller. To overcome the effects of cross-coupled interactions, a decoupler, $D(s) = [d_{ij}(s)]$, is designed first. By using the method proposed in [97], we have

$$d_{ji}(s) = \frac{G^{ij}(s)}{G^{ii}(s)} d_{ii}(s), \tag{6.33}$$

and $Q(s) = G(s)D(s)$ as

$$Q(s) = \text{diag}\{q_{ii}(s)\} = \text{diag}\left\{ \frac{|G(s)|}{G^{ii}(s)} d_{ii}(s) \right\},$$

where $G^{ij}(s)$ is cofactor corresponding to $g_{ij}(s)$ in $G(s)$. $q_{ii}(s)$ may be complicated to implement or even not rational and cannot be used to design controllers directly, so that model reduction techniques based on step tests [157] are applied to obtain rational and proper estimates of $q_{ii}(s)$, $\hat{q}_{ii}(s)$. With the PID tuning methods proposed in the above sections, single variable PID controllers, $k_{ii}(s)$, $i = 1, \cdots, m$, are designed for $\hat{q}_{ii}(s)$, $i = 1, \cdots, m$, and the multivariable controller $C(s)$, with

$$c_{ij}(s) = d_{ij}(s)k_{jj}(s), \tag{6.34}$$

is obtained. Suppose $\hat{C}(s) = [\hat{c}_{ij}(s)]$ is a multivariable PID controller. If $c_{ij}(s)$ in $C(s)$ is PID type, we choose $\hat{c}_{ij}(s) = c_{ij(s)}$. For $c_{ij}(s)$, which is not PID type, its estimate in form of PID, $\hat{c}_{ij}(s)$, is obtained by using model reduction techniques in [158]. The multivariable PID controller $\hat{C}(s)$ is then designed for $G(s)$.

Example 6.8. Consider a multivariable process,

$$G(s) = \begin{bmatrix} \dfrac{1}{s+1} & \dfrac{1}{s+2} \\ \dfrac{1}{s+3} & \dfrac{1}{s+1.5} \end{bmatrix}.$$

By choosing $d_{11}(s) = d_{22}(s) = 1$, the decoupler is designed as follows

$$D(s) = \begin{bmatrix} 1 & -\dfrac{s+1}{s+2} \\ -\dfrac{s+1.5}{s+3} & 1 \end{bmatrix},$$

according to (6.33). We have

$$Q(s) = \begin{bmatrix} \dfrac{2.5s+4.5}{(s+1)(s+2)(s+3)} & 0 \\ 0 & \dfrac{2.5s+4.5}{(s+1.5)(s+2)(s+3)} \end{bmatrix}.$$

One first-order time delay model of $Q(s)$ is obtained by using the method proposed in [157],

$$\hat{Q}(s) = \begin{bmatrix} \dfrac{0.7597e^{-0.286s}}{s+1.013} & 0 \\ 0 & \dfrac{0.7671e^{-0.288s}}{s+1.534} \end{bmatrix}.$$

Approximate Dominant Pole placement

For $\hat{q}_{11}(s) = 0.7597e^{-0.286s}/(s+1.013)$, suppose that the desired damping ratio is $\xi = 0.6$ and T_s is calculated from (6.1) as 7.13. The proposed single variable PID tuning method leads to

$$k_{11}(s) = 0.1621 \left(1 + \frac{1}{0.1718s} - 1.3291s \right).$$

For $\hat{q}_{22}(s) = 0.7671e^{-0.288s}/(s+1.534)$, suppose that the desired damping ratio is $\xi = 0.6$ and T_s is calculated from (6.1) as 5.52. The proposed single variable PID tuning method leads to

$$k_{22}(s) - 0.2599 \left(1 + \frac{1}{0.1488s} - 0.4296s \right).$$

$C(s)$ is calculated according to

$$C(s) = \begin{bmatrix} k_{11}(s) & k_{22}(s)d_{12}(s) \\ k_{11}(s)d_{21}(s) & k_{22}(s) \end{bmatrix}.$$

$c_{12}(s) = k_{22}(s)d_{12}(s)$ and $c_{21}(s) = k_{11}(s)d_{12}(s)$ are high-order controllers. By using the method in [158], we have

$$\hat{c}_{12}(s) = -0.5540 - \frac{0.8733}{s} + 0.1976s,$$

and

$$\hat{c}_{21}(s) = -0.2459 - \frac{0.4718}{s} + 0.1378s,$$

respectively. $\hat{C}(s)$ is

$$\hat{C}(s) = \begin{bmatrix} 0.1621 + \dfrac{0.9435}{s} - 0.2154s & -0.5540 - \dfrac{0.8733}{s} + 0.1976s \\ -0.2459 - \dfrac{0.4718}{s} + 0.1378s & 0.2599 + \dfrac{1.7466}{s} - 0.1117s \end{bmatrix}.$$

The step responses of the resultant multivariable PID control system to unit set-point changes are shown in Fig. 6.26. For the first loop, the settling time of the multivariable PID control system is 6.95 and the overshoot is 12.47% with the corresponding damping ratio of 0.55. For the second loop, the settling time is 5.11 and the overshoot is 10.68% with the corresponding damping ratio of 0.58. Step responses of the original control system with $C(s)$ as the controller are also given in Fig. 6.26 for comparison.

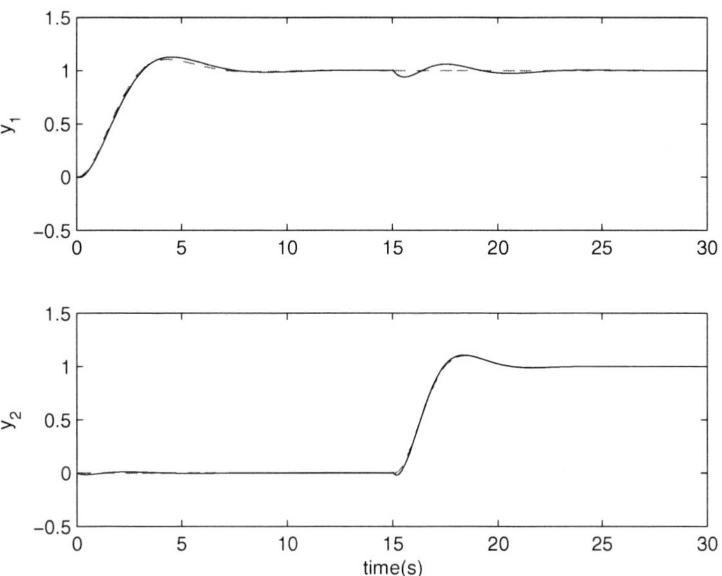

Fig. 6.26. Step response of Example 6.8. (Solid line: $\hat{C}(s)$; dash line: $C(s)$)

Guaranteed Dominant Pole placement

For $q_{11}(s) = (2.5s + 4.5)/[(s+1)(s+2)(s+3)]$, suppose that the desired damping ratio is $\xi = 0.6$ and $T_s = 7.13$. The dominant poles are $0.5610 \pm 0.7480i$. We take $m = 3$. Choosing $K_{p,11} = 1$, the following PID controller is obtained as

$$k_{11}(s) = 1 + \frac{1.5595}{s} + 0.5159s,$$

with the closed-loop poles as -4.3018, -1.8659 and $-0.5610 \pm 0.7480i$. For $q_{22}(s) = (2.5s + 4.5)/[(s+1.5)(s+2)(s+3)]$, suppose that the desired damping ratio is $\xi = 0.6$ and T_s is 7.13. The dominant poles are $0.5610 \pm 0.7480i$. Choosing $K_{p,22} = 1$, the following PID controller is obtained as

$$k_{22}(s) = 1 + \frac{2.0703}{s} + 0.9231s,$$

with the closed-loop poles as -5.8705, -1.8152 and $-0.5610 \pm 0.7480i$.
   $C(s)$ is calculated according to

$$C(s) = \begin{bmatrix} k_{11}(s) & k_{22}(s)d_{12}(s) \\ k_{11}(s)d_{21}(s) & k_{22}(s) \end{bmatrix}.$$

$c_{12}(s) = k_{22}(s)d_{12}(s)$ and $c_{21}(s) = k_{11}(s)d_{12}(s)$ are high-order controllers. By using the method in [158], their PID estimates are obtained and we have

$$\hat{C}(s) = \begin{bmatrix} 1 + \dfrac{1.5595}{s} + 0.5159s & -1.0029 - \dfrac{1.0352}{s} - 0.4668s \\ -0.7562 - \dfrac{0.7798}{s} - 0.3404s & 1 + \dfrac{2.0703}{s} + 0.9231s \end{bmatrix}.$$

The step responses of the resultant multivariable PID control system to unit set-point changes are shown in Fig. 6.27. Step responses of the original control system with the controller of $C(s)$ are also given in Fig. 6.27 for comparison.

Example 6.9. Consider the Vinate and Luyben plant,

$$G(s) = \begin{bmatrix} \dfrac{-0.2e^{-s}}{7s+1} & \dfrac{1.3e^{-0.3s}}{7s+1} \\ \dfrac{-2.8e^{-1.8s}}{9.5s+1} & \dfrac{4.3e^{-0.35s}}{9.2s+1} \end{bmatrix}.$$

By choosing $d_{11}(s) = 1$ and $d_{22}(s) = e^{-0.7s}$, the decoupler is designed as follows

$$D(s) = \begin{bmatrix} 1 & 6.5 \\ \dfrac{2.8(9.2s+1)e^{-1.45s}}{4.3(9.5s+1)} & e^{-0.7s} \end{bmatrix},$$

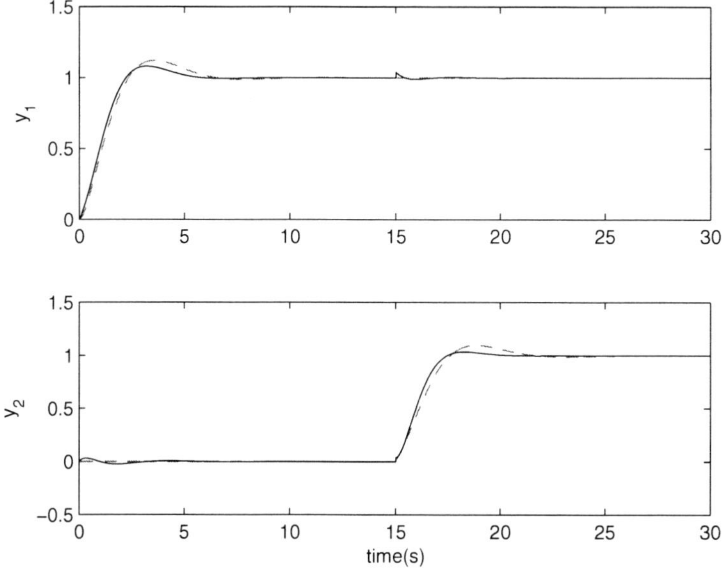

Fig. 6.27. Step Response for Example 6.8 (Solid line: $\hat{C}(s)$; dash line: $C(s)$)

according to (6.33). One first-order time delay model of $Q(s) = G(s)D(s)$ is obtained by using the method proposed in [157],

$$\hat{Q}(s) = \begin{bmatrix} \dfrac{0.08677\mathrm{e}^{-1.86s}}{s+0.1342} & 0 \\ 0 & \dfrac{-1.459\mathrm{e}^{-2.27s}}{s+0.105} \end{bmatrix}.$$

Approximate Dominant Pole Placement

For $\hat{q}_{11}(s) = 0.08677\mathrm{e}^{-1.86s}/(s+0.1342)$, suppose that the desired damping ratio is $\xi = 0.7$ and T_s is calculated from (6.1) as 47.48. The proposed single variable PID tuning method leads to

$$k_{11}(s) = 0.2612\left(1 + \frac{1}{1.9506s} - 7.1775s\right).$$

For $\hat{q}_{22}(s) = -1.459\mathrm{e}^{-2.27s}/(s+0.105)$, suppose that the desired damping ratio is $\xi = 0.7$ and T_s is calculated from (6.1) as 60.00. The proposed single variable PID tuning method leads to

$$k_{22}(s) = -0.0118\left(1 + \frac{1}{2.3800s} - 10.1746s\right).$$

After $C(s)$ is calculated from $D(s)$ and k_{ii}, $i = 1, 2$, according to (6.34), $\hat{C}(s)$ is obtained as

$$\hat{C}(s) = \begin{bmatrix} 0.2612 + \dfrac{0.1339}{s} - 1.8748s & -0.0767 - \dfrac{0.0322}{s} + 0.7804s \\[2ex] 0.1540 + \dfrac{0.0872}{s} - 1.1404s & -0.0072 - \dfrac{0.0050}{s} + 0.1264s \end{bmatrix}.$$

The step responses of the resultant multivariable PID control system to unit set-point changes are shown in Fig. 6.28. For the first loop, the settling time of the multivariable PID control system is 49.84 and the overshoot is 6.6% with the corresponding damping ratio of 0.65. For the second loop, the settling time is 67.76 and the overshoot is 7.34% with the corresponding damping ratio of 0.64. Step responses of the original control system with the controller of $C(s)$ are also given in Fig. 6.28 for comparison.

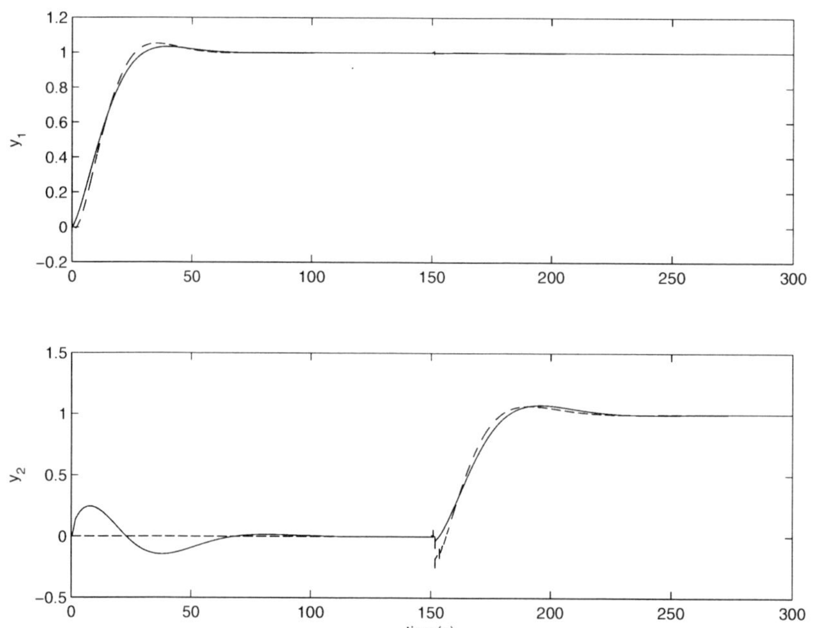

Fig. 6.28. Step response of Example 6.9 (Solid line: $\hat{C}(s)$; dash line: $C(s)$)

Guaranteed Dominant Pole Placement

For $\hat{q}_{11}(s) = 0.08677e^{-1.86s}/(s + 0.1342)$, suppose that the desired damping ratio is $\xi = 0.7$ and T_s is 47.48. The dominant poles are $-0.0842 \pm 0.0859i$. We take $m = 3$. The proposed method leads to $-1/K_{p,11} \in (-\infty, -0.569)$ for two assigned poles to dominate all others. We choose $K_{p,11} = 0.85$ and the following PID controller is obtained as

$$k_{11}(s) = 0.8500 + \frac{0.1803}{s} + 1.8096s.$$

For $\hat{q}_{22}(s) = -1.459\mathrm{e}^{-2.27s}/(s+0.105)$, suppose that the desired damping ratio is $\xi = 0.7$ and T_s is 60.00. The dominant poles are $-0.0668 \pm -0.0681\mathrm{i}$. We take $m = 3$. The proposed method leads to $1/K_{p,22} \in (-\infty, -11.86)$ for two assigned poles to dominate all others. We choose $K_{p,22} = -0.04$ and the following PID controller is obtained as

$$k_{22}(s) = -0.0400 - \frac{0.0067}{s} - 0.1031s.$$

After $C(s)$ is calculated from $D(s)$ and k_{ii}, $i = 1,2$, according to (6.34), $\hat{C}(s)$ is obtained as

$$\hat{C}(s) = \begin{bmatrix} 0.8500 + \dfrac{0.1803}{s} + 1.8096s & -0.2600 - \dfrac{0.0435}{s} - 0.6701s \\[2ex] 0.5197 + \dfrac{0.1174}{s} + 1.3221s & -0.0354 - \dfrac{0.0067}{s} - 0.0767s \end{bmatrix}.$$

The step responses of the resultant multivariable PID control system to unit set-point changes are shown in Fig. 6.29. Step responses of the original control system with the controller of $C(s)$ are also given in Fig. 6.29 for comparison.

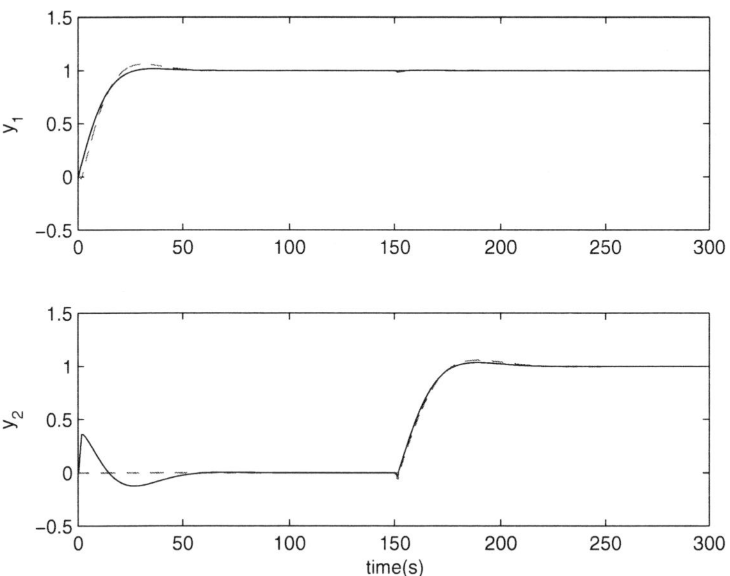

Fig. 6.29. Step Response for Example 6.9 (Solid line: $\hat{C}(s)$; dash line: $C(s)$)

Example 6.10. Consider the well-known Wood/Berry process,

$$G(s) = \begin{bmatrix} \dfrac{12.8\mathrm{e}^{-s}}{16.7s+1} & \dfrac{-18.9\mathrm{e}^{-3s}}{21.0s+1} \\[2ex] \dfrac{6.6\mathrm{e}^{-7s}}{10.9s+1} & \dfrac{-19.4\mathrm{e}^{-3s}}{14.4s+1} \end{bmatrix}.$$

By choosing $d_{11}(s) = d_{22}(s) = 1$, the decoupler is designed as follows

$$D(s) = \begin{bmatrix} 1 & \dfrac{(315.63s + 18.90)\mathrm{e}^{-2s}}{268.80s + 12.80} \\ \dfrac{(95.04s + 6.60)\mathrm{e}^{-4s}}{211.46s + 19.40} & 1 \end{bmatrix},$$

according to (6.33). One first-order time delay model of $Q(s) = G(s)D(s)$ is obtained as,

$$\hat{Q}(s) = \begin{bmatrix} \dfrac{6.374\mathrm{e}^{-1.065s}}{5.414s + 1} & 0 \\ 0 & \dfrac{-9.691\mathrm{e}^{-3.12s}}{7.942s + 1} \end{bmatrix}.$$

Approximate Dominant Pole Placement

For $\hat{q}_{11}(s) = 6.374\mathrm{e}^{-1.065s}/(5.414s + 1)$, suppose that the desired damping ratio is $\xi = 0.7$ and T_s is calculated from (6.1) as 32.35. The proposed single variable PID tuning method leads to

$$k_{11}(s) = 0.0204\left(1 + \frac{1}{1.0388s} - 9.4551s\right).$$

For $\hat{q}_{22}(s) = -9.691\mathrm{e}^{-3.12s}/(7.942s + 1)$, suppose that the desired damping ratio is $\xi = 0.7$ and T_s is calculated from (6.1) as 59.14. The proposed single variable PID tuning method leads to

$$k_{22}(s) = -0.0187\left(1 + \frac{1}{2.7246s} - 3.0100s\right).$$

After $C(s)$ is calculated from $D(s)$ and k_{ii}, $i = 1,2$, according to (6.34), $\hat{C}(s)$ is obtained as

$$\hat{C}(s) = \begin{bmatrix} 0.0204 + \dfrac{0.0196}{s} - 0.1929s & 0.0073 - \dfrac{0.0101}{s} - 0.4114s \\ 0.0287 + \dfrac{0.0067}{s} - 0.2643s & -0.0187 - \dfrac{0.0069}{s} + 0.0563s \end{bmatrix}.$$

The step responses of the resultant multivariable PID control system to unit set-point changes are shown in Fig. 6.30. Step responses of the original control system with the controller of $C(s)$ are also given in Fig. 6.30 for comparison. The original control system can achieve the desired performance approximately. The performance of the resultant multivariable PID control is not good as the original control system, but it is still acceptable.

Guaranteed Dominant Pole Placement

For $\hat{q}_{11}(s) = 6.374\mathrm{e}^{-1.065s}/(5.414s + 1)$, suppose that the desired damping ratio is $\xi = 0.7$ and T_s is 32.35. The dominant poles are $-0.1236 \pm 0.1261i$. We take $m = 3$. The proposed method leads to $-1/K_{p,11} \in (-\infty, -4.8489)$ for two assigned poles to dominate all others. We choose $K_{p,11} = 0.2$ and the following PID controller is obtained as

$$k_{11}(s) = 0.200 + \frac{0.0416}{s} + 0.5470s.$$

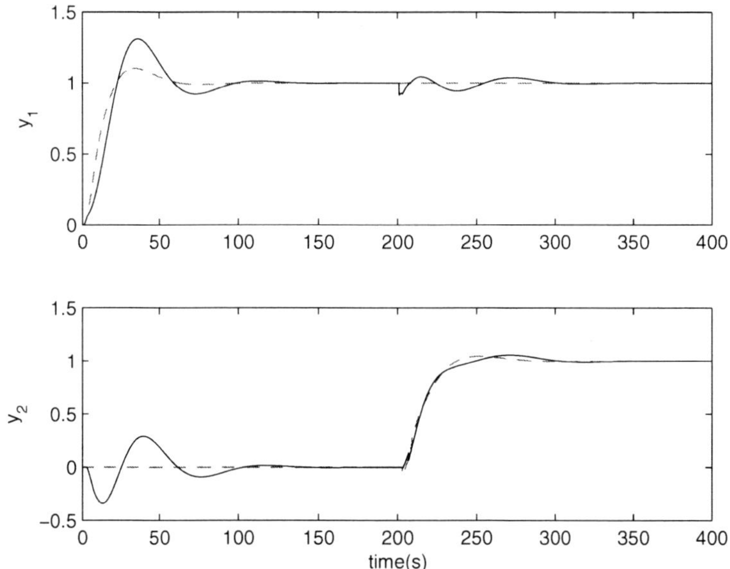

Fig. 6.30. Step response of Example 6.10 (Solid line: $\hat{C}(s)$; dash line: $C(s)$)

For $\hat{q}_{22}(s) = -9.691e^{-3.12s}/(7.942s + 1)$, suppose that the desired damping ratio is $\xi = 0.7$ and T_s is 59.14. The dominant poles are $0.0676 \pm 0.0690i$. We take $m = 3$. The proposed method leads to $1/K_{p,22} \in (-\infty, -12.08)$ for two assigned poles to dominate all others. We choose $K_{p,22} = -0.08$ and the following PID controller is obtained as

$$k_{22}(s) = -0.0800 - \frac{0.0110}{s} - 0.4143s.$$

After $C(s)$ is calculated from $D(s)$ and k_{ii}, $i = 1, 2$, according to (6.34), $\hat{C}(s)$ is obtained as

$$\hat{C}(s) = \begin{bmatrix} 0.2000 + \dfrac{0.0416}{s} + 0.5470s & -0.0682 - \dfrac{0.0162}{s} - 0.3707s \\ 0.1103 + \dfrac{0.0142}{s} - 0.0200s & -0.0800 - \dfrac{0.0110}{s} - 0.4143s \end{bmatrix}.$$

The step responses of the resultant multivariable PID control system to unit set-point changes are shown in Fig. 6.31. Step responses of the original control system with the controller of $C(s)$ are also given in Fig. 6.31 for comparison. The original control system can achieve the desired performance approximately. The performance of the resultant multivariable PID control is not good as the original control system, but it is still acceptable.

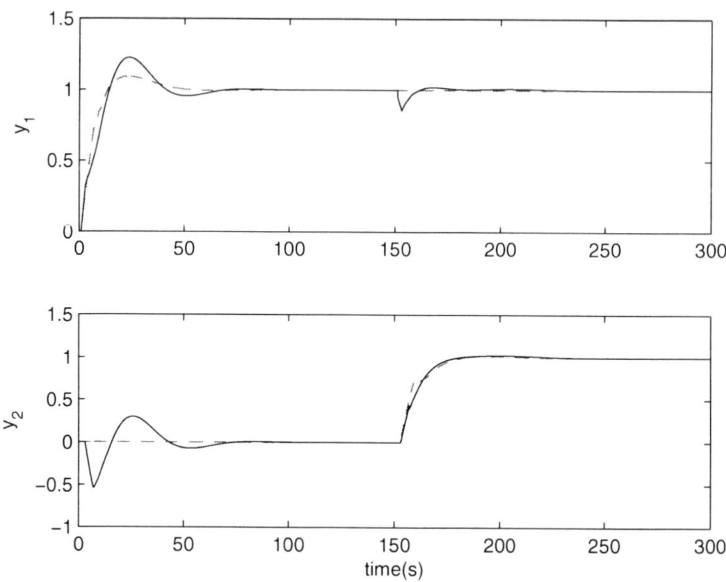

Fig. 6.31. Step Response for Example 6.10 (Solid line: $\hat{C}(s)$; dash line: $C(s)$)

6.5   Conclusion

In this chapter, two methods on approximate and guaranteed pole placement respectively are proposed.

For approximate pole placement, an analytical PID design method has been presented for continuous-time delay systems to achieve approximate pole placement with dominance. It greatly simplifies the continuous infinite spectrum assignment problem with a delay process to a 3rd-order pole placement problem in discrete domain for which the closed-form solution exists and is converted back to its continuous PID controller. The method works well for both monotonic and oscillatory processes of low or high order. Finally, the method is extended to multivariable cases successfully.

For guaranteed pole placement, two simple yet effective methods have been presented for guaranteed dominant pole placement by PID, based on Root locus and Nyquist plot, respectively. Each method is demonstrated with examples. Obviously, the methods are not limited to PID controllers. They can be extended to other controllers where one controller parameter is used as the variable gain and all other parameters are solved in terms of this gain to meet the fixed pole requirements.

7 Optimal Multivariable PID Control Based on LMI Approach

In the previous two chapters, design methods for multivariable PID control have been demonstrated based on IMC and dominant pole placement, respectively. They are in line with traditional design methods commonly used for process control and are broadly accepted in practice. On the other hand, most of the developments in the modern control theory have not been applied to PID control due to their complexity or inconvenience of use. It is our wish to bridge the modern control theory and the industrial control practice which could benifit from the advancement in the former. This chapter aims to make an initial step towards solving this problem. The basic idea is to transform the PID control to the equivalent static output feedback (SOF) control by augmenting to the process with some new state variables induced by the PID controller such that the well established results in SOF field can be employed to design a multivariable PID controller for various specifications such as stability, H_2/H_∞ performance and maximum output control.

7.1 Introduction

The static output feedback plays a very important role in control theory and applications. Recently, it has attracted considerable attention (see e.g. [159, 85, 86, 160, 161, 162, 163] and references therein). Yet, it is still left with some open problems. Unlike the state feedback case, a SOF gain which stabilizes the system is not easy to find. Linear matrix inequality (LMI) [89] is one of the most effective and efficient tools in controller design and a great deal of LMI-based design methods of SOF design have been proposed in the last decade [90, 164, 91, 165, 166, 167, 168, 169, 170, 171, 172]. Among these methods, an iterative linear matrix inequality (ILMI) method was proposed by Cao et al. [90] and later employed to solve some multivariable PID controller design problems [79, 80]. In this context, a new additional matrix-valued variable is introduced so that the involved stability condition becomes conservative (sufficient but far from necessary). The iterative algorithm in [90] tried to find a sequence of the additional variables such that the relevant sufficient condition is close to the necessary and sufficient one. The similar idea is used in the so-called substitutive LMI method in [172]. In both works, the additional matrix variables are updated at the current

Q.-G. Wang et al.: PID Control for Multivariable Processes, LNCIS 373, pp. 167–202, 2008.
springerlink.com
© Springer-Verlag Berlin Heidelberg 2008

iteration step using the decision variables (matrix-valued) obtained in the preceding step. With additional variables, the dimensions of the LMIs become higher. It is possible that the decision variables obtained in the preceding step can be used in the next one directly without introducing the additional variables and the dimensions of the LMIs need not be increased. In addition, we will develop some efficient way to get suitable initial values for some decision variable in the iterative procedure, which has not been dealt with in the existing approaches.

Notation: $\mathbb{R}^n$ denotes the n-dimensional real Euclidean space; I is an identity matrix with an appropriate dimension; the superscripts 'T' and '-1' stand for the matrix transpose and inverse, respectively; $W > 0$ ($W \geq 0$) means that W is real, symmetric and positive-definite (positive-semidefinite); $\|\cdot\|$ denotes either the Euclidean vector norm or the induced matrix 2-norm.

7.2 Transformation from PID Controllers to SOF Controllers

7.2.1 Transformation

Consider the linear time-invariant system

$$\dot{x} = Ax + Bu, \qquad y = Cx, \tag{7.1}$$

with the following PID controller

$$u = F_1 y + F_2 \int_0^t y \, dt + F_3 \frac{dy}{dt}, \tag{7.2}$$

where $x(t) \in \mathbb{R}^n$ is state variable, $u(t) \in \mathbb{R}^l$ is control inputs, $y(t) \in \mathbb{R}^m$ is outputs, A, B, C are matrices with appropriate dimensions, and $F_1, F_2, F_3 \in \mathbb{R}^{l \times m}$ are matrices to be designed. We assume that system (7.1) is one of the minimum state space realizations of some transfer matrix $H(s)$. Controller (7.2) is an ideal PID controller. Readers are referred to [2] for how to change the ideal PID controllers into practical ones and the relationships between the parameters of the two kinds of the controllers. Let $z_1 = x$, $z_2 = \int_0^t y \, dt$. Denote $z = [z_1^T, z_2^T]^T$. The variable z can be viewed as the state vector of a new system, whose dynamics is governed by

$$\dot{z}_1 = \dot{x} = A z_1 + Bu, \quad \dot{z}_2 = y = C z_1.$$

i.e.,

$$\dot{z} = \bar{A}z + \bar{B}u, \qquad \bar{A} = \begin{bmatrix} A & 0 \\ C & 0 \end{bmatrix}, \qquad \bar{B} = \begin{bmatrix} B \\ 0 \end{bmatrix}.$$

Combining equation (7.1) and the definition of z yields

$$y = C z_1 = \begin{bmatrix} C & 0 \end{bmatrix} z, \quad \int_0^t y \, dt = z_2 = \begin{bmatrix} 0 & I \end{bmatrix} z,$$

$$\frac{dy}{dt} = C\dot{x} = CAx + CBu = \begin{bmatrix} CA & 0 \end{bmatrix} z + CBu.$$

Denoting $\bar{C}_1 := \begin{bmatrix} C & 0 \end{bmatrix}$, $\bar{C}_2 := \begin{bmatrix} 0 & I \end{bmatrix}$, $\bar{C}_3 := \begin{bmatrix} CA & 0 \end{bmatrix}$, and $\bar{y}_i := \bar{C}_i z$, $i = 1, 2, 3$, we have

$$u = F_1 \bar{y}_1 + F_2 \bar{y}_2 + F_3 \bar{y}_3 + F_3 CBu. \tag{7.3}$$

Suppose the matrix $I - F_3 CB$ is invertible. Let

$$\begin{aligned}
\bar{y} &= \begin{bmatrix} \bar{y}_1^T & \bar{y}_2^T & \bar{y}_3^T \end{bmatrix}^T, \qquad \bar{C} = \begin{bmatrix} \bar{C}_1^T & \bar{C}_2^T & \bar{C}_3^T \end{bmatrix}^T, \\
\bar{F} &= \begin{bmatrix} \bar{F}_1 & \bar{F}_2 & \bar{F}_3 \end{bmatrix} \\
&= \begin{bmatrix} (I - F_3 CB)^{-1} F_1 & (I - F_3 CB)^{-1} F_2 & (I - F_3 CB)^{-1} F_3 \end{bmatrix}.
\end{aligned}$$

Thus the problem of PID controller design reduces to that of SOF controller design for the following system:

$$\dot{z} = \bar{A} z + \bar{B} u, \quad \bar{y} = \bar{C} z, \quad u = \bar{F} \bar{y}. \tag{7.4}$$

Once the composite matrix $\bar{F} = \begin{bmatrix} \bar{F}_1 & \bar{F}_2 & \bar{F}_3 \end{bmatrix}$ is found, the original PID gains can be recovered from

$$F_3 = \bar{F}_3 (I + CB\bar{F}_3)^{-1}, \quad F_2 = (I - F_3 CB)\bar{F}_2, \quad F_1 = (I - F_3 CB)\bar{F}_1. \tag{7.5}$$

The invertibility of matrix $I + CB\bar{F}_3$ is guaranteed by the following proposition.

Proposition 7.1. *Matrix $I - F_3 CB$ is invertible if and only if matrix $I + CB\bar{F}_3$ is invertible, where F_3 and $\bar{F}_3$ are related to each other by*

$$\bar{F}_3 = (I - F_3 CB)^{-1} F_3, \quad or \quad F_3 = \bar{F}_3 (I + CB\bar{F}_3)^{-1}.$$

Proof. Suppose matrix $I - F_3 CB$ is invertible. Then we have

$$\begin{aligned}
\Delta &:= \begin{bmatrix} I & -CB \\ F_3 & I - F_3 CB \end{bmatrix} \\
&= \begin{bmatrix} I & -CB(I - F_3 CB)^{-1} \\ 0 & I \end{bmatrix} \begin{bmatrix} I + CB\bar{F}_3 & 0 \\ 0 & I - F_3 CB \end{bmatrix} \\
&\quad \times \begin{bmatrix} I & 0 \\ (I - F_3 CB)^{-1} F_3 & I \end{bmatrix}. \tag{7.6}
\end{aligned}$$

On the other hand, we have

$$\begin{bmatrix} I & 0 \\ -F_3 & I \end{bmatrix} \begin{bmatrix} I & -CB \\ F_3 & I - F_3 CB \end{bmatrix} = \begin{bmatrix} I & -CB \\ 0 & I \end{bmatrix},$$

which implies that matrix Δ is invertible. From equation (7.6) we necessarily have that matrix $I + CB\bar{F}_3$ is also invertible. The converse claim can be proved similarly. $\square$

7.2.2 Well-Posedness of PID Control

In dynamic output feedback systems, the first question one would ask is whether the feedback interconnection makes sense. This is the well known well-posedness problem [173], which is defined as follows.

Definition 7.1 (Zhou et al., [173]). *A feedback system is said to be well-posed if all closed-loop transfer matrices are well-defined and proper.*

In our case, we have the following claim.

Proposition 7.2. *The closed-loop system is well-posed if and only if Condition 7.1 holds.*

Condition 7.1 *The matrix $I - F_3 CB$ is invertible.*

Proof. Sufficiency: If Condition 7.1 holds, we can solve u uniquely from (7.3). Thus the closed loop system can be expressed by the standard state space model (7.4). Therefore all the closed-loop transfer matrices for any outputs of the system are well-defined and proper.

Necessity: We prove it by contradiction. Suppose that $I - F_3 CB$ is singular. Then there is a vector $k \in \mathbb{R}^{1 \times l}$, $k \neq 0$, such that $k(I - F_3 CB) = 0$. Substituting this equation into (7.3) will lead to

$$
\begin{aligned}
0 &\equiv k(I - F_3 CB)u = k(F_1 \bar{y}_1 + F_2 \bar{y}_2 + F_3 \bar{y}_3) \\
&= (kF_1 \bar{C}_1 + kF_2 \bar{C}_2 + kF_3 \bar{C}_3)z =: \bar{k}z.
\end{aligned}
\tag{7.7}
$$

Notice that equation (7.7) is valid for all feedback matrices F_1 and F_2. Since $k \neq 0$, from the following equation

$$
\begin{aligned}
\bar{k} &= k \begin{bmatrix} F_1 & F_2 & F_3 \end{bmatrix} \begin{bmatrix} \bar{C}_1 \\ \bar{C}_2 \\ \bar{C}_3 \end{bmatrix} = k \begin{bmatrix} F_1 & F_2 & F_3 \end{bmatrix} \begin{bmatrix} C & 0 \\ 0 & I \\ CA & 0 \end{bmatrix} \\
&= \begin{bmatrix} kF_1 C + kF_3 CA & kF_2 \end{bmatrix},
\end{aligned}
$$

we can see that there exists an F_2 such that $\bar{k} \neq 0$. Now if we define $\bar{k}z$ as an output of the closed loop system, the transfer matrices from this output to the input u will have no definition. This is a contradiction and hence the proof is completed. $\square$

Equation (7.7) shows that if Condition 7.1 fails to hold, the state vector z will be confined in a reduced dimensional sub-space, which contradicts the definition of state variables. Notice that Condition 7.1 holds automatically if $CB = 0$.

There are two approaches to deal with Condition 7.1 in the design of feedback matrices. The first approach is to do *nothing* but post-checking whether $I + CB\bar{F}_3$ and hence $I - F_3 CB$ are invertible. This is based on the observation that the probability of the solved $\bar{F}_3$ which makes $I + CB\bar{F}_3$ singular is zero in the whole possible parameter space consisting of $\bar{F}_3$. The second approach is to add another LMI to the corresponding algorithms, which will be discussed later.

7.3 Feedback Stabilization with PID Controllers

7.3.1 Static Output Feedback Stabilization — Some Results Revisited

As noted in the previous section, our method is closely related to SOFS problem. Therefore some results on SOFS are first revisited to provide a preliminary. As is well known, SOFS problem is one of the most important open questions in control community, see, e.g., the survey by Syrmos, et al. [160]. Generally speaking, it seems that no analytic solutions to this problem exist and the only promising approach to follow is to exploit the special structure of particular problems. Yet this observation does not discourage ones from their enthusiasms to find numerical solutions to the problem. The efforts in this direction are reflected in the papers by Cao et al. [90], Crusius and Trofino [91], where the powerful technique of linear matrix inequalities is used, which makes the relevant solution algorithms very effective if the associated algorithms have solutions. Here we summarize the main result in Cao et al. [90] since it has a nice property that the approach there is independent of the particular state-space representation for the systems considered.

The objective of SOFS problem is to find a static output feedback controller

$$u = Fy, \tag{7.8}$$

where $F \in \mathbb{R}^{l \times m}$, such that the closed-loop system $\dot{x} = (A + BFC)x$ is asymptotically stable.

Lemma 7.1 (Cao et al. [90]). *System (7.1) is stabilizable via static output feedback if and only if there exist matrices $P > 0$ and F satisfying the following matrix inequality:*

$$A^T P + PA - PBB^T P + \left(B^T P + FC\right)^T \left(B^T P + FC\right) < 0. \tag{7.9}$$

Remark 7.1. It is easy to show that matrix inequality (7.9) is independent of a particular state-space realization of the systems considered. That is, if matrix inequality (7.9) has solutions to one minimum state space realization (A, B, C) of some plant $H(s)$, then it has solutions to any one of the minimum state space realization $(\bar{A}, \bar{B}, \bar{C})$ of $H(s)$ with the same feedback matrix F. It is a common practice that almost all industrialized PID controllers are designed based on frequency models of systems. Therefore we can pick up any one of the state space realizations of the system to study the possibility of output feedback stabilization if the frequency model of the system is in hand.

The negative sign of the term $-PBB^T P$ in matrix inequality (7.9) makes its solution very complicated. Suppose we can find a matrix Ψ which depends on P affinely and satisfies

$$\Psi \leq PBB^T P. \tag{7.10}$$

Then it is easy to show that system (7.1) can be stabilized by $u = Fy$ if the following inequality

$$A^T P + PA - \Psi + \left(B^T P + FC\right)^T \left(B^T P + FC\right) < 0 \tag{7.11}$$

has a solution for (P,F). By Schur complement [89], inequality (7.11) is equivalent to the following inequality

$$\begin{bmatrix} A^T P + PA - \Psi & (B^T P + FC)^T \\ B^T P + FC & -I \end{bmatrix} < 0. \tag{7.12}$$

Now matrix inequality (7.12) depends affinely on (P,F) once other parameters in Ψ are given.

In Cao et al. [90], Ψ is given by

$$\Psi = X^T BB^T P + P^T BB^T X - X^T BB^T X,$$

where $X > 0$. In this case, inequality (7.10) is always satisfied and the equal sign holds if and only if $X^T B = P^T B$. Once X is given, matrix inequality (7.12) can be solved very efficiently by LMI tool box in Matlab. Based upon the above consideration, the following iterative linear matrix inequality algorithm is proposed to solve SOFS problem [90].

Algorithm 7.1 (ILMI algorithm for SOFS). Given system's state space realization (A,B,C).

Step 1. Choose $Q_0 > 0$ and solve P for the Riccati equation

$$A^T P + PA - PBB^T P + Q_0 = 0, \quad P > 0.$$

Set $i = 1$ and $X_1 = P$.

Step 2. Solve the following optimization problem for P_i, F and α_i.

OP1: Minimize α_i subject to the following LMI constraints

$$\begin{bmatrix} \Sigma_{1i} & (B^T P_i + FC)^T \\ B^T P_i + FC & -I \end{bmatrix} < 0, \qquad P_i > 0 \tag{7.13}$$

where $\Sigma_{1i} = A^T P_i + P_i A - X_i BB^T P_i - P_i BB^T X_i + X_i BB^T X_i - \alpha_i P_i$. Denote by α_i^* the minimized value of α_i.

Step 3. If $\alpha_i^* \leq 0$, the matrix pair (P_i, F) solves SOFS problem. Stop. Otherwise go to Step 4.

Step 4. Solve the following optimization problem for P_i and F.

OP2: Minimize $\mathrm{tr}(P_i)$ subject to LMI constraints (7.13) with $\alpha_i = \alpha_i^*$, where tr stands for the trace of a square matrix. Denote by P_i^* the optimal P_i.

Step 5. If $\| X_i B - P_i^* B \| < \varepsilon$, where ε is a prescribed tolerance, go to Step 6; otherwise set $i := i + 1$, $X_i = P_i^*$, and go to Step 2.

Step 6. It cannot be decided by this algorithm whether SOFS problem is solvable. Stop.

For the properties of solution series α_i^* in optimization problem OP1 and P_i^* in optimization problem OP2 and the consideration on initial value selection, readers are referred to reference Cao et al. [90].

7.3.2   Feedback Stabilization with PID Controllers

Consider system (7.1) again, but now we use PID controller (7.2) instead of SOF controller (7.8). Our objective here is to design the feedback matrices F_1, F_2, F_3 such that system (7.1) is stabilized by the controller. Using the transformation given in Section 2, we can transform system (7.1) with controller (7.2) into system (7.4). It is evident that (7.1) is asymptotically stable if and only if (7.4) is asymptotically stable provided that Condition 7.1 holds. The stabilizing feedback matrices $(\bar{F}_1, \bar{F}_2, \bar{F}_3)$ in (7.4) can be found by solving $\bar{F}$ through the application of Algorithm 7.1 to system $(\bar{A}, \bar{B}, \bar{C})$.

As we have mentioned, there are two approaches to guarantee the well-posedness of PID control system. Here we discuss the second approach in detail. To guarantee the nonsingularity of the matrix $I + CB\bar{F}_3$, we add the following LMI

$$I + (CB\bar{F}_3)^T + CB\bar{F}_3 > 0 \tag{7.14}$$

to Algorithm 7.1, i.e., Steps 2 and 4 in Algorithm 7.1 are changed to the following

Step 2′.   Solve the optimization problem for P_i, $\bar{F}$ and α_i: Minimize α_i subject to the constraints LMIs (7.13) and (7.14) (OP 1).

Step 4′.   Solve the optimization problem for P_i and $\bar{F}$: Minimize $\mathrm{tr}(P_i)$ subject to the constraints LMIs (7.13) and (7.14) with $\alpha_i = \alpha_i^*$ (OP 2).

The inequality (7.14) comes from the observation that if (7.14) holds, we have

$$(I + CB\bar{F}_3)^T (I + CB\bar{F}_3) = I + (CB\bar{F}_3)^T + CB\bar{F}_3 + (CB\bar{F}_3)^T (CB\bar{F}_3) > 0.$$

Thus follows that $I + CB\bar{F}_3$ is nonsingular. Notice that LMI (7.14) is a very conservative condition.

What we suggest is to first try the first approach, i.e., post-checking if $I + CB\bar{F}_3$ is invertible without using the constraint (7.14). If it fails, use the above modified algorithm with constraint (7.14).

7.4 H_2 Suboptimal Control with PID Controllers

In this section, the design problem of PID controllers under H_2 performance specification is investigated. First we study SOF case and then we extend the result to PID case. Consider the system

$$\dot{x} = Ax + B_1 w + B_2 u, \quad y_s = C_s x, \quad y_r = C_r x, \tag{7.15}$$

where $x(t) \in \mathbb{R}^n$ is state variable, $u(t) \in \mathbb{R}^{l_1}$ is system inputs which are manipulatable by the controller, $w(t) \in \mathbb{R}^{l_2}$ is system inputs which, standing for some exogenous disturbances, are not manipulatable by the controller, $y_s(t) \in \mathbb{R}^{m_1}$ is the vector of sensed or measured outputs, $y_r(t) \in \mathbb{R}^{m_2}$ is the vector of regulated or controlled outputs, and A, B_1, B_2, C_s, C_r are matrices with appropriate dimensions. The static output feedback H_2 control (SOFH_2) problem is to find a control of the form

$$u = Fy_s, \tag{7.16}$$

such that the closed-loop transfer function, T_{wy_r}, from w to y_r is stable and

$$\left\|T_{wy_r}\right\|_2 < \gamma, \tag{7.17}$$

where γ is a positive number and $\|\cdot\|_2$ denotes 2-norm of system transfer matrix.

On substitution of (7.16) into (7.15), it is easy to find [173] that

$$\left\|T_{wy_r}\right\|_2^2 = \mathrm{tr}\left(C_r L_c C_r^T\right),$$

where L_c is the controllability Gramian of system (7.15) satisfying

$$(A + B_2 F C_s)L_c + L_c(A + B_2 F C_s)^T + B_1 B_1^T = 0,$$

supposing that $A + B_2 F C_s$ is Hurwitz. It is well known that for any positive definite matrix P satisfying

$$(A + B_2 F C_s)P + P(A + B_2 F C_s)^T + B_1 B_1^T < 0, \tag{7.18}$$

the relationship $P > L_c$ holds. Notice that the condition that $A + B_2 F C_s$ is Hurwitz is implied by inequality (7.18). Thus if

$$\mathrm{tr}(C_r P C_r^T) < \gamma^2, \tag{7.19}$$

the requirement (7.17) is satisfied. Now we try to develop an ILMI algorithm to solve inequalities (7.18)-(7.19).

Lemma 7.2. *For any fixed A, B_1, B_2, C_s and F, there exists a positive definite matrix P which solves inequality (7.18) if and only if the following inequality has a positive definite matrix solution*

$$AP + PA^T - PC_s^T C_s P + \left(B_2 F + P C_s^T\right)\left(B_2 F + P C_s^T\right)^T + B_1 B_1^T < 0. \tag{7.20}$$

Proof. The proof is similar to that of Theorem 1 of [90], where the term $B_1 B_1^T$ is absent. Due to this difference, the parameter ρ whose meaning is as in the proof of Theorem 1 of [90] should satisfy the condition that $\rho > 1$. Other steps are omitted.

Based upon the above arguments, the solution to $\mathrm{SOF}H_2$ problem can be summarized as:

Proposition 7.3. *The H_2 performance index (7.17) can be achieved by SOF controller (7.16) if the matrix inequalities (7.19) and (7.20) have solutions for (P, F).*

By the same reasoning as in Sect. 3, we can use the following algorithm to solve the feedback matrix F in Proposition 7.3.

Algorithm 7.2 (ILMI algorithm for $\mathrm{SOF}H_2$). Initial data: System's state space realization (A, B_1, B_2, C_s, C_r) and performance index γ.

Step 1. Choose $Q_0 > 0$ and solve P for the Riccati equation

$$AP + PA^T - PC_s^T C_s P + Q_0 = 0, \quad P > 0.$$

Set $i = 1$ and $X_1 = P$.

Step 2. Solve the following optimization problem for P_i, F and α_i.

OP1: Minimize α_i subject to the following LMI constraints

$$\begin{bmatrix} \Sigma_{2i} & B_2 F + P_i C_s^T \\ \left(B_2 F + P_i C_s^T\right)^T & -I \end{bmatrix} < 0 \qquad (7.21)$$

$$\mathrm{tr}\left(C_r P_i C_r^T\right) < \gamma^2 \qquad (7.22)$$

$$P_i > 0 \qquad (7.23)$$

where $\Sigma_{2i} = AP_i + P_i A^T + B_1 B_1^T - X_i C_s^T C_s P_i - P_i C_s^T C_s X_i + X_i C_s^T C_s X_i - \alpha_i P_i$. Denote by α_i^* the minimized value of α_i.

Step 3. If $\alpha_i^* \le 0$, the obtained matrix F solves SOFH_2 problem. Stop. Otherwise go to Step 4.

Step 4. Solve the following optimization problem for unknowns P_i, and F.

OP2: Minimize $\mathrm{tr}(P_i)$ subject to LMI constraints (7.21)–(7.23) with $\alpha_i = \alpha_i^*$. Denote by P_i^* the optimal P_i.

Step 5. If $\| X_i B - P_i^* B \| < \varepsilon$, where ε is a prescribed tolerance, go to Step 6; otherwise set $i := i+1$, $X_i = P_i^*$, and go to Step 2.

Step 6. It cannot be decided by this algorithm whether SOFH_2 problem is solvable. Stop.

Now consider system (7.15) and the performance specification (7.17) with PID controller

$$u = F_1 y_s + F_2 \int_0^t y_s \mathrm{d}t + F_3 \frac{\mathrm{d}y_s}{\mathrm{d}t}, \quad F_1, F_2, F_3 \in \mathbb{R}^{l_1 \times m_2} \qquad (7.24)$$

instead of static output feedback controller (7.16). We make the following assumption.

Assumption 7.1. *Suppose that each row vector in C_s and each column vector in B_1 are orthogonal, i.e., $C_s B_1 = 0$.*

The well-posedness of system (7.15) with dynamic feedback controller (7.24) implies that

Condition 7.2 *The matrix $I - F_3 C_s B_2$ is invertible.*

Define new matrices as

$$\bar{A} = \begin{bmatrix} A & 0 \\ C_s & 0 \end{bmatrix}, \quad \bar{B}_1 = \begin{bmatrix} B_1 \\ 0 \end{bmatrix}, \quad \bar{B}_2 = \begin{bmatrix} B_2 \\ 0 \end{bmatrix}$$

$$\bar{C}_{s1} = \begin{bmatrix} C_s & 0 \end{bmatrix}, \quad \bar{C}_{s2} := \begin{bmatrix} 0 & I \end{bmatrix}, \quad \bar{C}_{s3} := \begin{bmatrix} C_s A & 0 \end{bmatrix},$$

$$\bar{C}_s := \begin{bmatrix} \bar{C}_{s1}^T & \bar{C}_{s2}^T & \bar{C}_{s3}^T \end{bmatrix}^T, \quad \bar{C}_r := \begin{bmatrix} C_r & 0 \end{bmatrix},$$

$$\bar{F}_i := (I - F_3 C_s B_2)^{-1} F_i, \ i = 1,2,3, \quad \bar{F} := \begin{bmatrix} \bar{F}_1 & \bar{F}_2 & \bar{F}_3 \end{bmatrix}.$$

Then system (7.15) with PID controller (7.24) can be transformed into the SOF system:

$$
\begin{cases}
\dot{z} &= \bar{A}z + \bar{B}_1 w + \bar{B}_2 u, \ \text{with}\, z_1 = x, \ z_2 = \displaystyle\int_0^t y_s \mathrm{d}t. \\
\bar{y}_s &:= \bar{C}_s z, \quad \bar{y}_r := \bar{C}_r z, \\
u &= \bar{F}\bar{y}_s.
\end{cases}
\tag{7.25}
$$

It is evident that the relationship (e.g., transfer function) between w and y_r is preserved under the above transformation. Therefore the performance specification (7.17) can be achieved by PID controller (7.24) if and only if it can be achieved by SOF system (7.25) provided that Condition 7.2 holds. Thus the composite feedback matrix $\bar{F}$ in equation (7.25) can be found by applying Algorithm 7.2 to system (7.25).

7.5 H_∞ Suboptimal Control with PID Controllers

In this section, we investigate the design problem of PID controllers under H_∞ performance specification. We also begin with SOF case and then extend the result to PID case. Consider the system

$$
\dot{x} = Ax + B_1 w + B_2 u, \quad y_s = C_s x, \quad y_r = C_r x + Du,
\tag{7.26}
$$

where D is a constant matrix, and x, u, y_s, y_r, w, A, B_1, B_2, C_s, C_r are the same as (7.15). The static output feedback H_∞ suboptimal control (SOFH_∞) problem is to find a controller of the form (7.16) such that the closed-loop transfer function, T_{wy_r}, from w to y_r is stable and

$$
\|T_{wy_r}\|_\infty < v,
\tag{7.27}
$$

where $v > 0$ and $\|\cdot\|_\infty$ denotes H_∞-norm of system transfer matrix, see Zhou et al. [173].

From Bounded Real Lemma and Schur complement [89], it is not difficult to show that (7.16) is an H_∞ suboptimal controller if and only if there exist a positive definite matrix P such that

$$
\begin{bmatrix}
(A + B_2 FC_s)^T P + P(A + B_2 FC_s) & PB_1 & (C_r + DFC_s)^T \\
B_1^T P & -v^2 I & 0 \\
C_r + DFC_s & 0 & -I
\end{bmatrix} < 0,
$$

which is also equivalent to

$$
\begin{aligned}
(A + B_2 FC_s)^T P + P(A + B_2 FC_s) + v^{-2} PB_1 B_1^T P \\
+ (C_r + DFC_s)^T (C_r + DFC_s) < 0.
\end{aligned}
\tag{7.28}
$$

Inequality (7.28) is in the quadratic form of the unknowns (P, F). As in previous sections, we can develop an ILMI algorithm to solve it.

Algorithm 7.3 (ILMI algorithm for SOFH_∞). Initial data: System's state space realization $(A, B_1, B_2, C_s, C_r, D)$ and performance index v.

Step 1. Choose $Q_0 > 0$ and solve P for the Riccati equation

$$A^T P + PA - PB_2 B_2^T P + Q_0 = 0, \quad P > 0.$$

Set $i = 1$ and $X_1 = P$.

Step 2. Solve the following optimization problem for P_i, F and α_i.

OP1: Minimize α_i subject to the following LMI constraints

$$\begin{bmatrix} \Sigma_{3i} & P_i B_1 & (C_r + DFC_s)^T & (B_2^T P_i + FC_s)^T \\ B_1^T P_i & -v^2 I & 0 & 0 \\ C_r + DFC_s & 0 & -I & 0 \\ B_2^T P_i + FC_s & 0 & 0 & -I \end{bmatrix} < 0,$$
$$P_i > 0 \tag{7.29}$$

where $\Sigma_{3i} = A^T P_i + P_i A - X_i B_2 B_2^T P_i - P_i B_2 B_2^T X_i + X_i B_2 B_2^T X_i - \alpha_i P_i$. Denote by α_i^* the minimized value of α_i.

Step 3. If $\alpha_i^* \leq 0$, the obtained matrix F solves SOFH_∞ problem. Stop. Otherwise go to Step 4.

Step 4. Solve the following optimization problem for unknowns P_i and F.

OP2: Minimize $\mathrm{tr}(P_i)$ subject to LMI constraints (7.29) with $\alpha_i = \alpha_i^*$. Denote by P_i^* the optimal P_i.

Step 5. If $\| X_i B - P_i^* B \| < \varepsilon$, where ε is a prescribed tolerance, go to Step 6; otherwise set $i := i + 1$, $X_i = P_i^*$, and go to Step 2.

Step 6. It cannot be decided by this algorithm whether SOFH_∞ problem is solvable. Stop.

Now consider PID controller (7.24). Suppose Assumption 7.1 and Condition 7.2 hold. Using the same notations for $\bar{A}$, $\bar{B}_1$, $\bar{B}_2$, $\bar{C}_s$, $\bar{C}_r$ and $\bar{F}$ as those in Sect. 4.2, we can write the dynamics of the closed-loop system in the form

$$\dot{z} = \bar{A}z + \bar{B}_1 w + \bar{B}_2 u, \quad \bar{y}_s := \bar{C}_s z, \quad \bar{y}_r := \bar{C}_r z + Du, \quad u = \bar{F}\bar{y}_s. \tag{7.30}$$

Thus the feedback matrices $(\bar{F}_1, \bar{F}_2, \bar{F}_3)$ can be obtained by applying Algorithm 7.3 to system (7.30).

7.6 Maximum Output Control with PID Controllers

In this section, we investigate the design problem of PID controllers under the performance requirement that the system output is smaller than a specified value when the input signal is bounded. Consider system (7.26) with $x(0) = 0$. Note that w is viewed as an external command signal here instead of the exogenous disturbances in previous sections. Other notations for the system considered are the same as in Sect. 7.4. As before, we begin with SOF case first. The static output feedback maximum output control (SOFMOC) problem is to find a control of the form (7.16) such that the maximum regulated output, denoted by $Y_{r,\max} := \sup_{t \geq 0} \|y_r(t)\|$, of the closed-loop system under the command input w is smaller than a given positive number σ, i.e.,

$$Y_{r,\max} \leq \sigma. \tag{7.31}$$

To make sense for SOFMOC problem, we need the following assumption.

Assumption 7.2. *The reference input signal w is bounded uniformly over $[0, +\infty)$, i.e., $\|w(t)\| \leq 1$, $\forall t \in [0, +\infty)$.*

To guarantee performance specification (7.31) to be satisfied, following the same procedure as that in [89] we can obtain

Proposition 7.4. *If there exist matrices $P > 0$, F and numbers $\tau_2 \geq 0$, $\eta > 0$ such that the matrix inequalities*

$$\begin{bmatrix} P & (C_r + DFC_s)^T \\ (C_r + DFC_s) & \dfrac{\sigma^2}{\eta} I \end{bmatrix} > 0, \tag{7.32}$$

$$\begin{bmatrix} \Sigma_4 & PB_1 \\ B_1^T P & -\tau_2 \eta I \end{bmatrix} < 0. \tag{7.33}$$

hold, where $\Sigma_4 = (A + B_2 FC_s)^T P + P(A + B_2 FC_s) + \tau_2 P$, then the performance index (7.31) is satisfied.

In inequalities (7.32) and (7.33), we can always choose $\eta = 1$ theoretically without loss of generality. However, a larger η may lead to a faster convergence speed, according to our simulation experience. The most difficult problem is how to deal with the unknown variable τ_2. In the sequel, we will develop a method to find an "optimal" value for τ_2. According to Schur complement, (7.33) is equivalent to

$$(A + B_2 FC_s)^T P + P(A + B_2 FC_s) + \tau_2 P + \frac{1}{\tau_2 \eta} PB_1 B_1^T P < 0. \tag{7.34}$$

Let $\Phi = \tau_2 P + PB_1 B_1^T P / (\tau_2 \eta)$. Notice that only Φ in the left hand side of inequality (7.34) depends on τ_2. Since $P > 0$ and $PB_1 B_1^T P \geq 0$, it seems that a "minimum" Φ should exist. The rule for the choice of τ_2 is just to make Φ as small as possible. Here the meaning of the words "minimum" and "optimal" is in the sense of some kind of matrix norm. For the sake of easiness in numerical computation, we use Frobenius norm, denoted as $\| \cdot \|_F$. Suppose $P = \Gamma_1 \Gamma_1$ and $PB_1 B_1^T P = \Gamma_2 \Gamma_2$, where Γ_1 and Γ_2 are positive definite and positive semidefinite matrices, respectively. A simple algebra yields

$$\Phi = \left(\sqrt{\tau_2} \Gamma_1 - \sqrt{\frac{1}{\tau_2 \eta}} \Gamma_2 \right) \left(\sqrt{\tau_2} \Gamma_1 - \sqrt{\frac{1}{\tau_2 \eta}} \Gamma_2 \right) + \sqrt{\frac{1}{\eta}} (\Gamma_1 \Gamma_2 + \Gamma_2 \Gamma_1).$$

Thus it is clear that the choice of τ_2 should make $\sqrt{\tau_2}\Gamma_1$ and $\sqrt{1/(\tau_2\eta)}\Gamma_2$ as near as possible, i.e.,

$$\left\|\sqrt{\tau_2}\Gamma_1 - \sqrt{\frac{1}{\tau_2\eta}}\Gamma_2\right\|_F \longrightarrow \min.$$

Since

$$\left\|\sqrt{\tau_2}\Gamma_1 - \sqrt{\frac{1}{\tau_2\eta}}\Gamma_2\right\|_F^2 = \operatorname{tr}\left(\sqrt{\tau_2}\Gamma_1 - \sqrt{\frac{1}{\tau_2\eta}}\Gamma_2\right)\left(\sqrt{\tau_2}\Gamma_1 - \sqrt{\frac{1}{\tau_2\eta}}\Gamma_2\right)$$

$$= \operatorname{tr}(\tau_2 P + \frac{1}{\tau_2\eta}PB_1 B_1^T P) - \sqrt{\frac{1}{\eta}}\operatorname{tr}(\Gamma_1\Gamma_2 + \Gamma_2\Gamma_1),$$

the optimal τ_2 is thus given by

$$\tau_2 = \sqrt{\frac{\operatorname{tr}\left(PB_1 B_1^T P\right)}{\eta\operatorname{tr}(P)}}. \tag{7.35}$$

The above results are summarized in the following algorithm.

Algorithm 7.4 (ILMI algorithm for SOFMOC). Initial data: System's state space realization $(A, B_1, B_2, C_s, C_r, D)$, performance index σ, and a given positive number η.

Step 1. Choose $Q_0 > 0$ and solve P for the Riccati equation

$$A^T P + PA - PB_2 B_2^T P + Q_0 = 0, \quad P > 0.$$

Set $i = 1$ and $X_1 = P$. Calculate τ_2 according to

$$\tau_2 = \sqrt{\operatorname{tr}\left(PB_1 B_1^T P\right)/(\eta\operatorname{tr}(P))}.$$

Step 2. Solve the following optimization problem for P_i, F and α_i.
OP1: Minimize α_i subject to the following LMI constraints

$$\begin{bmatrix} \Sigma_{4i} & P_i B_1 & \left(B_2^T P_i + FC_s\right)^T \\ B_1^T P_i & -\tau_2\eta I & 0 \\ B_2^T P_i + FC_s & 0 & -I \end{bmatrix} < 0, \tag{7.36}$$

$$\begin{bmatrix} P_i & (C_r + DFC_s)^T \\ (C_r + DFC_s) & \dfrac{\sigma^2}{\eta}I \end{bmatrix} > 0, \tag{7.37}$$

$$P_i > 0, \tag{7.38}$$

where $\Sigma_{4i} = A^T P_i + P_i A - X_i B_2 B_2^T P_i - P_i B_2 B_2^T X_i + X_i B_2 B_2^T X_i + \tau_2 P_i - \alpha_i P_i$.
Denote by α_i^* the minimized value of α_i.

Step 3. If $\alpha_i^* \leq 0$, the obtained matrix F solves SOFMOC. Stop. Otherwise go to Step 4.

Step 4. Solve the following optimization problem for unknowns P_i, F.

OP2: Minimize $\mathrm{tr}(P_i)$ subject to LMI constraints (7.36)–(7.38) with $\alpha_i = \alpha_i^*$. Denote by P_i^* the optimal P_i.

Step 5. If $\| X_i B - P_i^* B \| < \varepsilon$, where ε is a prescribed tolerance, go to Step 6; otherwise set $i := i+1$, $X_i = P_i^*$, $\tau_2 = \sqrt{\mathrm{tr}\left(P_i^* B_1 B_1^T P_i^*\right) / \left(\eta\, \mathrm{tr}(P_i^*)\right)}$, and go to Step 2.

Step 6. It cannot be decided by this algorithm whether SOFMOC is solvable. Stop.

Notice that α_i may cease to decrease after some iterations. In this case, we can fix τ_2 at its value just when α_i begins not to decrease and then use the above algorithm to find the desired feedback matrices.

Now consider system (7.26) and the performance specification (7.31) with PID controller (7.24). Suppose Assumptions 7.1 and 7.2 and Condition 7.2 hold. Using the same notations for $\bar{A}$, $\bar{B}_1$, $\bar{B}_2$, $\bar{C}_s$, $\bar{C}_r$ and $\bar{F}$, as those in Sect. 7.4, we can write the dynamics of the closed-loop system in the form of (7.30). Thus the feedback matrices $(\bar{F}_1, \bar{F}_2, \bar{F}_3)$ can be calculated by applying Proposition 7.4 and Algorithm 7.4 to system (7.30).

7.7 A Numerical Example: Design of Aircraft Controllers

In this section we apply the algorithms developed in Sects. 7.3–7.6 to the design of aircraft controllers. The longitudinal dynamics of an aircraft trimmed at 25000 ft and 0.9 Mach are unstable and have two right half plane phugoid modes. One of the state space realization of its linearized model is as follows [74]:

$$\dot{x} = Ax + B_2 u, \quad y_s = C_s x,$$

where

$$A = \begin{bmatrix} -0.0266 & -36.6170 & -18.8970 & -32.0900 & 3.2509 & -0.7626 \\ 0.0001 & -1.8997 & 0.9831 & -0.0007 & -0.1708 & -0.0050 \\ 0.0123 & 11.7200 & -2.6316 & 0.0009 & -31.6040 & 22.3960 \\ 0 & 0 & 1.0000 & 0 & 0 & 0 \\ 0 & 0 & 0 & 0 & -30.0000 & 0 \\ 0 & 0 & 0 & 0 & 0 & -30.0000 \end{bmatrix},$$

$$B_2 = \begin{bmatrix} 0 & 0 \\ 0 & 0 \\ 0 & 0 \\ 0 & 0 \\ 30 & 0 \\ 0 & 30 \end{bmatrix}, \quad C_s = \begin{bmatrix} 0 & 1 & 0 & 0 & 0 & 0 \\ 0 & 0 & 0 & 1 & 0 & 0 \end{bmatrix}.$$

Table 7.1. SOF- and PID-Controllers and Their Performance

Problem	Feedback matrices	Poles (closed-loop)	Performance required	Actual performance	Iteration number	α
SOF (Stab.)	$F = \begin{bmatrix} 7.0158 & -4.3414 \\ 2.1396 & -4.4660 \end{bmatrix}$	$-0.0475 \pm j0.0853$ $-0.7576 \pm j10.7543$ -29.2613 -33.6825	Stability	Stable	20	-1.9×10^{-4}
PID (Stab.)	$F_1 = \begin{bmatrix} 10.1359 & -1.7947 \\ 6.9912 & -9.4140 \end{bmatrix}$ $F_2 = \begin{bmatrix} 0.3817 & -0.6939 \\ 0.6528 & -1.1978 \end{bmatrix}$ $F_3 = \begin{bmatrix} 2.6162 & -1.4722 \\ 0.8212 & -1.6284 \end{bmatrix}$	-0.0003 -0.0243 -0.0774 $-3.07 \pm j0.46$ $-17.38 \pm j34.73$ -37.0751	Stability	Stable	1	-4.4×10^{-4}
SOF (H_2)	$F = \begin{bmatrix} 15.8817 & 2.7088 \\ 20.9358 & 3.1029 \end{bmatrix}$	-0.0441 -1.6641 $-2.06 \pm j6.11$ -28.1157 -30.6106	$H_2 = 1600$	$H_2 = 1.598$	1	-1.7×10^{-3}
PID (H_2)	$F_1 = \begin{bmatrix} 140.9071 & 68.8308 \\ 171.9978 & 80.6037 \end{bmatrix}$ $F_2 = \begin{bmatrix} 115.4286 & 2.6966 \\ 141.6570 & 1.9412 \end{bmatrix}$ $F_3 = \begin{bmatrix} 138.0913 & 62.4469 \\ 168.4135 & 74.3155 \end{bmatrix}$	-0.0002 $-0.0042 \pm j0.0038$ $-0.0077 \pm j0.0066$ $-0.3315 \pm j0.3570$ -7.2870	$H_2 = 1600$	$H_2 = 0.096$	14	-1.5×10^{-2}
SOF (H_∞)	$F = \begin{bmatrix} 0.2838 & 0.0313 \\ -0.8725 & -0.0289 \end{bmatrix}$	$-0.1042 \pm j0.1536$ $-1.6525 \pm j4.3864$ -30.0027 -31.0378	$H_\infty < 5$	$H_\infty = 0.863$	26	-4.5×10^{-2}
PID (H_∞)	$F_1 = \begin{bmatrix} 422.17 & 221.64 \\ -188.84 & -104.44 \end{bmatrix}$ $F_2 = \begin{bmatrix} 0.3845 & -0.5019 \\ 0.1068 & -0.3373 \end{bmatrix}$ $F_3 = \begin{bmatrix} 48.03 & -31.85 \\ -19.31 & 8.80 \end{bmatrix}$	-0.0001 -0.0020 -0.0050 -0.7200 -19.9100 $-48.02 \pm j77.79$ -191.0000	$H_\infty < 5$	$H_\infty = 1.000$	3	-4.9×10^{-4}
SOF (MOC)	$F = \begin{bmatrix} -0.1874 & 0.0738 \\ -0.7973 & -0.0672 \end{bmatrix}$	-0.1180 $-1.1537 \pm j1.9003$ -1.5090 -30.0179 -30.6016	$\sigma_0 = 5$	$\sigma_0 = 0.089$	$30 + 217$	-5.2×10^{-4}
PID (MOC)	$F_1 = \begin{bmatrix} 38.2059 & 21.8277 \\ -39.2057 & -21.8279 \end{bmatrix}$ $F_2 = \begin{bmatrix} 0.0249 & 0.0203 \\ -0.0244 & -0.0202 \end{bmatrix}$ $F_3 = \begin{bmatrix} 5.3442 & -0.2300 \\ -5.3442 & 0.2300 \end{bmatrix}$	-1.5×10^{-5} -9.4×10^{-4} -0.0214 -0.7275 -11.4136 $-24.499 \pm j85.589$ -29.9743	$\sigma_0 = 5$	$\sigma_0 = 1.0$	2	-1.4×10^{-5}

The control variables are elevon and canard actuators. The output variables are angle of attack and attitude angle. The two unstable eigenvalues of A are $0.6899 \pm j0.2484$. In the cases of H_2, H_∞ and maximum output controls, we choose $B_1 = \begin{bmatrix} 0 & 0 & 0 & 0 & 30 & 0 \end{bmatrix}^T$, $D = \begin{bmatrix} 1 & 1 \end{bmatrix}$ and $C_r = \begin{bmatrix} 0 & 1 & 0 & 0 & 0 & 0 \end{bmatrix}$, respectively.

For our system, $C_s B_2 = 0$ and $C_s B_1 = 0$. Therefore Condition 7.2 and Assumption 7.1 always hold. Thus the closed loop system with PID controllers is always well-posed.

Using Algorithms 7.1–7.4, respectively, we can obtain SOF and PID controllers corresponding to different performance specifications accordingly. The results are summarized in Table 7.1.

Notice that the implementation of Algorithm 7.4 for MOC problem in SOF case should be divided into two phases when a smaller η is chosen, where it is $\eta = 10^4$. In the first phase, parameter τ_2 is updated using (7.35). After 30 iterations, when $\tau_2 = 0.0608$, α decreases slowly. Then we move to the second phase, i.e., fix τ_2 at $\tau_2 = 0.0608$ and use Algorithm 7.4 again (certainly τ_2 does not change in the corresponding steps) to find the feedback matrix F. After another 217 iterations, we get the solution.

As can be seen from Table 7.1, using Algorithms 7.2 and 7.4 to the design of H_2 suboptimal control and MOC problems respectively may yield quite conservative results. The reason for the case of H_2 problem is that the matrix P satisfying (7.20) may have a very large trace, which can be observed from the proof of Lemma 7.2 and conflicts with the requirement (7.19), and the reason for the case of MOC problem is due to the fact that Algorithm 7.4 actually guarantees $|y_r| < \sigma$ for all w satisfying $|w| < 1$, while the characteristics that $w = 1$ for all $t > 0$ has not been exploited.

7.8 Improvement of Convergence in Static Output Feedback

In this section, a new ILMI algorithm is proposed for SOF stabilization problem without introducing any additional variables, and assisted with a separate ILMI algorithm to find good initial variables. The algorithms for SOF stabilization are also extended to solve the SOF H_∞ control problem. They are applied to multivariable PID control. Numerical examples show the effectiveness and an improvement of the algorithms over the existing methods [174].

7.8.1 SOF Stabilization

Consider the following system:

$$\begin{cases} \dot{x}(t) & = & Ax(t) + B_1\omega(t) + B_2u(t), \\ z(t) & = & C_1x(t) + D_{11}\omega(t) + D_{12}u(t), \\ y(t) & = & C_2x(t) + D_{21}\omega(t), \end{cases} \tag{7.39}$$

where $x(t) \in \mathbb{R}^n$ is the state vector, $\omega(t) \in \mathbb{R}^r$ is the external signal, $u(t) \in \mathbb{R}^m$ is the controlled input, $z \in \mathbb{R}^q$ is the controlled output, and $y(t) \in \mathbb{R}^p$ is the measured output. $A, B_1, B_2, C_1, C_2, D_{11}, D_{12}$ and D_{21} are constant matrices with appropriate dimensions. The following assumption is made on system (7.39).

Assumption 7.3. (A, B_2) *is stabilizable and* (C_2, A) *is detectable.*

The SOF stabilization problem is to find a SOF controller

$$u(t) = Fy(t), \tag{7.40}$$

where $F \in \mathbb{R}^{m \times p}$ such that the closed-loop system with $\omega(t) = 0$ given by

$$\dot{x} = (A + B_2 F C_2)x(t), \tag{7.41}$$

is stable. As we all know, the closed-loop system (7.41) is stable if and only if there exists a $P = P^T > 0$ such that

$$P(A + B_2 F C_2) + (A + B_2 F C_2)^T P < 0. \tag{7.42}$$

Condition (7.42) is a BMI which is not a convex optimal problem. An ILMI method was proposed in [90], where a new variable X was introduced such that the stability condition becomes a sufficient one when $X \neq P$. The algorithm presented in [90] tried to find some X close to P by using an iterative method and the iterative procedure carries between P and X. On the other hand, a substitutive LMI formulation was proposed in [172], where some new variables such as L and M were introduced such that the stability condition also becomes sufficient one when $L \neq R^{-1}B_2 P$ or $M \neq F C_2$. The algorithm presented in [172] also tried to find some L and M close to $R^{-1}B_2 P$ and $F C_2$ respectively by using iterative method. It is clear that the introduced variables are calculated based on the information on P and F obtained in the preceding step in the iterative procedure. In fact, the information on P can be used directly without introducing additional variables. For example, when we derive a P in this step, we can employ it to derive the F in the next step. In the contrary, the F is also used to derive P, and so on. Therefore, these variables in the iterative procedures in [90] and [172] are unnecessary and the iteration can be carried out between P and F directly. Base on this idea, we propose a new algorithm as follows.

As mentioned in [90], if

$$P(A + B_2 F C_2) + (A + B_2 F C_2)^T P - \alpha P < 0 \tag{7.43}$$

holds, the closed-loop system matrix $A + B_2 F C_2$ has its eigenvalues in the strict left-hand side of the line $\alpha/2$ in the complex s-plane. If a $\alpha \leq 0$ satisfying (7.43) can be found, the SOF stabilization problem is solved.

The key point in our algorithm is to find an initial P. The P which satisfies (7.42) cannot be derived using LMI due to unknown F. By setting $V_1 = PB_2 F$, (7.42) becomes

$$PA + A^T P + V_1 C_2 + C_2^T V_1^T < 0. \tag{7.44}$$

However, (7.44) ignores B_2 so that this P does not take into account all the information. On the other hand, (7.42) is transformed to the following inequality by pro- and post-multiply $L = P^{-1}$,

$$(A + B_2 F C_2)L + L(A + B_2 F C_2)^T < 0. \tag{7.45}$$

By setting $V_2 = FC_2L$, (7.45) becomes

$$AL + LA^T + B_2V_2 + V_2^T B_2^T < 0. \tag{7.46}$$

According to the idea of the cone complementary linearization method [169], $L = P^{-1}$ yields $PL = I$, which is relaxed with the following LMI:

$$\begin{bmatrix} P & I \\ I & L \end{bmatrix} \geq 0 \tag{7.47}$$

and the linearized version of trace(PL) is minimized. Then, an iterative algorithm to find an initial P is stated as follows:

Algorithm 7.5

Step 1. Set $i = 1$ and $P_0 = I$ and $L_0 = I$.
Step 2. Derive a P_i and L_i by solving the following optimization problem for P_i, L_i, V_1 and V_2 :
OP1: Minimize trace$(P_iL_{i-1} + L_iP_{i-1})$ subject to the following LMI constraints

$$P_iA + A^T P_i + V_1C_2 + C_2^T V_1^T < 0, \tag{7.48}$$
$$AL_i + L_iA^T + B_2V_2 + V_2^T B_2^T < 0, \tag{7.49}$$
$$\begin{bmatrix} P_i & I \\ I & L_i \end{bmatrix} \geq 0. \tag{7.50}$$

Step 3. If trace$(P_iL_i) - n < \varepsilon_1$, a prescribed tolerance, an initial $P = P_i$ is found, stop.
Step 4. If the difference of two iterations satisfies trace(P_iL_i) − trace$(P_{i-1}L_{i-1})$
$< \varepsilon_2$, a prescribed tolerance, the initial P may not be found, stop.
Step 5. Set $i = i + 1$, goto Step 2.

If an initial P can not be found by Algorithm 1, the SOF control problem for system (7.39) with $\omega(t) = 0$ may not have solutions. On the other hand, after an initial P is found, an ILMI algorithm that stabilizes system (7.39) with $\omega(t) = 0$ using SOF is stated as follows:

Algorithm 7.6

Step 1. Set $i = 1$ and $P_1 = P$ as obtained from Algorithm 1.
Step 2. Solve the following optimization problem for F with given P_i:
OP1: Minimize α_i subject to the following LMI constraint

$$P_i(A + B_2FC_2) + (A + B_2FC_2)^T P_i - \alpha_iP_i < 0 \tag{7.51}$$

Step 3. If $\alpha_i \leq 0$, F is a stabilizating SOF gain, stop.
Step 4. Set $i = i + 1$. Solve the following optimization problem for P_i with given F:
OP2: Minimize α_i subject to the above LMI constraint (7.51).

Step 5. If $\alpha_i \leq 0$, F is a stabilization SOF gain, stop.
Step 6. Solve the following optimization problem for P_i with given F and α_i:
OP3: Minimize trace(P_i) subject to the above LMI constraint (7.51).
Step 7. If $\|P_i - P_{i-1}\|/\|P_i\| < \delta$, a prescribed tolerance, goto Step 8, else set $i = i+1$ and $P_i = P_{i-1}$, then goto Step 2.
Step 8. The system may not be stabilizable via SOF, stop.

Remark 7.2. The discussions on the iterative procedure and convergence of Algorithm 7.6 follow those in [90].

Remark 7.3. If $C_2 = I$, the SOF stabilization problem reduces to a state feedback problem. In fact, the initial P derived in Algorithm 7.5 is also the solution of state feedback stabilization problem since the state feedback gain F can be derived by $F = V_2 L^{-1} = V_2 P$. On the other hand, the F can also be obtained directly by OP1 in Algorithm 7.6 when the above initial P is given. Thus, the state feedback stabilization gain without conservativeness can also be derived using our Algorithms 7.5 and 7.6.

Remark 7.4. The algorithm presented here is different from those in [169]. The stopping criterion in [169] is given in terms of $\varepsilon_{\mathrm{eof}}$ which depends on selected α and β, where α and β should be sufficiently small. However, it is difficult to determine how small the values should be as they depend on a particular situation under consideration. If inappropriate α and β are selected, the algorithm may not be convergent. In our procedure, Algorithm 7.5 finds an initial P for a given ε_1, where the selection of ε_1 is not crucial as P obtained is not the final solution but will be elaborated in Algorithm 7.6. With the initial P, Algorithm 7.6 produces a static output feedback gain matrix F which can guarantee the stability of closed-loop system (7.41). The stopping condition is $\alpha < 0$, where α need not be specified a prior. Overall, our procedure is easier to use.

Example 7.1 ([90]). Consider the SOF stabilization problem of system (7.39) with $\omega(t) = 0$ and the following parameter matrices:

$$A = \begin{bmatrix} 0 & 1 \\ 1 & 0 \end{bmatrix}, \quad B_2 = \begin{bmatrix} 1 \\ 0 \end{bmatrix}, \quad C_2 = \begin{bmatrix} 1 & \beta \end{bmatrix}. \tag{7.52}$$

Table 7.2. SOF results (Example 7.1)

β	Method	Feedback gains	Poles	Itera no.	α
15	[90]	$F = -0.7369$	$-0.3684 \pm j3.1492$	15	-0.0377
	proposed	$F = -0.1333$	$-0.0667 \pm j0.9978$	2+1	-7.8046×10^{-4}
100	[90]	$F = -0.1748$	$-0.0874 \pm j4.0586$	366	-0.0017
	proposed	$F = -0.0200$	$-0.0100 \pm j1.0000$	2+1	-0.0012

The open-loop system is unstable since the eigenvalues are 1 and -1. The calculation results for SOF stabilization using the method in [90] and our Algorithms are listed in Table 7.2. It is noted that the iteration number in the form of $l+k$ in the table means that l is the iteration number of Algorithm 7.6 to find an initial P and k is the iteration number of Algorithm 7.6 to find a SOF gain. It can be seen that the convergence speed of Algorithm 7.6 is greatly faster than that in [90].

Example 7.2 ([163]). Consider the SOF stabilization problem of system (7.39) with $\omega(t) = 0$ and the following parameter matrices:

$$A = \begin{bmatrix} -4 & -2 & -8 & 5 & -1 & -8 & 4 \\ -9 & -7 & -6 & -3 & -2 & 2 & 6 \\ -7 & -3 & 7 & 5 & 2 & 10 & -1 \\ -6 & -3 & 8 & 1 & 2 & 3 & -7 \\ 0 & -5 & 6 & -3 & -4 & 6 & 1 \\ 2 & 8 & -4 & 6 & -9 & -2 & -4 \\ 5 & 8 & 3 & 1 & 9 & -6 & 3 \end{bmatrix}$$

$$B_2 = \begin{bmatrix} -3.9 & 2 & 0.1 & -2.5 & -1 & 2.5 & -1 \\ 0.5 & 0.5 & -1 & -0.5 & 1 & 2 & -0.05 \end{bmatrix}^T$$

$$C_2 = \begin{bmatrix} 3 & 6 & -5 & -2 & -1 & -7 & 5 \\ -1 & -4 & -7 & -1 & -6 & -5 & -3 \end{bmatrix}.$$

After more than 250 iterations, the algorithm in [163] converges. Algorithm 7.5 yields an initial P with only 4 iterations. Then, a SOF gain F is found as

$$F = \begin{bmatrix} -0.8871 & 4.9310 \\ -0.6576 & 0.9869 \end{bmatrix}$$

using Algorithm 7.6 with 2 iterations. In this case, $\alpha = -0.7481$ and the eigenvalues of the closed-loop system are $-7.8846 \pm j36.0334, -0.6354 \pm j12.2411, -0.3742, -4.9779 \pm j6.3825$. The convergence speed is faster than that in [163].

7.8.2 H_∞ Synthesis

An ILMI algorithm presented in the preceding section is now employed to solve the SOF H_∞ control problem. The objective of the SOF H_∞ synthesis is to to find a SOF controller (7.40) such that the transfer function of the closed-loop system satisfies H_∞ norm constraint

$$\|T_{z\omega}(s)\|_\infty < \gamma, \text{ for } \gamma > 0. \tag{7.53}$$

Equation (7.53) can be represented as a matrix inequality [90]:

$$\begin{bmatrix} PA_{\mathrm{cl}} + A_{\mathrm{cl}}^T P & PB_{\mathrm{cl}} & C_{\mathrm{cl}}^T \\ B_{\mathrm{cl}}^T P & -\gamma I & D_{\mathrm{cl}}^T \\ C_{\mathrm{cl}} & D_{\mathrm{cl}} & -\gamma I \end{bmatrix} < 0 \qquad (7.54)$$

where

$$\begin{aligned} A_{\mathrm{cl}} &= A + B_2 F C_2, \\ B_{\mathrm{cl}} &= B_1 + B_2 F D_{21}, \\ C_{\mathrm{cl}} &= C_1 + D_{12} F C_2, \\ D_{\mathrm{cl}} &= D_{11} + D_{12} F D_{21}. \end{aligned}$$

Similarly to Algorithm 7.6, we propose an algorithm to obtain the solution of matrix inequality (7.54) for a given $\gamma > 0$. It relies on, like Algorithm 7.5, an algorithm for finding an initial P for SOF H_∞ control problem:

Algorithm 7.7

Step 1. Set $i = 1$ and $P_0 = I$ and $L_0 = I$.
Step 2. Derive a P_i and L_i by solving the following optimization problem for P_i, L_i, V_1 and V_2:
OP1: Minimize $\mathrm{trace}(P_i L_{i-1} + L_i P_{i-1})$ subject to the LMI constraints (7.55), (7.56) and (7.57)

$$\begin{bmatrix} P_i A + A^T P_i + V_1 C_2 + C_2^T V_1^T & P_i B_1 + V_1 D_{21} & C_1^T + C_2^T F^T D_{12}^T \\ B_1^T P_i + D_2^T V_1^T & -\gamma I & D_{11}^T + D_{21}^T F^T D_{12}^T \\ C_1 + D_{12} F C_2 & D_{11} + D_{12} F D_{21} & -\gamma I \end{bmatrix} < 0$$

$$(7.55)$$

$$\begin{bmatrix} A L_i + L_i A^T + B_2 V_2 + V_2^T B_2^T & B_1 + B_2 F D_{21} & C_1^T L_i + V_2^T D_{12}^T \\ B_1^T + D_{21}^T F^T B_2^T & -\gamma I & D_{11}^T + D_{21}^T F^T D_{12}^T \\ L_i C_1 + D_{12} V_2 & D_{11} + D_{12} F D_{21} & -\gamma I \end{bmatrix} < 0$$

$$(7.56)$$

$$\begin{bmatrix} P_i & I \\ I & L_i \end{bmatrix} \geq 0. \qquad (7.57)$$

Step 3. If $\mathrm{trace}(P_i L_i) - n < \varepsilon_1$, a prescribed tolerance, an initial $P = P_i$ is found, stop.
Step 4. If the difference between the two iterations satisfies $\mathrm{trace}(P_i L_i) - \mathrm{trace}(P_{i-1} L_{i-1}) < \varepsilon_2$, a prescribed tolerance, the initial P may not be found, stop.
Step 5. Set $i = i + 1$, goto Step 2.

After an initial P is found, an ILMI algorithm for SOF H_∞ control problem for system (7.39) is stated as follows:

Algorithm 7.8

Step 1. Set $i = 1$ and $P = P_1$ as obtained from Algorithm 7.7.
Step 2. Solve the following optimization problem for F with given P_i:
OP1: Minimize α_i subject to the LMI constraint (7.58)

$$\begin{bmatrix} \Phi_{11} & P_i B_1 + P_i B_2 F D_{21} & C_1^T + C_2^T F^T D_{12}^T \\ B_1^T P_i + D_{21}^T F^T B_2^T P_i & -\gamma I & D_{11}^T + D_{21}^T F^T D_{12}^T \\ C_1 + D_{12} F C_2 & D_{11} + D_{12} F D_{21} & -\gamma I \end{bmatrix} < 0, \tag{7.58}$$

where
$$\Phi_{11} = P_i A + A^T P_i + P_i B_2 F C_2 + C_2^T F^T B_2^T P_i - \alpha P_i.$$

Step 3. If $\alpha_i \le 0$, F is a stabilization SOF H_∞ control gain for γ, stop.
Step 4. Set $i = i + 1$. Solve the following optimization problem for P_i with given F:
OP2: Minimize α_i subject to the above LMI constraint (7.58).
Step 5. If $\alpha_i \le 0$, F is a stabilizating SOF H_∞ control gain for γ, stop.
Step 6. Solve the following optimization problem for P_i with given F and α_i:
OP3: Minimize trace(P_i) subject to the above LMI constraint (7.58).
Step 7. If $\|P_i - P_{i-1}\|/\|P_i\| < \delta$, a prescribed tolerance, goto Step 8, else set $i = i + 1$ and $P_i = P_{i-1}$, then goto Step 2.
Step 8. It may not be decided by this algorithm whether SOF H_∞ control problem is solvable, stop.

Example 7.3 ([172]). Consider system (7.39) with the following parameter matrices:

$$A = \begin{bmatrix} -0.0366 & 0.0271 & 0.0188 & -0.4555 \\ 0.0482 & -1.01 & 0.0024 & -4.0208 \\ 0.1002 & 0.3681 & -0.707 & 1.42 \\ 0 & 0 & 1 & 0 \end{bmatrix},$$

$$B_1 = \begin{bmatrix} 1 & 0 & 0 \\ 0 & 0 & 0 \\ 0 & 0 & 0 \\ 0 & 0 & 0 \end{bmatrix}, \quad B_2 = \begin{bmatrix} 0.4422 & 0.1761 \\ 3.5446 & -7.5922 \\ -5.52 & 4.49 \\ 0 & 0 \end{bmatrix},$$

$$C_1 = \begin{bmatrix} 0 & 1 & 0 & 0 \\ 0 & 0 & 0 & 1 \end{bmatrix}, \quad C_2 = \begin{bmatrix} 1 & 0 & 0 & 0 \\ 0 & 1 & 0 & 0 \end{bmatrix},$$

$$D_{11} = \begin{bmatrix} 0.5 & 0 & 0 \\ 0 & 1 & 0 \end{bmatrix}, \quad D_{12} = \begin{bmatrix} 1 & 0 \\ 0 & 1 \end{bmatrix},$$

$$D_{21} = \begin{bmatrix} 0 & 0.1 & 0 \\ 0 & 0 & 0.1 \end{bmatrix}.$$

The SOF H_∞ norm obtained in [172] is 1.183. For $\gamma = 1.144$, the initial matrix P is obtained as

$$P_1 = \begin{bmatrix} 0.7130 & -0.4054 & -0.3283 & -1.1658 \\ -0.4054 & 2.2769 & 1.4029 & 1.2100 \\ -0.3283 & 1.4029 & 1.5941 & 1.0533 \\ -1.1658 & 1.2100 & 1.0533 & 3.2305 \end{bmatrix}$$

by Algorithm 7.7 after 24 iterations. Then the SOF H_∞ norm converges to 1.144 after 2 iterations using Algorithm 7.8. The resulting SOF gain is

$$F = \begin{bmatrix} 0.0976 & -3.8054 \\ -0.4191 & 4.6958 \end{bmatrix}.$$

In this case, $\alpha = -8.2776 \times 10^{-7}$ and the eigenvalues of the closed-loop system are $-50.1599, -0.2781, -0.2431 \pm j1.3534$.

7.8.3 PID Control

Motivated by its popularity in industry, let us consider now the following PID controller

$$u(t) = F_1 y(t) + F_2 \int_0^t y(\theta)\mathrm{d}\theta + F_3 \dot{y}(t) \tag{7.59}$$

instead of SOF controller (7.40), where $F_1, F_2, F_3 \in \mathbb{R}^{m \times p}$ are gain matrices to be designed. Without lose of generality, D_{21} is set as zero.

It is noted that the method is proposed in Sect. 7.2 to transform this PID controller design problem to a SOF control problem. In order to using the method proposed in Sects. 7.3–7.6, Condition 7.1 is needed: The matrix $I - F_3 C_2 B_2$ is invertible.

The invertibility of matrix $I + C_2 B_2 \bar{F}_3$ follows from Proposition 7.1 in Sect. 7.2.

Under Condition 7.1, one easily see that Algorithms 7.5 and 7.6 can be employed to derive the stabilizable PID control gains, F_1, F_2 and F_3. Similarly, Algorithms 7.7 and 7.8 can be used to derive the PID H_∞ controller for a given performance $\gamma > 0$.

Example 7.4 ([79]). One of the state space realization of the aircraft controller system model is as system (7.39) with the following parameters, as shown at the top of the next page:

$$A = \begin{bmatrix} -0.0266 & -36.6170 & -18.8970 & -32.0900 & 3.2509 & -0.7626 \\ 0.0001 & -1.8997 & 0.9831 & -0.0007 & -0.1708 & -0.0050 \\ 0.0123 & 11.7200 & -2.6316 & 0.0009 & -31.6040 & 22.3960 \\ 0 & 0 & 1.0000 & 0 & 0 & 0 \\ 0 & 0 & 0 & 0 & -30.0000 & 0 \\ 0 & 0 & 0 & 0 & 0 & -30.0000 \end{bmatrix},$$

$$B_1 = \begin{bmatrix} 0 \\ 0 \\ 0 \\ 0 \\ 30 \\ 0 \end{bmatrix}, \quad B_2 = \begin{bmatrix} 0 & 0 \\ 0 & 0 \\ 0 & 0 \\ 0 & 0 \\ 30 & 0 \\ 0 & 30 \end{bmatrix},$$

$$C_1 = \begin{bmatrix} 0 & 1 & 0 & 0 & 0 & 0 \end{bmatrix}, \quad C_2 = \begin{bmatrix} 0 & 1 & 0 & 0 & 0 & 0 \\ 0 & 0 & 0 & 1 & 0 & 0 \end{bmatrix},$$

$$D_{11} = [0], \quad D_{12} = \begin{bmatrix} 1 & 1 \end{bmatrix}, \quad D_{21} = \begin{bmatrix} 0 \\ 0 \end{bmatrix}.$$

The calculation results using Algorithms 7.7 and 7.8 and the method in [90] and Sects. 7.3–7.6 for SOF control and PID control are listed in Table 7.3. It can be seen that the convergence speed is faster than that in [90] and Sects. 7.3–7.6.

On the other hand, Remark 7.4 is demonstrated in this example. For example, when ε_1 in Algorithm 7.7, which is similar to those in [169], is set as 10^{-3}, 2 iterations obtain an initial P. However, based on this P, F is not a stabilization H_∞ control gain for given $\gamma = 1.001$. On the contrary, the corresponding F in Table 7.3 is derived after 14 iterations by using Algorithm 7.8.

7.9 Improvement by Descriptor Systems Approach

Recall a state-space system in (7.1) and (7.2) as

$$\dot{x}(t) = Ax(t) + Bu(t), \quad y(t) = Cx(t), \tag{7.60}$$

with PID controller of the form

$$u(t) = F_1 y(t) + F_2 \int_0^t y(\theta)\,\mathrm{d}\theta + F_3 \dot{y}(t), \tag{7.61}$$

Table 7.3. SOF and PID-controller and their performances for Example 7.4

Method	Problem	Feedback gains	Poles	Perf.	Itera. no.	α
[79,90]	SOF (Stab.)	$F = \begin{bmatrix} 7.0158 & -4.3414 \\ 2.1396 & -4.4660 \end{bmatrix}$	$-0.0475 \pm j0.0853$ $-0.7576 \pm j0.7543$ $-29.2613,\ -33.6825$	Stab.	20	-1.9×10^{-4}
Algorithm 7.8	SOF (Stab.)	$F = \begin{bmatrix} 0.6828 & 0.2729 \\ -0.1024 & -0.0348 \end{bmatrix}$	$-1.3274 \pm j4.6317$ $-0.7735,\ -0.0665$ $-31.0626,\ -30.0006$	Stab.	3+2	-0.1330
[79]	PID (Stab.)	$F_1 = \begin{bmatrix} 10.1359 & -1.7947 \\ 6.9912 & -9.4140 \end{bmatrix}$ $F_2 = \begin{bmatrix} 0.3817 & -0.6939 \\ 0.6528 & -1.1978 \end{bmatrix}$ $F_3 = \begin{bmatrix} 2.6162 & -1.4722 \\ 0.8212 & -1.6284 \end{bmatrix}$	$-0.0003,\ -0.0243$ $-0.0774,\ -37.0751$ $-3.07 \pm j0.46$ $-17.38 \pm j34.73$	Stab.	1	-4.4×10^{-4}
Algorithm 7.8	PID (Stab.)	$F_1 = \begin{bmatrix} 13.9831 & 2.0458 \\ 0.8758 & -1.3309 \end{bmatrix}$ $F_2 = \begin{bmatrix} 0.1710 & -0.3188 \\ 0.1550 & -0.2891 \end{bmatrix}$ $F_3 = \begin{bmatrix} 5.7954 & -4.5855 \\ 1.1572 & -1.9783 \end{bmatrix}$	-49.8699 $-19.4227 \pm j31.7602$ $-4.9964,\ -0.6490$ -1.3296×10^{-5} $-0.0275,\ -0.0388$	Stab.	2+3	-2.6390×10^{-5}
[79,90]	SOF (H_∞)	$F = \begin{bmatrix} 0.2838 & 0.0313 \\ -0.8725 & -0.0289 \end{bmatrix}$	$-0.1042 \pm j0.1536$ $-1.6525 \pm j4.3864$ $-30.0027,\ -31.0378$	$H_\infty < 5$	26	-4.5×10^{-2}
Algorithm 7.8	SOF (H_∞)	$F = \begin{bmatrix} 2.3982 & 0.2302 \\ -3.3982 & -0.2302 \end{bmatrix}$	$-0.0010 \pm j11.6634$ $-0.0547,\ -0.1260$ $-34.3691,\ -30.0061$	$H_\infty < 0.323$	50+18	-0.0020
[79]	PID (H_∞)	$F_1 = \begin{bmatrix} 442.17 & 221.64 \\ -188.84 & -104.44 \end{bmatrix}$ $F_2 = \begin{bmatrix} 0.3845 & -0.5019 \\ 0.1068 & -0.3373 \end{bmatrix}$ $F_3 = \begin{bmatrix} 48.03 & -31.85 \\ -19.31 & 8.80 \end{bmatrix}$	$-0.0001,\ -0.0020$ $-0.0050,\ -0.7200$ $-19.9100,\ -191.0000$ $-48.02 \pm j77.79$	$H_\infty < 5$	3	-4.9×10^{-4}
Algorithm 7.8	PID (H_∞)	$F_1 = \begin{bmatrix} 22.5780 & 5.9056 \\ -15.3968 & -4.9824 \end{bmatrix}$ $F_2 = \begin{bmatrix} 39.0276 & 20.4624 \\ -24.1171 & -13.3738 \end{bmatrix}$ $F_3 = \begin{bmatrix} 6.2065 & -4.6857 \\ -4.1822 & 2.9566 \end{bmatrix}$	$-22.9441 \pm j32.1154$ $-35.1083,\ -9.1967$ $-4.7860,\ -0.0249$ $-0.0637,\ -0.6649$	$H_\infty < 1.001$	2+14	-1.3142×10^{-4}

where $x \in \mathbb{R}^n$, $u \in \mathbb{R}^l$, and $y \in \mathbb{R}^m$ are state variable, control input and output, respectively; A, B, C are real matrices with appropriate dimensions; $F_i \in \mathbb{R}^{l \times m}$, $i = 1, 2, 3$, are matrices to be designed. In Sect. 7.2, system (7.60) with (7.61) is transformed into a SOF control system, and once a SOF gain $\bar{F} = [\bar{F}_1,\ \bar{F}_2,\ \bar{F}_3]$ is found, the original PID gains are recovered by (7.5) as

$$F_3 = \bar{F}_3(I + CB\bar{F}_3)^{-1}, \quad F_2 = (I - \bar{F}_3CB)\bar{F}_2, \quad F_1 = (I - \bar{F}_3CB)\bar{F}_1, \tag{7.62}$$

under Condition 7.1: The matrix $I - F_3CB$ is invertible.

It is seen that Condition 7.1 is crucial to the design problem. Although the parameter subspace consisting of F_3 that violates Condition 7.1 is zero measure, it still needs to remove the constraint which seriously affects not only the well-posedness of the SOF system but also the solution to the PID controller design. Another limitation is that all the system parameters A, B, C must be exactly known. If subjected to perturbations, the method in Sects. 7.3–7.6 will be invalid.

In this section, the problem of PID controller design is transformed to the static output feedback (SOF) controller design for a descriptor system, and two problems are considered: stabilization via PID control and robust stabilization with H_∞ performance. Under such a descriptor transformation, the disadvantages of the method in Sect. 7.2 can be overcome. In a more detail, the constraint of Condition 7.1 is removed, the PID gains are available immediately without additional computations once the SOF matrix is obtained, and systems with some parameters subject to perturbations can be treated. After the descriptor transformation, our goals for the considered problems are reduced to the corresponding SOF controller designs for descriptor systems. These SOF controller design problems, as in the same situation for standard systems [90, 79], have not been completely solved. In this section, the idea in [90] will be used and extended to descriptor systems for SOF stabilizing controller design. With some further developments based on the descriptor type bounded real lemma [101], we give a necessary and sufficient condition and a sufficient condition for the robust SOF stabilizing controller design with H_∞ performance.

7.9.1   Stabilization Via PID Control

Based on system (7.60), we introduce a new state variable

$$\bar{x}(t) = \left[x^T(t), \int_0^t x^T(\theta)\mathrm{d}\theta, \dot{x}^T(t) \right]^T,$$

and let the new output be $\bar{y}(t) = [y^T(t), \int_0^t y^T(\theta)\mathrm{d}\theta, \dot{y}^T(t)]^T$. Then system (7.60) with (7.61) is transformed into the following SOF control system:

$$\begin{aligned}
\bar{E}\dot{\bar{x}}(t) &= \bar{A}\bar{x}(t) + \bar{B}u(t), \\
\bar{y}(t) &= \bar{C}\bar{x}(t), \\
u(t) &= \bar{F}\bar{y}(t),
\end{aligned} \tag{7.63}$$

where

$$\bar{E} = \begin{bmatrix} I_n & 0 & 0 \\ 0 & I_n & 0 \\ 0 & 0 & 0 \end{bmatrix}, \quad \bar{A} = \begin{bmatrix} 0 & 0 & I_n \\ I_n & 0 & 0 \\ A & 0 & -I_n \end{bmatrix}, \quad \bar{B} = \begin{bmatrix} 0 \\ 0 \\ B \end{bmatrix},$$

$$\bar{C} = \begin{bmatrix} C & 0 & 0 \\ 0 & C & 0 \\ 0 & 0 & C \end{bmatrix}, \quad \bar{F} = \begin{bmatrix} F_1 & F_2 & F_3 \end{bmatrix}.$$

System of the form (7.63) is called a descriptor system due to $\mathrm{rank}(\bar{E}) < \dim(\bar{E})$ [100]. It is well-known that for a descriptor system Σ described by $E\dot{x}(t) = Ax(t)$ (or, the pair

(E,A)), the regularity condition $\det(sE - A) \neq 0$ guarantees the existence and uniqueness of solutions on $[0,\infty)$, the full degree condition $\deg(\det(sE - A)) = \mathrm{rank}E$ ensures that the system is impulse-free, i.e., there is no impulse behavior in the system, and the stability of the system means that spectrum condition $\sigma(E,A) := \{s : \det(sE - A) = 0\} \subset \mathbb{C}^-$ holds [100]. The system Σ is called *admissible* if it is regular, impulse-free and stable. Note that an impulse-free system may have initial jump for non-compatible initial conditions [175]. With these concepts in hand, the design of the original PID stabilizing controllers is converted to the design of SOF stabilizing controllers rendering the closed-loop descriptor system (7.63) admissible. It is seen that once the SOF gain $\bar{F}$ is obtained, the original PID gains F_i are available immediately due to the relation $\bar{F} = \begin{bmatrix} F_1 & F_2 & F_3 \end{bmatrix}$.

Next, we aim to design the SOF stabilizing controllers for system (7.63). The following lemma will be used.

Lemma 7.3 (Masubuchi et al., [101]). *The system described by $E\dot{x}(t) = Ax(t)$ is admissible if and only if there exists a matrix P such that*

$$E^T P = P^T E \geq 0, \quad A^T P + P^T A < 0.$$

With the aid of Lemma 7.3, we establish the following result which is an easy extension of that in [90].

Proposition 7.5. *System (7.63) is admissible if and only if there exist matrices P and $\bar{F}$ such that*

$$\bar{E}^T P = P^T \bar{E} \geq 0, \tag{7.64}$$
$$\bar{A}^T P + P^T \bar{A} - P^T \bar{B}\bar{B}^T P + (\bar{B}^T P + \bar{F}\bar{C})^T (\bar{B}^T P + \bar{F}\bar{C}) < 0. \tag{7.65}$$

Proof. By virtue of Lemma 7.3, system (7.63) is admissible if and only if there exist matrices Q and $\bar{F}$ such that inequalities

$$\bar{E}^T Q = Q^T \bar{E} \geq 0, \tag{7.66}$$
$$(\bar{A} + \bar{B}\bar{F}\bar{C})^T Q + Q^T (\bar{A} + \bar{B}\bar{F}\bar{C}) < 0 \tag{7.67}$$

hold. Using a procedure similar to the proof of Theorem 1 in [90], the equivalence of (7.66) and (7.67) to (7.64) and (7.65) can be verified. $\square$

Proposition 7.5 presents a necessary and sufficient condition in terms of matrix inequalities for the SOF stabilization of system (7.63), and thus for the original PID stabilization problem. The constraint of Condition 1 is removed. Indeed, Condition 1 is implied by (7.64) and (7.65). This is because (7.64) and (7.65) specify P in the following form:

$$P = \begin{bmatrix} P_{11} & 0 \\ P_{21} & P_{22} \end{bmatrix}, \quad 0 < P_{11} \in \mathbb{R}^{2n}, \ P_{22} \in \mathbb{R}^n \text{ is invertible.} \tag{7.68}$$

This further gives from (7.65) that

$$(-I_n + BF_3C)^T P_{22} + P_{22}^T(-I_n + BF_3C) < 0, \tag{7.69}$$

yielding that the matrix $-I_n + BF_3C$ and thus the matrix $I_l - F_3CB$ is invertible.

As in [90] and [79], we need to develop an iterative LMI algorithm to solve the quadratic matrix inequality (QMI) in (7.65). Note that the QMI in (7.65) is equivalent to a bilinear matrix inequality (BMI). This can be seen as follows. With a proof similar to that of Theorem 2 in [90], (7.65) is equivalent to

$$\bar{A}^T P + P^T \bar{A} - X^T \bar{B}\bar{B}^T P - P^T \bar{B}\bar{B}^T X + X^T \bar{B}\bar{B}^T X$$
$$+ (\bar{B}^T P + \bar{F}\bar{C})^T (\bar{B}^T P + \bar{F}\bar{C}) < 0 \tag{7.70}$$

for some matrix X. By Schur complement, (7.70) is equivalent to

$$\begin{bmatrix} \bar{A}^T P + P^T \bar{A} - X^T \bar{B}\bar{B}^T P - P^T \bar{B}\bar{B}^T X & X^T \bar{B} & (\bar{B}^T P + \bar{F}\bar{C})^T \\ \bar{B}^T X & -I & 0 \\ \bar{B}^T P + \bar{F}\bar{C} & 0 & -I \end{bmatrix} < 0$$

which is in the form of BMI.

The following is the algorithm in descriptor version which is similar to those in [90] and [79], and the explanation is given later.

Algorithm 7.9 (Iterative LMI algorithm)

Step 1. Set $i = 1$ and $X_1 = I_{3n}$.
Step 2. Solve the following optimization problem for P_i, $\bar{F}$ and α_i.
OP1: Minimize α_i subject to the following LMI constraints

$$\bar{E}^T P_i = P_i^T \bar{E} \geq 0, \tag{7.71}$$

$$\begin{bmatrix} \Sigma_i & (\bar{B}^T P_i + \bar{F}\bar{C})^T \\ \bar{B}^T P_i + \bar{F}\bar{C} & -I \end{bmatrix} < 0, \tag{7.72}$$

where $\Sigma_i = \bar{A}^T P_i + P_i^T \bar{A} - X_i^T \bar{B}\bar{B}^T P_i - P_i^T \bar{B}\bar{B}^T X_i + X_i^T \bar{B}\bar{B}^T X_i - 2\alpha_i \bar{E}^T P_i$. Denote by α_i^* the minimized value of α_i.
Step 3. If $\alpha_i^* \leq 0$, $\bar{F}$ is a stabilizing SOF gain. Stop. Otherwise, go to Step 4.
Step 4. Solve the following optimization problem for P_i, $\bar{F}$ and α_i.
OP2: Minimize trace($\bar{E}^T P_i$) subject to LMI constraints (7.71) and (7.72) with $\alpha_i = \alpha_i^*$. Denote by P_i^* the optimal P_i.
Step 5. If $\|\bar{B}^T X_i - \bar{B}^T P_i^*\| < \varepsilon$, where ε is a prescribed tolerance, go to Step 6; otherwise, set $i := i + 1$, $X_i = P_i^*$ and go to Step 2.
Step 6. The SOF problem cannot be solved by this algorithm. Stop.

The above algorithm has the same disadvantage as those in [90] and [79], i.e., it is based on a sufficient condition. Regarding Algorithm 7.9, we give some remarks below. For more details in the development of such an algorithm, please refer to [90] and [79].

Remark 7.5. It should be pointed out that Algorithm 7.1 (as well as Algorithms 7.2–7.4) in Sects. 7.3–7.6 should be revised so as not merely confined to the minimum state-space realizations. This is because the augmented SOF control system may not correspond to a minimum state-space realization of a transfer matrix although the original PID control system does. If so, the Riccati equation therein in Step 1 may not have a solution. For instance,

$$(A,B,C) = \left(\begin{bmatrix} 0 & 1 \\ 1 & 0 \end{bmatrix}, \begin{bmatrix} 0 \\ 1 \end{bmatrix}, \begin{bmatrix} 0 & 1 \end{bmatrix} \right)$$

is a minimum state-space realization of $H(s) = s/(s^2 - 1)$. However, with respect to parameters for the augmented SOF control system, the following Riccati equation has no solutions.

$$\tilde{A}^T P + P \tilde{A} - P \tilde{B} \tilde{B}^T P + I_3 = 0,$$

where

$$\tilde{A} = \begin{bmatrix} A & 0 \\ C & 0 \end{bmatrix} \in \mathbb{R}^{3 \times 3}, \quad \tilde{B} = \begin{bmatrix} B \\ 0 \end{bmatrix} \in \mathbb{R}^3.$$

So, all the algorithms in Sects. 7.3–7.6 should be fixed by removing the initial data selection of system's state-space realization, and just setting $X_1 = I$ in Step 1. The above concerns also apply to our Algorithm 7.9 for descriptor systems. Hence, we set the initial value $X_1 = I_{3n}$.

Remark 7.6. It can be shown that the LMIs in (7.71) and (7.72) is feasible by adjusting scalar α_i. The feasibility of LMIs in (7.71) and (7.72) implies that system (7.63) is regular, impulse-free and has all its roots in the left-hand side of $\Re(s) = \alpha_i$. In fact, (7.71) and (7.72) gives

$$\bar{E}^T P_i = P_i^T \bar{E} \geq 0,$$
$$\bar{A}^T P_i + P_i^T \bar{A} - P_i^T \bar{B} \bar{B}^T P_i - 2\alpha_i \bar{E}^T P_i + \left(\bar{B}^T P_i + \bar{F} \bar{C} \right)^T \left(\bar{B}^T P_i + \bar{F} \bar{C} \right) < 0,$$

due to $X_i^T \bar{B} \bar{B}^T P_i + P_i^T \bar{B} \bar{B}^T X_i - X_i^T \bar{B} \bar{B}^T X_i \leq P_i^T \bar{B} \bar{B}^T P_i$. Hence, by virtue of Proposition 7.5, we have

$$\bar{E}^T P_i = P_i^T \bar{E} \geq 0,$$
$$\left(-\alpha_i \bar{E} + \bar{A} + \bar{B} \bar{F} \bar{C} \right)^T P_i + P_i^T \left(-\alpha_i \bar{E} + \bar{A} + \bar{B} \bar{F} \bar{C} \right) < 0.$$

Remark 7.7. (7.71) and (7.72) can be combined to a single LMI. Let $E_L = [0, I_n]$ which is a maximum left annihilator of $\bar{E}$. Then the conditions in (7.71) and (7.72) are equivalent to the following LMI:

$$\begin{bmatrix} \bar{\Sigma}_i & \left(\bar{B}^T \left(Z_i \bar{E} + E_L^T Y_i \right) + \bar{F} \bar{C} \right)^T \\ \bar{B}^T \left(Z_i \bar{E} + E_L^T Y_i \right) + \bar{F} \bar{C} & -I \end{bmatrix} < 0, \tag{7.73}$$

for additional matrices $Z_i > 0$ and $Y_i \in \mathbb{R}^{n \times 3n}$, where $\bar{\Sigma}_i$ is as in Σ_i by changing P_i to $Z_i \bar{E} + E_L^T Y_i$. Indeed, (7.73) implies (7.71) and (7.72) by letting $P_i = Z_i \bar{E} + E_L^T Y_i$;

and conversely, (7.71) and (7.72) ensure (7.73) by noting that P_i has the same form as in (7.68) and can be decomposed as $Z_i \bar{E} + E_L^T Y_i$ with $Z_i = \mathrm{diag}\{P_{i11}, I_n\} > 0$ and $Y_i = [P_{i21}, P_{i22}]$.

So far, we have shown the improvements for PID stabilization problem using descriptor approach. We remark that this approach also applies to the designs of PID controllers under H_∞ and H_2 performances, respectively. The corresponding results can be obtained based on the bounded real lemma for descriptor systems [101] and the developments of H_2 control for descriptor systems [176]. These resulting improvements over those in Sects. 7.3–7.6 also remove the corresponding invertibility constraints. For conciseness, we do not address the details here, and in the next section we take the H_∞ case for example to show the robust PID controller design.

7.9.2 Robust Stabilization with H_∞ Performance

As stated in Sect. 7.9 that another limitation of the method in Sects. 7.3–7.6 lies in that all the system parameters A, B, C must be exactly known. In this section, we show that the descriptor approach proposed in this note also applies to uncertain systems. Consider the following system:

$$
\begin{aligned}
(I_n + \delta E)\dot{x}(t) &= (A + \delta A)x(t) + (B + \delta B)u(t) + (B_1 + \delta B_1)w(t), \\
y(t) &= Cx(t), \\
z(t) &= C_1 x(t) + Du(t),
\end{aligned}
\tag{7.74}
$$

with PID controller of the form (7.61), where $x \in \mathbb{R}^n$, $y \in \mathbb{R}^m$, $z \in \mathbb{R}^{m_1}$ $u \in \mathbb{R}^l$, and $w \in \mathbb{R}^{l_1}$ are state variable, measured output, controlled output, control input and exogenous disturbance, respectively; A, B, B_1, C, C_1, D are real matrices with appropriate dimensions; $F_i \in \mathbb{R}^{l \times m}$, $i = 1, 2, 3$, are matrices to be designed; $\delta E, \delta A, \delta B, \delta B_1$ stand for parameter perturbations in the following norm-bounded uncertainty form:

$$
\begin{bmatrix} \delta E & \delta A & \delta B & \delta B_1 \end{bmatrix} = M \Delta \begin{bmatrix} N_e & N_1 & N_2 & N_3 \end{bmatrix},
\tag{7.75}
$$

where M, N_e, N_1, N_2, N_3 are known constant real matrices with appropriate dimensions, the uncertainty matrix Δ satisfies

$$
\Delta^T \Delta \leq I.
\tag{7.76}
$$

We allow the derivative term to have uncertainties since our approach can deal with such a type of uncertainty. For the special case that no perturbation appears in the derivative term, just set $N_e = 0$. The purpose in this section is to design matrices F_i, $i = 1, 2, 3$, such that the the closed-loop transfer function, T_{wz}, from w to z is stable with H_∞ performance $\gamma > 0$ (i.e., $\|T_{wz}\|_\infty < \gamma$).

By using the same augmented variables as in Sect. 7.9.1, i.e.,

$$
\begin{aligned}
\bar{x}(t) &= [x^T(t), \int_0^t x^T(\theta)\mathrm{d}\theta, \dot{x}^T(t)]^T, \\
\bar{y}(t) &= [y^T(t), \int_0^t y^T(\theta)\mathrm{d}\theta, \dot{y}^T(t)]^T,
\end{aligned}
$$

we transform the uncertain system (7.74) with (7.61) into the following uncertain SOF control system:

$$\begin{aligned}
\bar{E}\dot{\bar{x}}(t) &= (\bar{A}+\bar{M}\Delta\bar{N}_1)\bar{x}(t)+(\bar{B}+\bar{M}\Delta\bar{N}_2)u(t)+(\bar{B}_1+\bar{M}\Delta\bar{N}_3)w(t),\\
\bar{y}(t) &= \bar{C}\bar{x}(t),\\
z(t) &= \bar{C}_1 x(t)+\bar{D}u(t),\\
u(t) &= \bar{F}\bar{y}(t),
\end{aligned} \tag{7.77}$$

where

$$\bar{B}_1=\begin{bmatrix}0\\0\\B_1\end{bmatrix},\quad \bar{C}_1=\begin{bmatrix}C_1 & 0 & 0\end{bmatrix},\quad \bar{D}=D,\quad \bar{M}=\begin{bmatrix}0\\0\\M\end{bmatrix},$$

$$\bar{N}_1=\begin{bmatrix}N_1 & 0 & -N_e\end{bmatrix},\quad \bar{N}_2=N_2,\quad \bar{N}_3=N_3,$$

and the other parameters remain the same as in system (7.63). It is easy to verify that the transfer function from w to z remains the same under the above descriptor transformation. Hence, the design of robust PID controller of the form (7.61) which stabilizes uncertain system (7.74) with H_∞ performance γ is converted to the design of a single SOF gain $\bar{F}$ which makes the closed-loop system (7.77) admissible with H_∞ performance γ for all allowable perturbations. Once the robust SOF gain $\bar{F}$ is obtained, the original PID gains F_i is available by $\bar{F}=\begin{bmatrix}F_1 & F_2 & F_3\end{bmatrix}$. In the rest of this section, a suitable robust SOF controller is designed in the sense of quadratical admissibility with H_∞ performance γ, which is defined based on the following bounded real lemma for descriptor systems.

Lemma 7.4 (Bounded real lemma, [101]). *The descriptor system*

$$\begin{aligned}
E\dot{x}(t) &= Ax(t)+Bw(t),\\
z(t) &= Cx(t),
\end{aligned} \tag{7.78}$$

is admissible with H_∞ performance γ if and only if there exists a matrix P such that

$$E^T P = P^T E \geq 0,$$
$$A^T P + P^T A + \gamma^{-2}P^T BB^T P + C^T C < 0.$$

Definition 7.2. *The uncertain descriptor system*

$$\begin{aligned}
E\dot{x}(t) &= A(\delta)x(t)+B(\delta)w(t),\\
z(t) &= C(\delta)x(t),
\end{aligned}$$

is quadratically admissible with H_∞ performance γ (QAH_∞-γ) if there exists a matrix P such that

$$E^T P = P^T E \geq 0,$$
$$A^T(\delta)P + P^T A(\delta) + \gamma^{-2}P^T B(\delta)B^T(\delta)P + C^T(\delta)C(\delta) < 0,$$

for all allowable perturbations δ.

Obviously, the QAH$_\infty$-γ concept is an extension of the quadratic stability for uncertain standard systems. The following is a necessary and sufficient condition in terms of matrix inequalities for system (7.77).

Proposition 7.6. *System (7.77) is QAH$_\infty$-γ under SOF gain $\bar{F}$ if and only if there exist a matrix P and scalars $\rho > 0$ and $\omega > 0$ such that*

$$\bar{E}^T P = P^T \bar{E} \geq 0, \tag{7.79}$$

$$R := \gamma^2 I - \rho \bar{N}_3^T \bar{N}_3 > 0, \tag{7.80}$$

$$\begin{bmatrix} \Omega + P^T S P & \star & \star & \star & \star \\ \left(\bar{B} + \bar{B}_1 R^{-1} \bar{N}_3^T \bar{N}_2\right)^T P + \bar{F}\bar{C} & -I & 0 & 0 & 0 \\ \bar{N}_3^T \left(\bar{N}_1 + \bar{N}_2 \bar{F}\bar{C}\right) & 0 & -\omega^{-1}\rho^{-2} R & 0 & 0 \\ \bar{C}_1 + \bar{D}\bar{F}\bar{C} & 0 & 0 & -\omega^{-1} I & 0 \\ \bar{N}_1 + \bar{N}_2 \bar{F}\bar{C} & 0 & 0 & 0 & -(\omega\rho)^{-1} I \end{bmatrix}$$
$$< 0, \tag{7.81}$$

where $\star$ denotes the symmetric block, and

$$\begin{aligned} \Omega &:= P^T\left(\bar{A} + \bar{B}_1 R^{-1}\bar{N}_3^T \bar{N}_1\right) + \left(\bar{A} + \bar{B}_1 R^{-1}\bar{N}_3^T \bar{N}_1\right)^T P, \\ S &:= \omega^{-1}\bar{B}_1 R^{-1}\bar{B}_1^T + (\omega\rho)^{-1}\bar{M}\bar{M}^T \\ &\quad - \left(\bar{B} + \bar{B}_1 R^{-1}\bar{N}_3^T \bar{N}_2\right)\left(\bar{B} + \bar{B}_1 R^{-1}\bar{N}_3^T \bar{N}_2\right)^T. \end{aligned}$$

Proof. By Definition 7.2 and Schur complement, system (7.77) is QAH$_\infty$-γ under $\bar{F}$ if and only if there exists a matrix Q such that

$$\bar{E}^T Q = Q^T \bar{E} \geq 0, \tag{7.82}$$

$$\Psi + \begin{bmatrix} Q^T \bar{M} \\ 0 \end{bmatrix} \Delta \begin{bmatrix} \bar{N}_1 + \bar{N}_2 \bar{F}\bar{C} & \bar{N}_3 \end{bmatrix}$$
$$+ \begin{bmatrix} \left(\bar{N}_1 + \bar{N}_2 \bar{F}\bar{C}\right)^T \\ \bar{N}_3^T \end{bmatrix} \Delta^T \begin{bmatrix} \bar{M}^T Q & 0 \end{bmatrix} < 0, \tag{7.83}$$

hold for all allowable Δ, where

$$\Psi = $$
$$\begin{bmatrix} \left(\bar{A} + \bar{B}\bar{F}\bar{C}\right)^T Q + Q^T\left(\bar{A} + \bar{B}\bar{F}\bar{C}\right) + \left(\bar{C}_1 + \bar{D}\bar{F}\bar{C}\right)^T\left(\bar{C}_1 + \bar{D}\bar{F}\bar{C}\right) & Q^T \bar{B}_1 \\ \bar{B}_1^T Q & -\gamma^2 I \end{bmatrix}.$$

Using the well-known fact that $Y + U\Delta V + V^T \Delta^T U^T < 0$ with $\Delta^T \Delta \leq I$ is equivalent to $Y + \rho^{-1} U U^T + \rho V^T V < 0$ for some $\rho > 0$, we have that (7.83) holds if and only if there exists $\rho > 0$ such that

$$\Psi + \rho^{-1} \begin{bmatrix} Q^T \bar{M} \bar{M}^T Q & 0 \\ 0 & 0 \end{bmatrix} + \rho \begin{bmatrix} \left(\bar{N}_1 + \bar{N}_2 \bar{F} \bar{C}\right)^T \left(\bar{N}_1 + \bar{N}_2 \bar{F} \bar{C}\right) & \star \\ \bar{N}_3^T \left(\bar{N}_1 + \bar{N}_2 \bar{F} \bar{C}\right) & \bar{N}_3^T \bar{N}_3 \end{bmatrix} < 0,$$

which, by Schur complement again and after simple manipulations, is equivalent to that $R > 0$ and

$$\begin{aligned}
Q^T &\left[\left(\bar{A} + \bar{B}_1 R^{-1} \bar{N}_3^T \bar{N}_1\right) + \left(\bar{B} + \bar{B}_1 R^{-1} \bar{N}_3^T \bar{N}_2\right) \bar{F} \bar{C}\right] \\
&+ \left[\left(\bar{A} + \bar{B}_1 R^{-1} \bar{N}_3^T \bar{N}_1\right) + \left(\bar{B} + \bar{B}_1 R^{-1} \bar{N}_3^T \bar{N}_2\right) \bar{F} \bar{C}\right]^T Q \\
&+ Q^T \left(\bar{B}_1 R^{-1} \bar{B}_1^T + \rho^{-1} \bar{M} \bar{M}^T\right) Q \\
&+ \rho^2 \left(\bar{N}_1 + \bar{N}_2 \bar{F} \bar{C}\right)^T \bar{N}_3 R^{-1} \bar{N}_3^T \left(\bar{N}_1 + \bar{N}_2 \bar{F} \bar{C}\right) \\
&+ \left(\bar{C}_1 + \bar{D} \bar{F} \bar{C}\right)^T \left(\bar{C}_1 + \bar{D} \bar{F} \bar{C}\right) + \rho \left(\bar{N}_1 + \bar{N}_2 \bar{F} \bar{C}\right)^T \left(\bar{N}_1 + \bar{N}_2 \bar{F} \bar{C}\right) \\
&< 0.
\end{aligned} \tag{7.84}$$

A method similar to the proof of Proposition 7.5 or that of Theorem 1 in [90] leads to that (7.84) with (7.82) holds if and only if there exist matrix P and a scalar $\omega > 0$ such that (7.79) and

$$\begin{aligned}
\Omega &+ P^T S P + \left[\left(\bar{B} + \bar{B}_1 R^{-1} \bar{N}_3^T \bar{N}_2\right)^T P + \bar{F} \bar{C}\right]^T \\
&\times \left[\left(\bar{B} + \bar{B}_1 R^{-1} \bar{N}_3^T \bar{N}_2\right)^T P + \bar{F} \bar{C}\right] \\
&+ \omega \rho^2 \left(\bar{N}_1 + \bar{N}_2 \bar{F} \bar{C}\right)^T \bar{N}_3 R^{-1} \bar{N}_3^T \left(\bar{N}_1 + \bar{N}_2 \bar{F} \bar{C}\right) \\
&+ \omega \left(\bar{C}_1 + \bar{D} \bar{F} \bar{C}\right)^T \left(\bar{C}_1 + \bar{D} \bar{F} \bar{C}\right) + \omega \rho \left(\bar{N}_1 + \bar{N}_2 \bar{F} \bar{C}\right)^T \left(\bar{N}_1 + \bar{N}_2 \bar{F} \bar{C}\right) \\
&< 0,
\end{aligned} \tag{7.85}$$

hold. It is evident that (7.85) is equivalent to (7.81) by Schur complement. This completes the proof. $\qquad \square$

Inequality (7.81) is in the quadratic matrix inequality form as (7.65). Adopting the idea for those algorithms in [90] and [79], an iterative LMI algorithm similar to Algorithm 2.1 can be developed to solve the inequalities in (7.79)–(7.81). The details are omitted here.

From Proposition 7.6, if it is possible that $R > 0$ and $S \geq 0$ hold for some $\rho > 0$ and $\omega > 0$, then inequality (7.81) can be reduced to a strict LMI. However, the resulting condition becomes a sufficient one only, which is as follows.

Corollary 7.1. *The uncertain system of the form (7.77) is QAH_∞-γ under SOF gain $\bar{F}$ if there exist a matrix P and a scalar $\rho > 0$ such that the following LMIs hold.*

$$\bar{E}^T P = P^T \bar{E} \geq 0,$$
$$R > 0,$$
$$S \geq 0,$$

$$\begin{bmatrix} \Omega & \star & \star & \star & \star & \star \\ S^{1/2}P & -I & 0 & 0 & 0 & 0 \\ \left(\bar{B}+\bar{B}_1 R^{-1}\bar{N}_3^T \bar{N}_2\right)^T P + \bar{F}\bar{C} & 0 & -I & 0 & 0 & 0 \\ \bar{N}_3^T \left(\bar{N}_1 + \bar{N}_2\bar{F}\bar{C}\right) & 0 & 0 & -\omega^{-1}\rho^{-2}R & 0 & 0 \\ \bar{C}_1 + \bar{D}\bar{F}\bar{C} & 0 & 0 & 0 & -\omega^{-1}I & 0 \\ \bar{N}_1 + \bar{N}_2\bar{F}\bar{C} & 0 & 0 & 0 & 0 & -(\omega\rho)^{-1}I \end{bmatrix} < 0,$$

where the notations remain the same as in Proposition 7.6.

Proof. It is easy by applying Schur complement to (7.81).   □

The conservativeness of Corollary 7.1 comes from the requirement of $S \geq 0$. Besides, the resulting LMI conditions imply that the triple $(\bar{E}, \bar{A}, \bar{B}_1)$ must be stabilizable.

Example 7.5. Consider system (7.60) with $A = 1$, $B = 1$ and $C = 1$, and the PID controller of the form (7.61). If we choose the method in Sects. 7.3–7.6, the system is transformed into the following SOF control system:

$$\dot{\bar{x}}(t) = \bar{A}\bar{x}(t) + \bar{B}u(t), \quad \bar{y}(t) = \bar{C}\bar{x}(t), \quad u(t) = \bar{F}\bar{y}(t), \tag{7.86}$$

where

$$\bar{A} = \begin{bmatrix} 1 & 0 \\ 1 & 0 \end{bmatrix}, \quad \bar{B} = \begin{bmatrix} 1 \\ 0 \end{bmatrix}, \quad \bar{C} = \begin{bmatrix} 1 & 0 \\ 0 & 1 \\ 1 & 0 \end{bmatrix}, \quad \bar{F} = \begin{bmatrix} \bar{F}_1 & \bar{F}_2 & \bar{F}_3 \end{bmatrix},$$

$$\bar{F}_1 = (I - F_3CB)^{-1}F_1, \quad \bar{F}_2 = (I - F_3CB)^{-1}F_2, \quad \bar{F}_3 = (I - F_3CB)^{-1}F_3,$$

and once $\bar{F}$ is obtained, the original PID gains are recovered by (7.62). To this end, we compute that

$$\bar{A}^T P + P\bar{A} - P\bar{B}\bar{B}^T P + (\bar{B}^T P + \bar{F}\bar{C})^T (\bar{B}^T P + \bar{F}\bar{C}) < 0 \tag{7.87}$$

has a set of solutions as

$$\bar{F} = \begin{bmatrix} -2 & -1 & -1 \end{bmatrix}, \quad P = \begin{bmatrix} 3 & 1 \\ 1 & 2 \end{bmatrix} > 0. \tag{7.88}$$

However, this $\bar{F}$ violates the invertibility of $(I + CB\bar{F}_3)$ since $I + CB\bar{F}_3 = 0$. Hence, the original PID gains can not be obtained by this set of solution, and another set of solution has to be searched to meet the invertibility requirement.

This situation will not occur by using our descriptor method. Under our approach, the resulting SOF control system is formulated as in (7.63) where

$$\bar{E} = \begin{bmatrix} 1 & 0 & 0 \\ 0 & 1 & 0 \\ 0 & 0 & 0 \end{bmatrix}, \quad \bar{A} = \begin{bmatrix} 0 & 0 & 1 \\ 1 & 0 & 0 \\ 1 & 0 & -1 \end{bmatrix}, \quad \bar{B} = \begin{bmatrix} 0 \\ 0 \\ 1 \end{bmatrix},$$

$$\bar{C} = I_3, \quad \bar{F} = \begin{bmatrix} F_1 & F_2 & F_3 \end{bmatrix}.$$

It is seen that once $\bar{F}$ is obtained, the original PID gains F_i are available immediately without any requirement. It is easy to compute that (7.64) and (7.65) are satisfied under

$$\bar{F} = \begin{bmatrix} -2.5 & -1 & -0.2 \end{bmatrix}, \quad P = \begin{bmatrix} 5 & 1 & 0 \\ 1 & 2 & 0 \\ 3 & 1 & 1 \end{bmatrix}. \tag{7.89}$$

Then, we conclude immediately that the original stabilizing PID gains are given by $F_1 = -2.5$, $F_2 = -1$ and $F_3 = -0.2$.

Furthermore, if matrices A, B together with the derivative term are subjected to perturbations, the method in Sects. 7.3–7.6 is no longer valid. In such an uncertainty case, our descriptor method provided in this section is applicable to design a robust stabilizing PID controller. Due to space limitation, the details are omitted.

7.10 Conclusions

In this chapter, algorithms based on ILMI technique have been developed to design the feedback matrices of multivariable PID controllers which guarantee the stability of the closed loop systems, H_2 or H_∞ performance specifications, or maximum output control requirement, respectively. Several numerical examples on the design of PID controllers have been presented to illustrate the feasibility of the proposed methods.

Compared to most of the prevalent PID controller design methods, advantages of our methods are: (i) Stability of closed loop systems is guaranteed; (ii) No specific requirement on either system structure or system order is imposed. The disadvantages of our methods are: (i) System parameters must be known for the algorithms given in Sects. 7.3-7.8; (ii) The iterative algorithms developed here are based on sufficient criteria for the corresponding problems. Therefore when the algorithms are terminated without a solution, it remains unclear whether or not the problem is feasible.

Notice also that the performance parameters γ, v and σ in Sects. 7.4–7.6 are given. We have not tried to optimize these parameters in the algorithms developed here. In practical applications, a second iteration for optimizing these parameters would be needed.

To overcome the disadvantages of the method in Sects. 7.3–7.6, the descriptor approach is proposed, which could not only improve the existing results by removing the invertibility constraints but also well treat a class of uncertain systems with norm-bounded perturbations. However, one disadvantage remains as in [79], i.e., the iterative algorithm to solve the corresponding quadratic matrix inequality is developed based on sufficient conditions.

Finally, new ILMI algorithms for SOF stabilization and H_∞ control are proposed in this chapter, which avoid introduction of the additional variables, leading to lower dimensions of the LMIs than [90] and [172]. In particular, an algorithm to derive an initial value for some decision variable of the involved ILMI is also given. The algorithms are also applied to design the multivariable PID controller. Numerical examples show that the proposed algorithms produces better result and/or faster convergence than the existing ones.

8 Multivariable PID Control for Synchronization

Chapters 7 presents the new design methods for MIMO PID control in the framework of LMI and optimal control. In this chapter, some of these design methods are also applied to solve a practical problem — to achieve fast master-slave synchronization of Lur'e systems with multivariable PD/PID control.

8.1 Introduction

Since the seminal work of Pecora and Carroll [177], the topic of chaos synchronization has attracted great interest in both theoretical studies [178, 179, 180, 181, 182, 183] and practical applications [184,185,186,187,188]. Synchronization phenomena in nonidentical systems or the coupled systems with different order have been investigated [189]. A number of master-slave synchronization schemes for Lur'e systems have been proposed [183, 190, 182, 191]. Wen et al. [181] studied robust synchronization of chaotic systems under output feedback control with multiple random delays. Some linear-state-feedback synchronization control methods [192, 193, 194, 195] are reported. For example, in the reference [194], it is proved that global asymptotic synchronization can be attained via a linear output error feedback approach when the feedback gain chosen as a function of a free parameter is large enough. Yassen [196] investigated chaos synchronization by using adaptive control. Chaos synchronization has also been addressed using observers with linear output feedback [197], PI observers [190,198,199] and nonlinear observers [200,201,202]. Femat et al. proposed a Laplace domain controller and its applications to design PII^2 controller [203,204]. Their results enable one to observe the different synchronization phenomena of chaotic systems with different order and model [180]. Jiang and Zheng [192] proposed a linear-state-feedback synchronization criterion based on LMI. To our best knowledge, no work is reported in the literature on chaos synchronization via full PID control. This may result from the fact in PID control studies that many prevalent PID controller design methods were established on basis of frequency response methods. The intrinsic characteristic of synchronization implies that the state-space approach is preferable for serving our purpose.

Q.-G. Wang et al.: PID Control for Multivariable Processes, LNCIS 373, pp. 203–217, 2008.
springerlink.com
© Springer-Verlag Berlin Heidelberg 2008

Hua and Guan [199] transformed chaotic systems with PI controller into an augmented proportional control system, as reported earlier by Zheng et al. [79], and proposed a synchronization criterion using the LMI technique. However, their methodology is not applicable to chaotic systems with PD or PID controller. It is well known in the area of PID control that the integral control is mainly employed to improve the steady state tracking accuracy while the derivative control to enhance stability and speed the system response [2]. Obviously, it is desirable to enhance stability and speed synchronization response as concerning the chaos synchronization. This implies that the derivative control is desirable to increase synchronization speed for Lur'e systems. If there had existed appropriate design methods of PID controller, PID control for synchronization should have prevailed.

The objective of this chapter is to propose a multivariable PD/PID controller design to achieve fast master-slave synchronization of Lur'e systems [205]. Due to the fact that measuring all the state variables of a system is inconvenient or even impossible in many practical situation [206], output feedback control is considered. The free-weighting matrix approach [109, 110, 111, 207] and the S-procedure [89] are employed to establish the synchronization strategy. It is shown that our corollary covers the existing result in the case of proportional control alone. Numerical results demonstrate the improvement of speeding synchronization response with the aid of the the derivative action, as compared to the results of the same chaotic system based on PI control [199].

8.2 Problem Formulation

Consider a general master-slave type of coupled Lur'e systems with PD/PID controller:

$$\mathcal{M} \ : \ \begin{cases} \dot{x}(t) &= Ax(t) + B\sigma\left(C^T x(t)\right) \\ y(t) &= Hx(t) \end{cases}$$

$$\mathcal{S} \ : \ \begin{cases} \dot{\hat{x}}(t) &= A\hat{x}(t) + B\sigma\left(C^T \hat{x}(t)\right) + u(t) \\ \hat{y}(t) &= H\hat{x}(t) \end{cases} \tag{8.1}$$

$$\mathcal{C} \ : \ \begin{cases} u(t) &= K_p\left(y(t) - \hat{y}(t)\right) + K_d\left(\dot{y}(t) - \dot{\hat{y}}(t)\right) \\ u(t) &= K_p(y(t) - \hat{y}(t)) + K_i \int_0^t (y(\theta) - \hat{y}(\theta))\mathrm{d}\theta + K_d\left(\dot{y}(t) - \dot{\hat{y}}(t)\right), \end{cases}$$

with master system $\mathcal{M}$, slave system $\mathcal{S}$ and controller $\mathcal{C}$. The master and slave systems are Lur'e systems with control input $u \in \mathbb{R}^n$, state vectors $x, \hat{x} \in \mathbb{R}^n$, and matrices $A \in \mathbb{R}^{n \times n}$, $B \in \mathbb{R}^{n \times n_h}$, $C \in \mathbb{R}^{n \times n_h}$. The matrix $H \in \mathbb{R}^{l \times n}$ implies use of the output feedback and the outputs of subsystems are $y, \hat{y} \in \mathbb{R}^l$, respectively. The nonlinearity $\sigma(\cdot) = [\sigma_1, \sigma_2, \cdots, \sigma_{n_h}]^T$ satisfies a sector condition with $\sigma_j(\cdot)$, $j = 1, 2, \cdots, n_h$, belonging to sectors $[0, k_j]$, i.e.,

$$\sigma_j(\xi)(\sigma_j(\xi) - k_j\xi) \leq 0, \forall \xi, \text{ for } j = 1, 2, \cdots, n_h. \tag{8.2}$$

PD Control

For the PD controller, K_p and K_d are the proportional and derivative gain matrices, respectively. Let the synchronization error of system (8.1) be $e(t) = x(t) - \hat{x}(t)$. The error dynamic system is given by

$$
\begin{cases}
\dot{e}(t) &= Ae(t) + B\eta\left(C^T e(t), \hat{x}(t)\right) - u(t) \\
u(t) &= K_p H e(t) + K_d H \dot{e}(t)
\end{cases}
\tag{8.3}
$$

PID Control

For the PID controller, K_p, K_i and K_d are the proportional, integral and derivative gain matrices, respectively. Let the synchronization error of system (8.1) be $e(t) = x(t) - \hat{x}(t)$. The error dynamic system is given by

$$
\begin{cases}
\dot{e}(t) &= Ae(t) + B\eta\left(C^T e(t), \hat{x}(t)\right) - u(t) \\
u(t) &= K_p H e(t) + K_i \int_0^t H e(\theta)\mathrm{d}\theta + K_d H \dot{e}(t)
\end{cases}
\tag{8.4}
$$

where $\eta\left(C^T e(t), \hat{x}(t)\right) = \sigma\left(C^T e + C^T \hat{x}\right) - \sigma\left(C^T \hat{x}\right)$. Let $C^T = \left[c_1, \cdots, c_{n_h}\right]^T$ with $c_j \in \mathbb{R}^n$, $j = 1, 2, \cdots, n_h$. Assume that the nonlinearity $\eta\left(C^T e(t), \hat{x}(t)\right)$ belongs to sector $[0, k_j]$,

$$
0 \leq \frac{\eta_j\left(c_j^T e, \hat{x}\right)}{c_j^T e} = \frac{\sigma\left(c_j^T e + c_j^T \hat{x}\right) - \sigma\left(c_j^T \hat{x}\right)}{c_j^T e} \leq k_j, \forall e, \hat{x}, j = 1, 2, \cdots, n_h.
\tag{8.5}
$$

It follows from (8.5) that

$$
\eta_j\left(c_j^T e, \hat{x}\right)\left(\eta_j\left(c_j^T e, \hat{x}\right) - k_j c_j^T e\right) \leq 0, \forall e, \hat{x}, j = 1, 2, \cdots, n_h.
\tag{8.6}
$$

In this section, we will study the design and robust synthesis of PD/PID controller to globally synchronize the master system $\mathcal{M}$ to the slave system $\mathcal{S}$, i.e., $e(t) \to 0$ as $t \to \infty$.

8.3 A Strategy for Master-Slave Synchronization

8.3.1 PD Control

Most of the prevalent PID controller design methods established on basis of frequency response methods. The intrinsic characteristic of synchronization implies that the state-space approach is preferable for serving our purpose. By transforming the problem of PD controller design to that of proportional controller design for state-space analysis, we obtain the following global synchronization strategy based on the free-weighting matrix approach [109, 110] and the S-procedure [89].

Theorem 8.1. *The equilibrium point $e = 0$ of the error system (8.3) is globally asymptotically stable if there exist $P = P^T > 0$, $\Lambda = \mathrm{diag}(\lambda_1, \lambda_2, \cdots, \lambda_{n_h}) \geq 0$, $W = \mathrm{diag}(w_1, w_2, \cdots, w_{n_h}) \geq 0$, $S = \mathrm{diag}(s_1, s_2, \cdots, s_{n_h}) \geq 0$, and any appropriately dimensional matrices N_j, $j = 1, 2, 3$ such that the following condition (8.7) is feasible,*

$$\Phi = \begin{bmatrix} \Phi_{11} & \Phi_{12} & \Phi_{13} & \Phi_{14} \\ \Phi_{12}^T & \Phi_{22} & \Phi_{23} & \Phi_{24} \\ \Phi_{13}^T & \Phi_{23}^T & -2W & \Phi_{34} \\ \Phi_{14}^T & \Phi_{24}^T & \Phi_{34}^T & -2S \end{bmatrix} < 0, \tag{8.7}$$

where

$$\begin{aligned} \Phi_{11} &= N_1(A - K_pH) + (N_1(A - K_pH))^T, \\ \Phi_{12} &= P - N_1(K_dH + I) + (N_2(A - K_pH))^T, \\ \Phi_{13} &= CKW + (N_3(A - K_pH))^T, \\ \Phi_{14} &= CKS + N_1B, \\ \Phi_{22} &= -(N_2(K_dH + I)) - (N_2(K_dH + I))^T, \\ \Phi_{23} &= C\Lambda K - (N_3(K_dH + I))^T, \\ \Phi_{24} &= N_2B, \\ \Phi_{34} &= N_3B, \\ K &= \mathrm{diag}\{k_1, k_2, \cdots, k_{n_h}\}. \end{aligned}$$

Proof. Based on system (8.3), we introduce a new state vector $z(t) = [z_1, z_2]^T$ where

$$\begin{aligned} z_1(t) &= e(t), \tag{8.8} \\ z_2(t) &= \dot{e}(t). \tag{8.9} \end{aligned}$$

Note that

$$\begin{aligned} 0 &= \dot{e}(t) - z_2(t) \\ &= (A - K_pH)e - z_2 + B\eta - K_dH\dot{e} \\ &= (A - K_pH)z_1 - (K_dH + I)z_2 + B\eta. \tag{8.10} \end{aligned}$$

Construct the following Lyapunov functional:

$$V(t) = z_1^T(t)Pz_1(t) + 2\sum_{j=1}^{n_h} k_j\lambda_j \int_0^{c_j^T z_1} \sigma_j(s)ds, \tag{8.11}$$

where $P = P^T > 0$ and $\Lambda = \mathrm{diag}(\lambda_1, \lambda_2, \cdots, \lambda_{n_h}) \geq 0$ are to be determined. It is clear from (8.10) that for any appropriately dimensional matrices N_j, $j = 1, 2, 3$, the following relationship holds,

$$\begin{aligned} 0 = &\left[z_1^T(t)N_1 + z_2^T(t)N_2 + \sigma^T\left(C^Tz_1(t)\right)N_3\right] \\ &\times \left[(A - K_pH)z_1 - (K_dH + I)z_2 + B\eta\right], \tag{8.12} \end{aligned}$$

where N_j, $j = 1,2,3$, are considered as free-weighting matrices [109, 110]. Furthermore, for any $W = \mathrm{diag}(w_1, w_2, \cdots, w_{n_h}) \geq 0$ and $S = \mathrm{diag}(s_1, s_2, \cdots, s_{n_h}) \geq 0$, it follows from (8.2) and (8.6) that

$$
\begin{aligned}
0 \;\leq\; & -2 \sum_{j=1}^{n_h} \left[w_j \sigma_j \left(c_j^T z_1 \right) \left(\sigma_j \left(c_j^T z_1 \right) - k_j c_j^T z_1 \right) + s_j \eta_j \left(\eta_j - k_j c_j^T z_1 \right) \right] \\
=\; & 2 z_1^T(t) C K W \sigma \left(C^T z_1(t) \right) - 2 \sigma^T \left(C^T z_1(t) \right) W \sigma \left(C^T z_1(t) \right) \\
& + 2 z_1^T(t) C K S \eta \left(C^T z_1(t), \hat{x}(t) \right) \\
& - 2 \eta^T \left(C^T z_1(t), \hat{x}(t) \right) S \eta \left(C^T z_1(t), \hat{x}(t) \right) .
\end{aligned}
\tag{8.13}
$$

Inequality (8.13) is a standard application of the S-procedure [89]. Taking the time derivative of $V(t)$ and adding the terms on the right hand side of (8.12) and (8.13) into $\dot{V}(t)$, one obtains

$$
\begin{aligned}
\dot{V}(t) \;=\; & 2 z_1^T(t) P z_2(t) + 2 \sum_{j=1}^{n_h} k_j \lambda_j \sigma_j \left(c_j^T z_1 \right) c_j^T z_2(t) \\
\leq\; & 2 z_1^T(t) P z_2(t) + 2 \sigma^T \left(C^T z_1(t) \right) K \Lambda C^T z_2(t) \\
& + 2 \left[z_1^T(t) N_1 + z_2^T(t) N_2 + \sigma^T \left(C^T z_1(t) \right) N_3 \right] \\
& \times \left[(A - K_p H) z_1 - (K_d H + I) z_2 + B \eta \right] \\
& + 2 \left[z_1^T(t) C K W \sigma \left(C^T z_1(t) \right) \right. \\
& \quad - \sigma^T \left(C^T z_1(t) \right) W \sigma \left(C^T z_1(t) \right) \\
& \quad + z_1^T(t) C K S \eta^T \left(C^T z_1(t), \hat{x}(t) \right) \\
& \quad \left. - \eta^T \left(C^T z_1(t), \hat{x}(t) \right) S \eta \left(C^T z_1(t), \hat{x}(t) \right) \right] \\
=\; & \xi^T \Phi \xi,
\end{aligned}
\tag{8.14}
$$

where

$$
\xi(t) = \left[z_1(t), z_2(t), \sigma \left(C^T z_1(t) \right), \eta \left(C^T z_1(t), \hat{x}(t) \right) \right]^T .
$$

Thus, $\dot{V}(t) < -\varepsilon \| z_1(t) \|^2 = -\varepsilon \| e(t) \|^2$ for a sufficiently small ε if $\Phi < 0$, which ensures the asymptotic stability of equilibrium point $e = 0$. This completes the proof. $\qquad \square$

The following corollary 8.1 gives a LMI-based methodology to find the control parameters K_p and K_d.

Corollary 8.1. *For two given scalars δ_1 and δ_2, the equilibrium point $e = 0$ of the error system (8.3) is globally asymptotically stable if there exist $P = P^T > 0$, $\Lambda = \mathrm{diag}(\lambda_1, \lambda_2, \cdots, \lambda_{n_h}) \geq 0$, $W = \mathrm{diag}(w_1, w_2, \cdots, w_{n_h}) \geq 0$, $S = \mathrm{diag}(s_1, s_2, \cdots, s_{n_h}) \geq 0$, and any appropriately dimensional matrices N and M_j, $j = 1,2$ such that the following LMI (8.15) is feasible,*

$$
\Phi = \begin{bmatrix}
\Phi_{11} & \Phi_{12} & \Phi_{13} & \Phi_{14} \\
\Phi_{12}^T & \Phi_{22} & \Phi_{23} & \Phi_{24} \\
\Phi_{13}^T & \Phi_{23}^T & -2W & \Phi_{34} \\
\Phi_{14}^T & \Phi_{24}^T & \Phi_{34}^T & -2S
\end{bmatrix} < 0,
\tag{8.15}
$$

where

$$\begin{aligned}
\Phi_{11} &= \delta_1\left(NA - M_1H + (NA - M_1H)^T\right), \\
\Phi_{12} &= P - \delta_1(M_2H + N) + (NA - M_1H)^T, \\
\Phi_{13} &= CKW + \delta_2(NA - M_1H)^T, \\
\Phi_{14} &= CKS + \delta_1NB, \\
\Phi_{22} &= -M_2H - N - (M_2H + N)^T, \\
\Phi_{23} &= C\Lambda K - \delta_2(M_2H + N)^T, \\
\Phi_{24} &= NB, \\
\Phi_{34} &= \delta_2NB, \\
K &= \mathrm{diag}\{k_1, k_2, \cdots, k_{n_h}\}.
\end{aligned}$$

Moreover, the coefficients of PD controller are given by $K_p = N^{-1}M_1$ and $K_d = N^{-1}M_2$, respectively.

Proof. By setting $N = N_2$, $M_1 = NK_p$, $M_2 = NK_d$, $N_1 = \delta_1N$ and $N_3 = \delta_2N$ in the condition (8.7), one obtains LMI (8.15), which is solvable for the variables P, Λ, W, S, N, M_j, $j = 1,2$, by using the LMI technique. Note that $\Phi < 0$ implies that N is nonsingular. Thus, $K_p = N^{-1}M_1$ and $K_d = N^{-1}M_2$. $\qquad\square$

With a minor modification to Theorem 8.1, one obtains the following result on synchronization via proportional control alone.

Corollary 8.2. *Consider system (8.1) with proportional control only: $u(t) = K_p(y(t) - \hat{y}(t))$. The equilibrium point $e = 0$ of the corresponding error system (8.3) is globally asymptotically stable if there exist $P = P^T > 0$, $\Lambda = \mathrm{diag}(\lambda_1, \lambda_2, \cdots, \lambda_{n_h}) \geq 0$, $W = \mathrm{diag}(w_1, w_2, \cdots, w_{n_h}) \geq 0$, $S = \mathrm{diag}(s_1, s_2, \cdots, s_{n_h}) \geq 0$, and any appropriately dimensional matrices N_j, $j = 1,2,3$, such that the following the condition (8.16) is feasible,*

$$\Phi = \begin{bmatrix}
\Phi_{11} & \Phi_{12} & \Phi_{13} & \Phi_{14} \\
\Phi_{12}^T & \Phi_{22} & \Phi_{23} & \Phi_{24} \\
\Phi_{13}^T & \Phi_{23}^T & -2W & 0 \\
\Phi_{14}^T & \Phi_{24}^T & 0 & -2S
\end{bmatrix} < 0, \tag{8.16}$$

where

$$\begin{aligned}
\Phi_{11} &= N_1(A - K_pH) + (N_1(A - K_pH))^T, \\
\Phi_{12} &= P - N_1 + (N_2(A - K_pH))^T, \\
\Phi_{13} &= CKW + (N_3(A - K_pH))^T, \\
\Phi_{14} &= CKS + N_1B,
\end{aligned}$$

$$
\begin{aligned}
\Phi_{22} &= -N_2 - N_2^T, \\
\Phi_{23} &= C\Lambda K - N_3^T, \\
\Phi_{24} &= N_2 B, \\
\Phi_{34} &= N_3 B, \\
K &= \mathrm{diag}\{k_1, k_2, \cdots, k_{n_h}\}.
\end{aligned}
$$

It should be noted that our results belong to output feedback control which is commonly used in industry control. Curran et al. [183] reported a synchronization condition for state feedback control $K_p(x(t) - \hat{x}(t))$, which is repeated as follows:

Lemma 8.1. *For system (8.1) with state feedback control $K_p(x(t) - \hat{x}(t))$ alone, the corresponding error system (8.3) has a globally asymptotically stable equilibrium point $e = 0$ if there exist $P = P^T > 0$, $\Lambda = \mathrm{diag}(\lambda_1, \lambda_2, \cdots, \lambda_{n_h}) \geq 0$, $W = \mathrm{diag}(w_1, w_2, \cdots, w_{n_h}) \geq 0$, and $S = \mathrm{diag}(s_1, s_2, \cdots, s_{n_h}) \geq 0$ such that the following the condition (8.17) is feasible,*

$$
\tilde{\Phi} =
\begin{bmatrix}
\tilde{\Phi}_{11} & \tilde{\Phi}_{12} & \tilde{\Phi}_{13} \\
\tilde{\Phi}_{12}^T & -2W & \tilde{\Phi}_{23} \\
\tilde{\Phi}_{13}^T & \tilde{\Phi}_{23}^T & -2S
\end{bmatrix}
< 0, \tag{8.17}
$$

where

$$
\begin{aligned}
\tilde{\Phi}_{11} &= (A - K_p)^T P + P(A - K_p), \\
\tilde{\Phi}_{12} &= (A - K_p)^T C\Lambda K + CKW, \\
\tilde{\Phi}_{13} &= PB + CSK, \\
\tilde{\Phi}_{23} &= K\Lambda C^T B, \\
K &= \mathrm{diag}\{k_1, k_2, \cdots, k_{n_h}\}.
\end{aligned}
$$

We now show that if H is an identity matrix corresponding to the state feedback case, Corollary 8.2 is equivalent to Lemma 8.1. Let $H = I$ where I is an identity matrix. The matrix Φ in the condition (8.16), under the elementary operations for matrix Φ: left-multiplication of the second row by $(A - K_p)^T$ and B^T and adding them to the first row and the fourth row, respectively; right-multiplication the second column by $(A - K_p)$ and B and adding them into the first column and the fourth column, respectively; and the interchange of the first and second rows and the one of the first and second columns, becomes

$$
\Phi =
\begin{bmatrix}
\hat{\Phi}_{11} & \hat{\Phi}_{12} & \hat{\Phi}_{13} & \hat{\Phi}_{14} \\
\hat{\Phi}_{12}^T & \tilde{\Phi}_{11} & \tilde{\Phi}_{12} & \tilde{\Phi}_{13} \\
\hat{\Phi}_{13}^T & \tilde{\Phi}_{12}^T & -2W & \tilde{\Phi}_{23} \\
\hat{\Phi}_{14}^T & \tilde{\Phi}_{13}^T & \tilde{\Phi}_{23}^T & -2S
\end{bmatrix}
< 0, \tag{8.18}
$$

where $\tilde{\Phi}_{ij}$ are defined in (8.17) and

$$
\begin{aligned}
\hat{\Phi}_{11} &= -N_2 - N_2^T, \\
\hat{\Phi}_{12} &= P - N_2^T(A - K_p) - N_1, \\
\hat{\Phi}_{13} &= C\Lambda K - N_3, \\
\hat{\Phi}_{14} &= -N_2^T B.
\end{aligned}
$$

It is clear that the condition (8.17) is feasible if the condition (8.18) is feasible. On the other hand, by setting $H = I$, $N_1 = P$, $N_2 = -\alpha I$ and $N_3 = C\Lambda K$ where α is a sufficiently small positive scalar, we can rewrite matrix Φ in the condition (8.18) as

$$
\Phi = \begin{bmatrix}
\check{\Phi}_{11} & \check{\Phi}_{12} & \check{\Phi}_{13} & \check{\Phi}_{14} \\
\check{\Phi}_{12}^T & \tilde{\Phi}_{11} & \tilde{\Phi}_{12} & \tilde{\Phi}_{13} \\
\check{\Phi}_{13}^T & \tilde{\Phi}_{12}^T & -2W & \tilde{\Phi}_{23} \\
\check{\Phi}_{14}^T & \tilde{\Phi}_{13}^T & \tilde{\Phi}_{23}^T & -2S
\end{bmatrix} < 0, \tag{8.19}
$$

where

$$
\begin{aligned}
\check{\Phi}_{11} &= 2I\alpha, \\
\check{\Phi}_{12} &= \alpha(A - K_p), \\
\check{\Phi}_{13} &= 0, \\
\check{\Phi}_{14} &= \alpha B.
\end{aligned}
$$

Based on the Schur complement [89], the condition (8.19) is feasible if the condition (8.17) is feasible. Therefore, Corollary 8.2 covers the synchronization condition proposed by Curran et al. [183] as a special case, in which the synchronization of Lur'e systems was achieved via proportional control alone.

8.3.2 PID Control

In order to apply the LMI technique [89] for PID controller design, we transform system (8.1) to a system in descriptor form [80]. Introduce a new state variable

$$
z(t) = \begin{bmatrix} z_1(t) \\ z_2(t) \\ z_3(t) \end{bmatrix} = \begin{bmatrix} e(t) \\ \int_0^t He(t)\,d\theta \\ \dot{e}(t) \end{bmatrix}. \tag{8.20}
$$

Then system (8.1) with PID controller is transformed into the following SOF control system in the descriptor form,

$$
\begin{aligned}
E\dot{z}(t) &= \tilde{A}z(t) + \tilde{B}\eta(t) + \hat{B}u(t), \\
y_e(t) &= \tilde{H}z(t), \\
u(t) &= [K_p, K_i, K_d]y_e(t),
\end{aligned} \tag{8.21}
$$

where

$$
E = \begin{bmatrix} I_n & 0 & 0 \\ 0 & I_l & 0 \\ 0 & 0 & 0 \end{bmatrix}, \quad \tilde{A} = \begin{bmatrix} 0 & 0 & I_n \\ H & 0 & 0 \\ A & 0 & I_n \end{bmatrix},
$$

$$
\bar{B} = \begin{bmatrix} 0 \\ 0 \\ -B \end{bmatrix}, \quad \hat{B} = \begin{bmatrix} 0 \\ 0 \\ -I_n \end{bmatrix}, \quad \tilde{H} = \begin{bmatrix} H & 0 & 0 \\ 0 & I_l & 0 \\ 0 & 0 & H \end{bmatrix}.
$$

Let us now investigate robust synchronization criterion which is applicable for the derivation of the control parameters K_p, K_i and K_d. Construct the following Lyapunov function:

$$
V(t) = \begin{bmatrix} z_1^T & z_2^T \end{bmatrix} \begin{bmatrix} P_{11} & P_{12} \\ P_{12}^T & P_{22} \end{bmatrix} \begin{bmatrix} z_1 \\ z_2 \end{bmatrix} + 2 \sum_{j=1}^{n_h} \lambda_j \int_0^{c_j^T z_1} \sigma_j(s) \mathrm{d}s, \tag{8.22}
$$

where $P_{12} \in \mathbb{R}^{n \times l}$, $P_{11} \in \mathbb{R}^{n \times n}$, $P_{22} \in \mathbb{R}^{l \times l}$, $\begin{bmatrix} P_{11} & P_{12} \\ P_{12}^T & P_{22} \end{bmatrix} > 0$ and $\Lambda = \mathrm{diag}(\lambda_1, \lambda_2, \cdots,$ $\lambda_{n_h}) \geq 0$ are to be determined. For any $W = \mathrm{diag}(w_1, w_2, \cdots, w_{n_h}) \geq 0$ and $S = \mathrm{diag}(s_1, s_2, \cdots, s_{n_h}) \geq 0$, it follows from (8.2) and (8.5) that

$$
\begin{aligned}
0 \;\leq\; & \tilde{S}(t) \\
= \; & -2 \sum_{j=1}^{n_h} \left[w_j \sigma_j \left(\sigma_j - k_j c_j^T z_1 \right) + s_j \eta_j \left(\eta_j - k_j c_j^T z_1 \right) \right] \\
= \; & 2 \left(z_1^T CKW\sigma - \sigma^T W \sigma + z_1^T CKS\eta - \eta^T S\eta \right), \tag{8.23}
\end{aligned}
$$

where $K = \mathrm{diag}(k_1, k_2, \cdots, k_{n_h})$. Inequality (8.23) is a standard application of the S-procedure [89]. Note that

$$
0 = \dot{e}(t) - z_3(t) = (A - K_p H)z_1 - K_i z_2 - (I + K_d H)z_3 + B\eta. \tag{8.24}
$$

It is clear from (8.24) that for two given scalars δ_1 and δ_2 and any appropriately dimensional matrix N, the following relationship holds,

$$
\begin{aligned}
0 \;=\; & \tilde{\mathscr{L}}(t) \\
= \; & 2 \left(z_1^T \delta_1 N + z_2^T H \delta_2 N + z_3^T N \right) \\
& \times \left[(A - K_p H)z_1 - K_i z_2 - (I + K_d H)z_3 + B\eta \right], \tag{8.25}
\end{aligned}
$$

where N, $\delta_j N$, $j = 1, 2$, are considered as free-weighting matrices [109,110]. Let $\xi^T(t) = [z_1^T, z_2^T, z_3^T, \sigma^T, \eta^T]$, $M_1 = NK_p$, $M_2 = NK_i$, $M_3 = NK_d$. Taking the time derivative of

$V(t)$ and adding the terms on the right hand side of (8.23) and (8.25) into $\dot{V}(t)$, one obtains

$$
\begin{aligned}
\dot{V}(t) &\leq \tilde{S}(t) + \tilde{\mathcal{L}}(t) + 2\begin{bmatrix} z_1^T & z_2^T \end{bmatrix}\begin{bmatrix} P_{11} & P_{12} \\ P_{12}^T & P_{22} \end{bmatrix}\begin{bmatrix} z_3 \\ Hz_1 \end{bmatrix} + 2\sigma^T \Lambda C^T z_3 \\
&= \xi^T(t)\Psi\xi(t),
\end{aligned}
\tag{8.26}
$$

where $\Psi = (\psi_{ij})$, $i, j = 1, \cdots, 5$, is a symmetric matrix with

$$
\begin{aligned}
\psi_{11} &= P_{12}H + \delta_1 NA - \delta_1 M_1 H + (P_{12}H + \delta_1 NA - \delta_1 M_1 H)^T, \\
\psi_{12} &= H^T P_{22}^T - \delta_1 M_2 + (\delta_2 HNA - \delta_2 HM_1 H)^T, \\
\psi_{13} &= P_{11} - \delta_1 N - \delta_1 M_3 H + (NA - M_1 H)^T, \\
\psi_{14} &= CGW, \\
\psi_{15} &= \delta_1 NB + CGS, \\
\psi_{22} &= -\delta_2 HM_2 - (\delta_2 HM_2)^T, \\
\psi_{23} &= P_{12}^T - \delta_2 HN - \delta_2 HM_3 H - M_2^T, \\
\psi_{24} &= 0, \\
\psi_{25} &= \delta_2 HNB, \\
\psi_{33} &= -N - M_3 H - (N + M_3 H)^T, \\
\psi_{34} &= C\Lambda, \\
\psi_{35} &= NB, \\
\psi_{44} &= -2W, \\
\psi_{45} &= 0, \\
\psi_{55} &= -2S.
\end{aligned}
$$

Thus, $\dot{V}(t) < -\varepsilon \|e(t)\|^2$ for a sufficiently small ε if

$$
\Psi < 0,
\tag{8.27}
$$

which ensures the asymptotic stability of equilibrium point $e = 0$. Note that for two given scalars δ_1 and δ_2, LMI (8.27) is solvable for the variables P_{11}, P_{12}, P_{22}, Λ, W, S, N and M_j, $j = 1, 2, 3$, by using the LMI technique. Thus, one can obtain the gains $K_p = N^{-1}M_1$, $K_i = N^{-1}M_2$ and $K_d = N^{-1}M_3$.

Remark 8.1. It should be noted that if one takes $z_2(t) = \int_0^t e(\theta)d\theta$ as doing in [80], the solution set of LMI (8.27) becomes empty if matrix H is not of full column rank. This property is strongly conservative for output feedback control. However, if the new state variable is taken the form in (8.20), i.e., $z_2(t) = \int_0^t He(t)d\theta$, the aforesaid conservativeness is fully overcome. Moreover, in comparison to the design of PID control in [80], the dimension of the descriptor system (8.21) is reduced by $(m - l)$.

8.4 Examples

Example 8.1. The Chua's chaotic circuit system [199],

$$\begin{cases} \dot{x} = -3.2x + 10y - 2.95(|x+1| - |x-1|), \\ \dot{y} = x - y + z, \\ \dot{z} = -14.87y, \end{cases} \tag{8.28}$$

can be represented in Lur'e form by

$$A = \begin{bmatrix} -3.2 & 10 & 0 \\ 1 & -1 & 1 \\ 0 & -14.87 & 0 \end{bmatrix}, \quad B = \begin{bmatrix} 5.9 \\ 0 \\ 0 \end{bmatrix}, \quad C = \begin{bmatrix} 1 & 0 & 0 \end{bmatrix}, \tag{8.29}$$

and $\sigma(\xi) = (1/2)(|\xi+1| - |\xi-1|)$ belongs to sector $[0,k]$ with $k = 1$. Suppose that the output matrix $H = \begin{bmatrix} 1 & 0 & 0 \\ 0 & 1 & 0 \end{bmatrix}$.

Setting $\delta_1 = 3.4$ and $\delta_2 = 0$, we compute LMI (8.15) and obtain a set of solutions as

$$M_1 = \begin{bmatrix} -2770 & -9157 \\ -6445 & -21235 \\ -2757 & -6117 \end{bmatrix}, \quad M_2 = \begin{bmatrix} -13 & 2806.6 \\ -44.7 & 6281 \\ -812.3 & -1856.1 \end{bmatrix},$$

and

$$N = \begin{bmatrix} -0.8 & 2761.8 & -812.3 \\ 0 & 6293.2 & -1850.9 \\ 0 & -5.2 & 12.2 \end{bmatrix}.$$

Thus, it follows from Corollary 8.1 that

$$K_p = \begin{bmatrix} -73.9813 & -203.4694 \\ -77.2121 & -172.5968 \\ -259.0389 & -575.3563 \end{bmatrix}, \quad K_d = \begin{bmatrix} -8.3400 & -62.7661 \\ -22.4153 & -50.0598 \\ -76.1879 & -173.5968 \end{bmatrix}. \tag{8.30}$$

With the gains K_p and K_d given in (8.30), the master-slave synchronization of chaotic system (8.28) under PD control is simulated. In Fig. 8.1(a), the time history of the error system (8.3) shows the fact that the synchronization is achieved quickly and implies

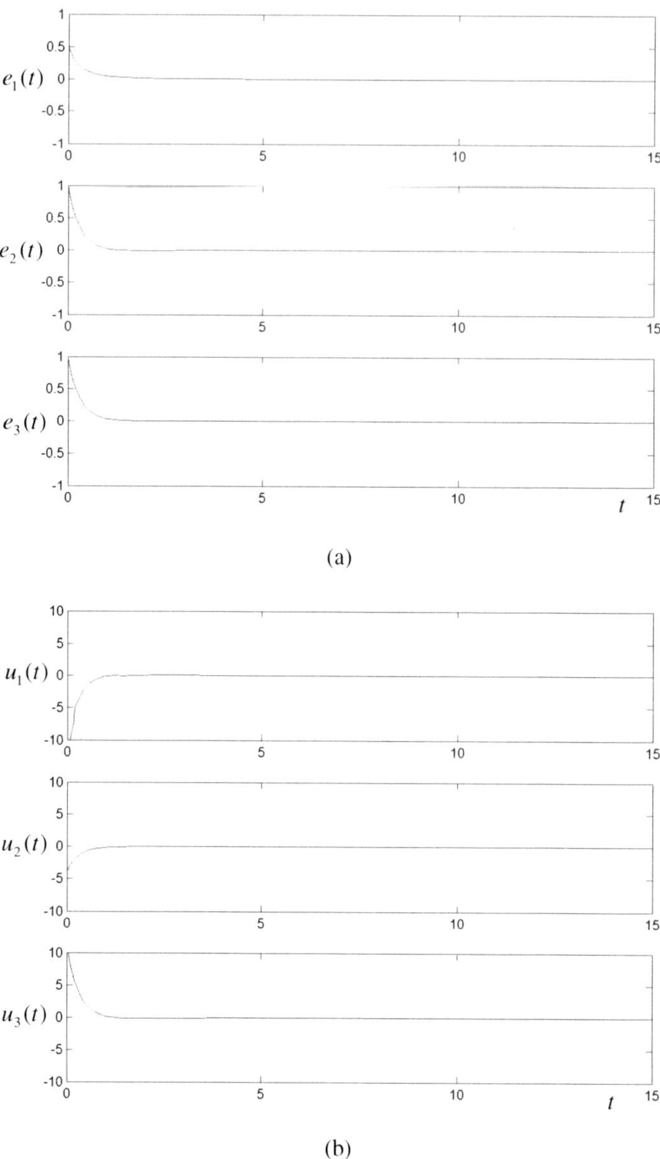

Fig. 8.1. Synchronization results under PD control: (a) error signal $e(t)$; (b) control signal $u(t)$; Time unit of t: second

the improvement of speeding synchronization response by using PD controller, as compared to the results of the same chaotic system based on PI control [199]. Fig. 8.2(b) exhibits the control signal.

Example 8.2. To illustrate the merits and effectiveness of our results, consider the paradigm in nonlinear physics — Chua's circuit [191]:

$$\begin{cases} \dot{x} = a(y - h(x)), \\ \dot{y} = x - y + z, \\ \dot{z} = -by, \end{cases} \tag{8.31}$$

with nonlinear characteristic

$$h(x) = m_1 x + \frac{1}{2}(m_0 - m_1)(|x+c| - |x-c|), \tag{8.32}$$

and parameters $a = 9$, $b = 14.28$, $c = 1$, $m_0 = -(1/7)$, $m_1 = 2/7$. The system can be represented in Lur'e form by

$$A = \begin{bmatrix} -18/7 & 9 & 0 \\ 1 & -1 & 1 \\ 0 & -14.28 & 0 \end{bmatrix}, \quad B = \begin{bmatrix} 27/7 \\ 0 \\ 0 \end{bmatrix}, \quad C = \begin{bmatrix} 1 \\ 0 \\ 0 \end{bmatrix},$$

and $\sigma(v) = (1/2)(|v+c| - |v-c|)$ belongs to sector $[0, k_1]$ with $k_1 = 1$, i.e., $\sigma(v)(\sigma(v) - k_1 v) \leq 0$ and $n_h = 1$. Suppose that the output matrix $H = \begin{bmatrix} 1 & 0 & 0 \\ 0 & 1 & 0 \end{bmatrix}$, which is is not of full column rank. Let $\delta_1 = \delta_2 = 2.4$. It is easy to compute LMI (8.27) and the original PID gains are given by

$$K_p = \begin{bmatrix} -3750.6 & -4846.4 \\ 66.8 & 87.5 \\ 161.4 & 198.0 \end{bmatrix}, \quad K_i = \begin{bmatrix} -3730.9 & -4865.3 \\ 65.8 & 88.5 \\ 161.4 & 212.3 \end{bmatrix},$$

$$K_d = \begin{bmatrix} -1566.9 & -2020.3 \\ 27.4 & 35.8 \\ 67.2 & 88.5 \end{bmatrix}. \tag{8.33}$$

With the gains, the master-slave synchronization of chaotic system (8.1) under PID control is simulated. In Fig. 8.2(a), the time history of the error system (8.4) shows that the synchronization is achieved quickly. Fig. 8.2(b) exhibits the control signal.

We finally discuss how noises affect the error dynamics. Assume that for Chua's circuit (8.31) and (8.32), there exist random noises in the output signals of the master system $\mathscr{M}$ and the slave system $\mathscr{S}$ as follows:

$$y(t) = Hx(t) + \breve{B}\gamma_1 \varepsilon(t), \quad \hat{y}(t) = H\hat{x}(t) + \breve{B}\gamma_2 \varepsilon(t), \tag{8.34}$$

where $\breve{B} = [1, 1]^T$, the positive constants γ_1 and γ_2 represent the noise levels (magnitudes), and $\varepsilon(t)$ stands for the uniformly distributed random signal, bounded by $|\varepsilon(t)| \leq 1$.

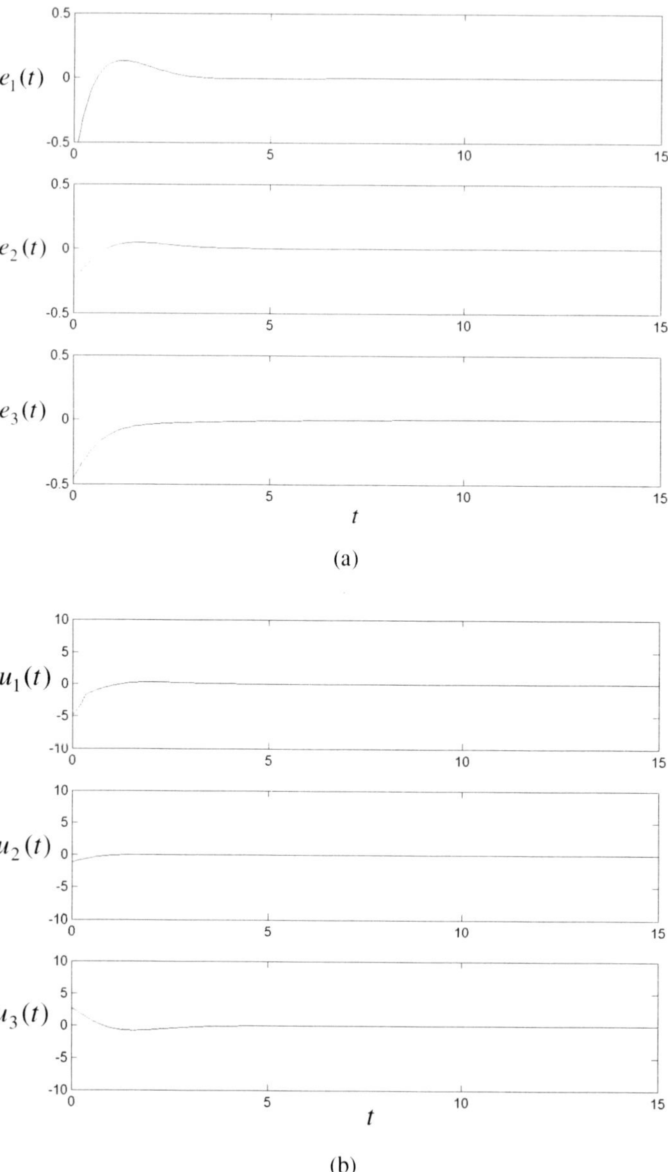

Fig. 8.2. Synchronization results under PID control: (a) the time history of the synchronization error $e(t)$; (b) the time history of the control signal $u(t)$; Time unit of t: second

It follows from Fig. 8.2 that with the gains K_p, K_i and K_d given in (8.33), the synchronization in the noiseless case can be achieved before $t = 5$. Let $\gamma = \gamma_1 - \gamma_2 = 0.01, 0.1, 1$ and 10, respectively. During $t \in [5, 100]$, the results of the variances of $|e(t)|$ (briefly, var($|e(t)|$)) via the noise level γ are shown by the log-log plot in

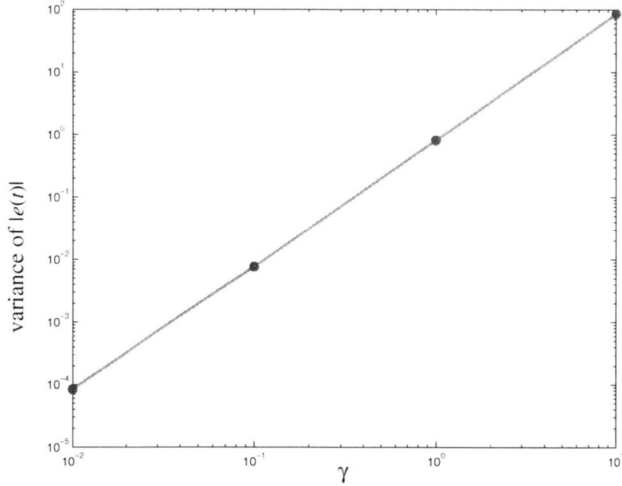

Fig. 8.3. The log-log plot of the variances of $|e(t)|$ via the noise level γ

Fig. 8.3. It implies that the error during synchronization will not be asymptotically stable, but remain bounded in presence of noise. When $0 < \gamma < 1$, we have $\mathrm{var}(|e(t)|) < \gamma$. However, $\mathrm{var}(|e(t)|)$ exponentially increases and $\mathrm{var}(|e(t)|) > \gamma$ if $\gamma > 1$.

8.5 Conclusion

The fast master-slave synchronization of Lur'e systems via PD control has been studied in this Chapter. A synchronization strategy is established with the help of the free-weighting matrix approach and the S-procedure and illustrated by an example. Our result covers the existing one in the case of proportional control and implies contribution of derivative control to speed synchronization response.

We have also investigated the master-slave synchronization of Lur'e system under PID control. The LMI technique, the free-weighting matrix approach and the S-procedure are used to derive the robust synchronization criterion, by which the merits of derivative and integral parts of the controller are applicable for speeding synchronization and disturbance attenuation, respectively. In comparison with the existing literature on designing PID controller based on LMI, the improvement of the solvability is achieved.

9 Multivariable Process Identification

Chapters 5–7 present several new PID controller design methods for MIMO systems but all these methods are model-based. That's why system identification plays an important role in the control engineering. This chapter focuses on system identification for both SISO and MIMO systems. Several identification algorithms are presented. The biggest improvement to the existing methods is that our methods can give accurate system model even in presence of time delay, unknown initial conditions and disturbances.

9.1 Introduction

Identification and control of single variable processes have been well studied [2, 208]. Since time delay is present in many industrial processes and has a significant effect on control system performance, there has been continuing interest in estimation of time delay and identification of continuous-time delay systems in general [209, 210, 211, 212]. Process identification requires some tests to excite the process dynamics. Typical test signals used in process identifications include step, pulse, relay, pseudo-random binary sequence (PRBS) and sinusoidal functions. Among them, step, pulse and relay experiments are more popular for their simplicity. Relay feedback test is demonstrated in [154]. This is a kind of closed-loop test and was pioneered by Åström and Hägglund [2]. It needs almost no prior knowledge of the process. In the early stage of development, only stationary response of relay feedback system is used to estimate the process frequency response at the oscillation frequency. More recent development on identification of the process frequency response can yield multiple points from a single relay test with the help of the fast Fourier transform (FFT) technique [213, 214, 215, 216]. With the estimated frequency response, a transfer function model with time delay can be obtained by some fitting techniques and enables tuning and implementation of model-based controllers.

Another important branch of methodology for identifications of continuous-time systems is the integration approach. It was first proposed by Dianessis [217]. In [218], the effect of deterministic disturbances at system input and output is included in the analysis. A similar integral-equation approach has been derived by Golubev and Wang [219] from a frequency-domain error criterion. From their works, efficiency and robustness

Q.-G. Wang et al.: PID Control for Multivariable Processes, LNCIS 373, pp. 219–249, 2008.
springerlink.com
© Springer-Verlag Berlin Heidelberg 2008

of integral equation methods have been shown. It was Wang and Zhang [157] who first proposed to apply integration method to identify continuous-time delay systems from step tests. Their method takes advantage of explicit formulas for multi-integration of a step input and devises a linear regression equation containing delay parameter. The least-squares method is then applied to identify the regression parameters, from which the full model parameters including time delay are recovered. This method is robust and the identification results are satisfactory even without filtering of the measured output, which is corrupted by noise. However, Wang's integration method requires that the tests start from zero initial conditions and there is no significant disturbance during the test. Hwang and Lai [220] proposed a identification algorithm, which uses pulse signals as the input. Two regression equations are obtained from the two edges of the pulse signal, respectively. Then the estimation and/or the elimination of the non-zero initial conditions and disturbances become possible. Their regression parameter vectors involve all parameters together in either of the two steps and some of them are very complicated functions of process parameters and initial conditions. Recently, based on novel integration techniques, robust identification methods have been proposed for single variable time delay processes in presence of nonzero initial conditions and dynamic disturbances [28, 156, 220, 221, 222]. In [220] and [221], identifications from pulse tests were proposed. In [222], another continuous-time identification method of process models with time delay and nonzero initial conditions is proposed. The problem is solved through the linear filtering method. However, this method needs an iterative procedure for the time delay estimation. In [28], by treating a relay test as a sequence of step tests, a linear integral filter is adopted to devise the algorithm, and a full process model including time delay is identified. This method needs the output measurement before the relay test, and also considers, like many previous identification methods, the constant disturbance only. An improved general method was developed in [156]. Note that [157, 220, 222] assume open-loop tests, while [28] addresses a closed-loop test.

In practice, most industrial processes are multivariable in nature [59]. To achieve performance requirement, modern advanced controllers based on process models are implemented [223] and identifications of multivariable processes are in great demand [224,225]. In the context of continuous process identification, many methods have been proposed for the multivariable case, for example, [218], [226] and [227]. An important issue with continuous process identification is time delay. Its estimation needs special attention.

In this chapter, a new integral identification method is proposed for both SISO continuous-time delay system and multivariable processes with multiple time delays. The identification test can be of open-loop such as pseudo random binary signals and pulse tests, or of closed-loop type such as relay tests. No prior process data is needed. The initial conditions are allowed to be unknown or nonzero. The disturbance can be of general form, but not limited to the static one. A new regression equation is derived taking into account nature of the underlying test signal. The equation has more linearly independent functions and thus enables to identify a full process model with time delay as well as combined effects of unknown initial condition and disturbance without any iteration. All the parameters including time delay in the regression equation are estimated in one step. The method shows great robustness against noise in output measurements but requires no filtering of noisy data. The effectiveness of the proposed method is demonstrated through simulations and real time implementation.

9.2 Identification of SISO Processes

9.2.1 Second-Order Modelling

This section focuses on the modelling of second-order systems. It serves for motivation of the general method to be described in the next section and for recommended use in applications since such a second-order model essentially covers most practical industrial processes. Consider a second-order continuous-time delay system,

$$y^{(2)} + a_1 y^{(1)}(t) + a_0 y(t) = b_1 u^{(1)}(t-d) + b_0 u(t-d) + l(t), \tag{9.1}$$

where $y(t)$ and $u(t)$ are the output and input of the process, respectively; d is the time delay; and $l(t)$ is an unknown disturbance or a bias to the process. The task is to estimate the model parameters, a_1, a_0, b_1, b_0 and d from one test. The test input under consideration is supposed to be in the form of

$$u(t) = \sum_{j=0}^{N} u_j(t) = \sum_{j=0}^{N} h_j \mathbf{1}(t - t_j), \tag{9.2}$$

where $\mathbf{1}(t)$ is the unit step function, $N \geq 1$, and $u_j(t)$ is a step input with magnitude of h_j and applied at $t = t_j$. This form covers many types of signals including open-loop tests such as PRBS, rectangular pulses with magnitude of h and duration of T,

$$u(t) = h\mathbf{1}(t) - h\mathbf{1}(t - T), \tag{9.3}$$

and rectangular doublet pulses,

$$u(t) = h\mathbf{1}(t) - 2h\mathbf{1}\left(t - \frac{T}{2}\right) + h\mathbf{1}(t - T),$$

as well as close-loop tests such as relay tests, see one example in Section 9.2.1. The relay function is described as

$$u(t) = \begin{cases} u_+, & \text{if } e(t) > \varepsilon_+, \text{ or } e(t) \geq \varepsilon_- \text{ and } u(t_-) = u_+, \\ u_-, & \text{if } e(t) < \varepsilon_-, \text{ or } e(t) \leq \varepsilon_+ \text{ and } u(t_-) = u_-, \end{cases} \tag{9.4}$$

where ε_+, $\varepsilon_- \in \mathbb{R}$ with $\varepsilon_- < \varepsilon_+$ indicating hysteresis; $u_-, u_+ \in \mathbb{R}$ and $u_- \neq u_+$.

A multiple integration operator on $f(t)$ is defined as follows,

$$\begin{cases} P_0 f(t) = f(t), \\ P_j f(t) = \displaystyle\int_0^t \int_0^{\tau_{j-1}} \cdots \int_0^{\tau_1} f(\tau_0) \, \mathrm{d}\tau_0 \mathrm{d}\tau_1 \cdots \mathrm{d}\tau_{j-1}, \; j \geq 1. \end{cases} \tag{9.5}$$

Applying P_2 to (9.1) yields

$$P_2 y^{(2)}(t) + a_1 P_2 y^{(1)}(t) + a_0 P_2 y(t)$$
$$= b_1 P_2 u^{(1)}(t-d) + b_0 P_2 u(t-d) + P_2 l(t). \tag{9.6}$$

For the left-hand side, we have

$$P_2 y^{(2)}(t) = y(t) - y(0) - y^{(1)}(0)t, \tag{9.7}$$

$$P_2 y^{(1)}(t) = \int_0^t y(\tau_0)\,d\tau_0 - y(0)t, \tag{9.8}$$

and

$$P_2 y(t) = \int_0^t \int_0^{\tau_1} y(\tau_0)\,d\tau_0\,d\tau_1. \tag{9.9}$$

For the right hand side, it is straightforward to verify that

$$P_2 \mathbf{1}(t-d) = \frac{(t-d)^2}{2!}\mathbf{1}(t-d),$$

and

$$P_2 \mathbf{1}^{(1)}(t-d) = (t-d)\mathbf{1}(t-d).$$

It then follows that

$$P_2 u(t-d) = \sum_{j=0}^{N} P_2 u_j(t-d) = \sum_{j=0}^{N} \frac{h_j(t-t_j-d)^2}{2!}\mathbf{1}(t-t_j-d), \tag{9.10}$$

and

$$P_2 u^{(1)}(t-d) = \sum_{j=0}^{N} P_2 u_j^{(1)}(t-d) = \sum_{j=0}^{N} h_j(t-t_j-d)\mathbf{1}(t-t_j-d). \tag{9.11}$$

Choose t to meet

$$t_k + d \le t < t_{k+1} + d, \tag{9.12}$$

where t_k and t_{k+1} are the kth and $(k+1)$th input switch instants, respectively. Equations (9.10) and (9.11) become

$$P_2 u(t-d) = \sum_{j=0}^{k} \frac{h_j(t-t_j-d)^2}{2!}, \tag{9.13}$$

and

$$P_2 u^{(1)}(t-d) = \sum_{j=0}^{k} h_j(t-t_j-d). \tag{9.14}$$

Suppose that there holds

$$P_2 l(t) = \sum_{j=0}^{Q} \beta_j t^j, \tag{9.15}$$

where Q is an integer. Equation (9.15) stands for the multiple integrations of the generalized disturbances [220] more than a static disturbance for which $l(t) = c\mathbf{1}(t)$, $P_2 l(t) = ct^2/2$ and $Q = 2$.

Substituting (9.7), (9.8), (9.9), (9.13) ,(9.14) and (9.15) into (9.6) gives

$$y(t) - y(0) - y^{(1)}(0)t + a_1 \left(\int_0^t y(\tau_0)\mathrm{d}\tau_0 - y(0)t \right) + a_0 \int_0^t \int_0^{\tau_1} y(\tau_0)\mathrm{d}\tau_0\mathrm{d}\tau_1$$

$$= b_1 \sum_{j=0}^{k} h_j(t - t_j - d) + b_0 \sum_{j=0}^{k} \frac{h_j(t - t_j - d)^2}{2!} + \sum_{j=0}^{Q} \beta_j t^j. \tag{9.16}$$

Equation (9.16) can then be rearranged as follows,

$$\phi^T(t,t_k)\theta = \gamma(t), \quad t_k + d \le t < t_{k+1} + d, \tag{9.17}$$

where

$$\phi(t,t_k) = \begin{bmatrix} -\int_0^t y(\tau_0)\mathrm{d}\tau_0 \\ -\int_0^t \int_0^{\tau_1} y(\tau_0)\mathrm{d}\tau_0\mathrm{d}\tau_1 \\ \sum_{j=0}^{k} h_j \\ \sum_{j=0}^{k} h_j(t - t_j) \\ \sum_{j=0}^{k} h_j(t - t_j)^2 \\ 1 \\ t \\ t^2 \\ \vdots \\ t^Q \end{bmatrix}, \quad \theta = \begin{bmatrix} a_1 \\ a_0 \\ \theta_0 \\ \theta_1 \\ \theta_2 \\ \alpha_0 \\ \alpha_1 \\ \alpha_2 \\ \vdots \\ \alpha_Q \end{bmatrix} = \begin{bmatrix} a_1 \\ a_0 \\ \dfrac{b_0 d^2}{2} - b_1 d \\ b_1 - b_0 d \\ \dfrac{b_0}{2} \\ \beta_0 + y(0) \\ \beta_1 + y^{(1)}(0) + a_1 y(0) \\ \beta_2 \\ \vdots \\ \beta_Q \end{bmatrix},$$

$$\gamma(t) = y(t).$$

The parameters α_i, $i = 0, 1, \cdots, Q$, are used to account for the effects of the aforementioned nonzero initial conditions and the disturbance. Choose $t = t_{ki}$, $i = 0, 1, 2, \cdots, M_k$, to meet $t_k + d \le t_{ki} < t_{k+1} + d$. One invokes (9.17) for t_{ki}:

$$\Psi_k \theta = \Gamma_k, \tag{9.18}$$

where $\Psi_k = \left[\phi(t_{k0}, t_k), \cdots, \phi(t_{kM_k}, t_k) \right]^T$ and $\Gamma_k = \left[\gamma(t_{k0}), \cdots, \gamma(t_{kM_k}) \right]^T$. From the $N+1$ input switches of one test, Γ_k and $\Psi_k, k = 0, \cdots, N$, are obtained and combined to

$$\Psi\theta = \Gamma,$$

where $\Psi = \left[\Psi_0^T, \cdots, \Psi_N^T \right]^T$ and $\Gamma = \left[\Gamma_0^T \cdots \Gamma_N^T \right]^T$. The ordinary least-squares method can be applied to find the solution

$$\hat{\theta} = \left[\hat{a}_1, \hat{a}_0, \hat{\theta}_0, \hat{\theta}_1, \hat{\theta}_2, \hat{\alpha}_0, \hat{\alpha}_1, \ldots, \hat{\alpha}_P \right]^T = \left(\Psi^T \Psi \right)^{-1} \Psi^T \Gamma.$$

In the presence of noise in the measurement of the process output, the instrumental variable (IV) method is adopted to guarantee the identification consistency. For our case, the instrumental variable $Z(t_{ki})$ is chosen as

$$Z(t_{ki}) = \left[(t_{ki})^{-(M_{id}-1)} \quad \cdots \quad (t_{ki})^{-1} \quad 1 \quad t_{ki} \quad \cdots \quad (t_{ki})^{2n+2+Q-M_{id}} \right],$$

where M_{id} is the quotient of $(2n+2+Q)/2$; n is order of the model, and $n = 2$ for second-order modeling.

After θ is estimated, its first 2 elements directly yield the parameters a_1 and a_0, and the others produce b_1, b_0 and d via

$$\begin{cases} b_0 = 2\hat{\theta}_2, \\ d = \dfrac{-\hat{\theta}_1 \pm \sqrt{\hat{\theta}_1^2 - 4\hat{\theta}_0\hat{\theta}_2}}{2\hat{\theta}_2}, \\ b_1 = \hat{\theta}_1 + b_0 d. \end{cases} \tag{9.19}$$

Selection of $t = t_{ki}$ depends on d, while d is to be identified and unknown. It is possible to estimate a range of d. Let d be in the range, $[d_{\min}, d_{\max}]$. $d_{\min}$ may be set as the time from the input signal injection to the point when the output response still remains unchanged from the past trend, while $d_{\max}$ is the time from the input signal injection to the point when the output response has changed from the past trend well beyond the noise band. Besides, such a range can be estimated with purely numerical method [220]. With $d_{\min} < d < d_{\max}$, we can then choose $t_k + d_{\max} \le t_{ki} < t_{k+1} + d_{\min}$. One difference between this method and the one by Ahamed et al. [222] is this choice of t. We implicitly assume some priori knowledge of time delay, while Ahamed et al. [222] find d by iteration.

It is easy to extend our method to identify the model parameters from the test which has the input in the form of:

$$u(t) = \sum_{j=0}^{N} h_j(t - t_j)\mathbf{1}(t - t_j),$$

where t_j is an input switch time. It is straightforward to find that

$$P_n(t - d)\mathbf{1}(t - d) = \frac{(t - d)^{n+1}}{(n + 1)!}\mathbf{1}(t - d).$$

Following the above development procedure, one will obtain an identification method similar to the proposed one. Because this kind of test signals are not widely used, the identification based on such inputs is not discussed in details in this chapter.

Example 9.1. Consider a continuous-time delay process,

$$y^{(2)}(t) + 2y^{(1)}(t) + y(t) = u(t - 5) + l,$$

subject to $y(0) = -1.5$, $y^{(1)}(0) = -1.5$ and $l = 0.2$. The relay test in (9.4) is applied at $t = 0$ with $u_+ = 1$, $u_- = -1$, $\varepsilon_+ = 0.4$ and $\varepsilon_- = -0.4$. The process input and output are

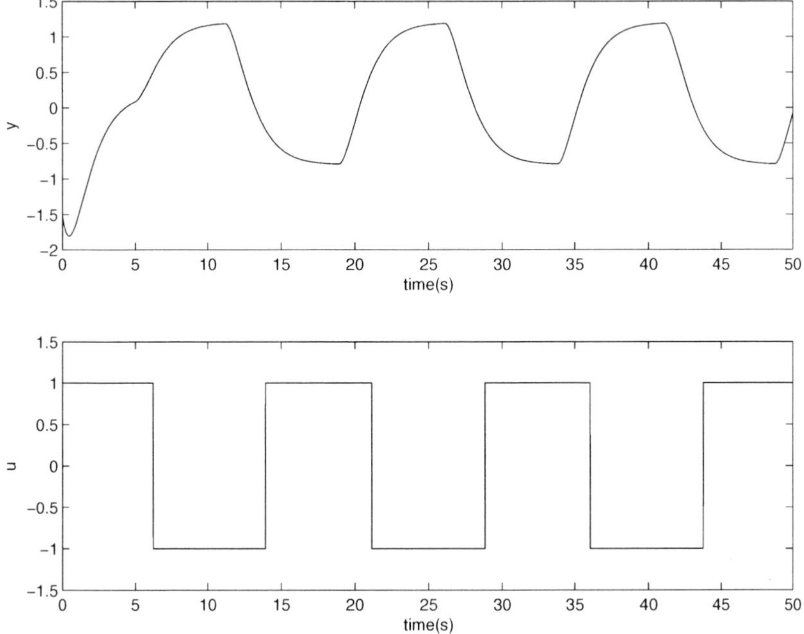

Fig. 9.1. Relay experiment for Example 9.1

shown in Fig. 9.1. Suppose $3.5 < d < 6.5$. The proposed method with $m = 0$ and $Q = 2$ leads to

$$\theta = \begin{bmatrix} 2.0202 & 1.0203 & 12.7793 & -5.1066 & 0.5102 & -1.4793 & -4.6046 & 0.1020 \end{bmatrix}^T.$$

The model is recovered as

$$y^{(2)}(t) + 2.02y^{(1)}(t) + 1.02y(t) = 1.02u(t - 5). \tag{9.20}$$

Suppose that the identification error is measured by the worst case error,

$$\text{ERR} = \max \left| \frac{\hat{G}(j\omega_i) - G(j\omega_i)}{G(j\omega_i)} \right|, i = 1, \cdots, M, \tag{9.21}$$

where $\hat{G}(j\omega_i)$ and $G(j\omega_i)$ are the estimated response and the actual ones, respectively. Only $\omega_i \in [0, \omega_c]$, where ω_c is the phase crossover frequency of the process, are considered. For this example, ERR $= 0.62\%$, which is due to computational errors.

For the same process, a pulse in (9.3) is applied at $t = 0$ with $h = 1$ and $T = 10$. The process input and output are shown in Fig. 9.2. The proposed method with $m = 0$ and $Q = 2$ leads to the same identification result as in (9.20).

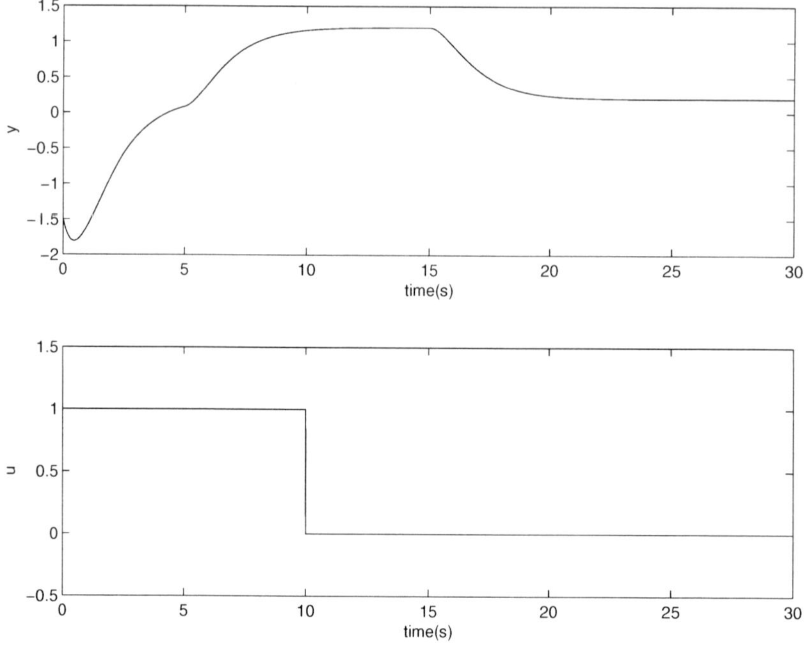

Fig. 9.2. Pulse test for Example 9.1

We then consider a changing disturbance. The changing disturbance is simulated by letting $\mathbf{1}(t)$ pass through the transfer function of $0.2/(15s+1)$. The proposed method with $m=0$ and $Q=3$ leads to

$$y^{(2)}(t) + 1.943y^{(1)}(t) + 0.9831y(t) = 0.9753u(t-4.98),$$

with ERR $= 0.8\%$.

To simulate practical conditions, white noise is added to corrupt the output. The noise-to-signal ratio defined by

$$\text{NSR} = \frac{\text{mean(abs(noise))}}{\text{mean(abs(signal))}},$$

is used to represent the noise level. A relay test in (9.4) is applied at $t=0$ with $u_+ = 1$, $u_- = -1$, $\varepsilon_+ = 0.8$ and $\varepsilon_- = -0.8$. The output is corrupted by noise of NSR $= 5\%$, 10%, 20%, 30% and 40%, respectively. The proposed method is applied without low-pass filtering and the identification errors are 0.91%, 1.12%, 4.59%, 8.47% and 18.97%, respectively.

In Table 9.1, the identification results for a number of second order processes [222] are given and compared with those in [222]. The NSR for all cases are 10%. These identification results are from 500 Monte Carlo simulations. The parameters shown are the means of 500 Monte Carlo simulations and the numbers in the parentheses are the

Table 9.1. Identification results for different second order processes

True models	Identified models	Identification method
$\dfrac{1.25\mathrm{e}^{-0.234s}}{0.25s^2+0.7s+1}$	$\dfrac{1.25(\pm0.02)\mathrm{e}^{-0.239(\pm0.042)s}}{0.25(\pm0.029)s^2+0.697(\pm0.02)s+1}$	Ahmed's
	$\dfrac{1.255(\pm0.0631)\mathrm{e}^{-0.22(\pm0.034)s}}{0.262(\pm0.028)s^2+0.716(\pm0.041)s+1}$	Proposed
$\dfrac{2\mathrm{e}^{-4.1s}}{100s^2+25s+1}$	$\dfrac{2(\pm0.04)\mathrm{e}^{-4.13(\pm0.742)s}}{99.4(\pm19.7)s^2+25(\pm0.67)s+1}$	Ahmed's
	$\dfrac{2.01(\pm0.091)\mathrm{e}^{-4.08(\pm0.119)s}}{100.7(\pm3.481)s^2+25.1(\pm1.183)s+1}$	Proposed
$\dfrac{(-4s+1)\mathrm{e}^{-0.615s}}{9s^2+2.4s+1}$	$\dfrac{(-4(\pm0.0913)s+1(\pm0.06))\mathrm{e}^{-0.6157(\pm0.07)s}}{8.99(\pm0.15)s^2+2.41(\pm0.15)s+1}$	Ahmed's
	$\dfrac{(-4.03(\pm0.08)s+1.03(\pm0.0757))\mathrm{e}^{-0.617(\pm0.0294)s}}{9.11(\pm0.2841)s^2+2.43(\pm0.1265)s+1}$	Proposed

estimated standard deviation of these estimates. The proposed method produces satisfactory identification results similar to [222], but the model parameters are recovered in one step without iterations. In Table 9.1, a non-minimum phase (NMP) process is also considered. Ahamed et al. [222] takes special procedure for identification of NMP processes. In contrast, the proposed method treats the identification of the NMP processes and that of minimum phase processes in the same way.

Remark 9.1. Our regression equation in (9.17) is different from that used by the previous integral identification methods, such as two-step algorithm in [220] where

$$\phi^T(t)\theta = \gamma(t),$$

where

$$\phi^T(t) \;=\; \begin{bmatrix} -y(t) & -P_1 y(t) & h & ht & \cdots & ht^Q \end{bmatrix},$$

$$\theta \;=\; \begin{bmatrix} a_2 & a_1 & \bar{\theta}_1 & \bar{\theta}_2 & \cdots & \bar{\theta}_Q \end{bmatrix},$$

$$\gamma(t) \;=\; P_2 y(t).$$

$\bar{\theta}_i$ are combinations of the model parameters, b_j, d, non-zero initial conditions and the disturbance. In our new regression equation, new elements, $\sum_{j=0}^{k} h_j(t-t_j)^i$, $i=0,1,2$, are added into $\phi(t,t_k)$. They are not only mutually independent but also independent with t^i, $i=0,1,2$. $\theta_i, i=0,1,2$ in θ are related to b_j and d, while $\alpha_i, i=0,\cdots,Q$ account for the effects of the nonzero initial conditions and disturbance. This enables estimation of all the regression parameters in one step.

Remark 9.2. In [28], the output measurement before the relay test is required and the input should be kept constant so as to eliminate the effect of the unknown initial conditions. Like many previous identification methods, Wang et al. [28] consider the static disturbance only. In contrast, the proposed method makes no use of process input and output before the test. It can be carried out under complex disturbances by including α_i,

$i = 0, \cdots, Q$, which account for the combined effects of the nonzero initial conditions and disturbance in the regression equations.

Remark 9.3. In [222], the filter transfer function as

$$F(s) = \frac{\beta^n}{s(s+\lambda)^n},$$

is applied. One has to choose the parameter λ, which is nontrivial [228]. Moreover, this method needs an iterative procedure for the time delay estimation and takes special procedure for identification of NMP processes. These problems are not present in the proposed method.

9.2.2 *n*-th Order Modeling

Consider an nth-order continuous-time delay system,

$$\begin{aligned}
y^{(n)}(t) + \cdots + a_1 y^{(1)}(t) + a_0 y(t) \\
= b_m u^{(m)}(t-d) + \cdots + b_1 u^{(1)}(t-d) + b_0 u(t-d) + l(t),
\end{aligned} \tag{9.22}$$

where $m < n$. Integrating (9.22) with (9.5) for n times, we have

$$P_n y^{(n)}(t) + \sum_{l=0}^{n-1} a_l P_n y^{(l)}(t) = \sum_{j=0}^{m} b_j P_n u^{(j)}(t-d) + P_n l(t). \tag{9.23}$$

It can be readily shown that

$$P_n \mathbf{1}^{(l)}(t-d) = \frac{(t-d)^{n-l}}{(n-l)!} \mathbf{1}(t-d), \quad l = 0, 1, \ldots, m,$$

and

$$P_n u^{(l)}(t-d) = \sum_{j=0}^{N} \frac{h_j (t-t_j-d)^{n-l}}{(n-l)!} \mathbf{1}(t-t_j-d), \quad l = 0, 1, \cdots, m.$$

Choose t to meet (9.12), and we have

$$P_n u^{(l)}(t-d) = \sum_{j=0}^{k} \frac{h_j (t-t_j-d)^{n-l}}{(n-l)!}, \quad l = 0, 1, \cdots, m. \tag{9.24}$$

The multiple integral of $l(t)$ is supposed to be

$$P_n l(t) = \sum_{j=0}^{Q} \beta_j t^j. \tag{9.25}$$

Equation (9.23) can be rearranged as

$$\phi^T(t,t_k)\theta = \gamma(t), \tag{9.26}$$

where

$$
\phi(t,t_k) = \begin{bmatrix} -\int_0^t y(\tau_0)\mathrm{d}\tau_0 \\ \vdots \\ -\int_0^t \int_0^{\tau_{n-1}} \cdots \int_0^{\tau_1} y(\tau_0)\mathrm{d}\tau_0\mathrm{d}\tau_1\cdots\mathrm{d}\tau_{n-1} \\ \sum_{j=0}^{k} h_j \\ \sum_{j=0}^{k} h_j(t-t_j) \\ \vdots \\ \sum_{j=0}^{k} h_j(t-t_j)^n \\ 1 \\ t \\ \vdots \\ t^Q \end{bmatrix}, \quad \theta = \begin{bmatrix} a_{n-1} \\ \vdots \\ a_0 \\ \theta_0 \\ \theta_1 \\ \vdots \\ \theta_n \\ \alpha_0 \\ \alpha_1 \\ \vdots \\ \alpha_Q \end{bmatrix},
$$

$$
\gamma(t) = y(t).
$$

The parameters α_i, $i = 1,\cdots,Q$, are used to account for the effects of the aforementioned nonzero initial conditions and the disturbance. Note that the first n elements of θ are the model parameters a_i, $i = 0,\cdots,n-1$, while θ_i, $i = 0,\cdots,n$ are combinations of the model parameters b_j, $j = 0,\cdots,m$, and d is given by

$$
\theta_i = \sum_{j=\max(n-m,i)}^{n} \frac{(-d)^{j-i}b_{n-j}}{(j-i)!i!}, \quad i = 0,1,\cdots,n. \tag{9.27}
$$

Choose $t = t_{ki}$, $i = 0,1,2,\cdots,M_k$, to meet $t_k + d \leq t_{ki} < t_{k+1} + d$. One invokes (9.17) for t_{ki}:

$$
\Psi_k\theta = \Gamma_k, \tag{9.28}
$$

where $\Psi_k = \left[\phi(t_{k0},t_k),\cdots,\phi(t_{kM_k},t_k)\right]^T$ and $\Gamma_k = \left[\gamma(t_{k0}),\cdots,\gamma(t_{kM_k})\right]^T$. From the $N+1$ input switches of one test, Γ_k and Ψ_k, $k = 0,\ldots,N$, are obtained and combined to

$$
\Psi\theta = \Gamma,
$$

where $\Psi = \left[\Psi_0^T,\cdots,\Psi_N^T\right]^T$ and $\Gamma = \left[\Gamma_0^T,\cdots,\Gamma_N^T\right]^T$. Once θ is estimated by applying the least-squares method or IV method, the model parameters can be recovered. From (9.27) for $i = 0,\cdots,m+1$, we can recover d from the following algebraic equation:

$$
\sum_{j=0}^{m+1} \frac{(n-m-1+j)!\theta_{n-1-m+j}d^j}{j!} = 0. \tag{9.29}
$$

Once d is obtained, the parameters b_j, $j = 0, \cdots, m$ are then calculated as

$$b_j = \sum_{i=0}^{j} \frac{(n-j+i)!\theta_{n-j+i}d^i}{i!}, j = 0, 1, \cdots, m. \tag{9.30}$$

The above approach produces $(m+1)$ possible solutions for time delay d and thus $(m+1)$ possible models. We follow the method in [157] and [220] to choose the appropriate model.

Example 9.2. Consider a continuous-time delay process, $G(s) = e^{-2s}/(s+1)^4$ or

$$y^{(4)}(t) + 4y^{(3)}(t) + 6y^{(2)}(t) + 4y^{(1)}(t) + y(t) = u(t-2) + l(t),$$

subject to $y^{(3)}(0) = y^{(2)}(0) = y^{(1)}(0) = y(0) = -0.5$. A changing disturbance is simulated by letting $\mathbf{1}(t)$ pass by $0.2/(20s+1)$. A relay experiment is performed at $t = 0$, with $u_+ = 1$, $u_- = -1$, $\varepsilon_+ = 0.4$ and $\varepsilon_- = -0.4$. The proposed method with $n = 2$, $m = 0$ and $Q = 3$ leads to

$$y^{(2)}(t) + 1.037y^{(1)}(t) + 0.3748y(t) = 0.3561u(t-3.04),$$

with ERR $= 5.15\%$. The proposed method with $n = 3$, $m = 0$ and $Q = 4$ leads to

$$y^{(3)}(t) + 2.038y^{(2)}(t) + 1.689y^{(1)}(t) + 0.4606y(t) = 0.475u(t-2.4),$$

with ERR $= 4.01\%$. The effectiveness of the proposed method is evident.

9.3 Identification of MIMO Processes

9.3.1 TITO Processes

To introduce our method with simplicity and clarity, let us consider a TITO continuous-time delay process first,

$$\begin{bmatrix} Y_1(s) \\ Y_2(s) \end{bmatrix} = \begin{bmatrix} G_{11}(s) & G_{12}(s) \\ G_{21}(s) & G_{22}(s) \end{bmatrix} \begin{bmatrix} U_1(s) \\ U_2(s) \end{bmatrix},$$

where $Y_1(s)$ and $Y_2(s)$ are the Laplace transforms of two outputs, $y_1(t)$ and $y_2(t)$, $U_1(s)$ and $U_2(s)$ are the Laplace transforms of two inputs, $u_1(t)$ and $u_2(t)$, and $G_{i,j}(s) = \alpha_{ij}(s)e^{-d_{ij}s}/\beta_{ij}(s)$, $i = 1,2$ and $j = 1,2$. The given TITO process may be decomposed into 2 two-input and single-output sub-processes, which can be described as

$$\begin{aligned} Y_i(s) &= \begin{bmatrix} G_{i1}(s) & G_{i2}(s) \end{bmatrix} U(s), \\ &= \begin{bmatrix} \dfrac{\alpha_{i1}(s)}{\beta_{i1}(s)}e^{-d_{i1}s} & \dfrac{\alpha_{i2}(s)}{\beta_{i2}(s)}e^{-d_{i2}s} \end{bmatrix} U(s), \quad i = 1,2. \end{aligned}$$

Let the common denominator of G_{i1} and G_{i2} be $\beta_i^*(s)$. We have

$$\beta_i^*(s)Y_i(s) = \begin{bmatrix} \alpha_{i1}^*(s)e^{-d_{i1}s} & \alpha_{i2}^*(s)e^{-d_{i2}s} \end{bmatrix} U(s), \quad i = 1,2.$$

The equivalent differential equations are

$$y_i^{(n_i)}(t) + \sum_{k=0}^{n_i-1} a_{i,k} y_i^{(k)}(t) = \sum_{j=1}^{2} \sum_{k=0}^{m_{ij}} b_{ij,k} u_j^{(k)}(t-d_{ij}) + w_i(t), \quad i = 1,2, \tag{9.31}$$

where $w_i(t)$ account for the unknown disturbances and biases. Our task is to identify $a_{i,k}$, $b_{ij,k}$ and d_{ij} from some tests on the process. During the identification test, two separate sets of piecewise step signals are applied on two inputs at $t = 0$, respectively. The test signals under consideration are

$$u_1(t) = \sum_{k=0}^{K_1} h_{1,k} \mathbf{1}(t - t_{1,k}),$$

where $\mathbf{1}(t)$ is the unit step, $K_1 \geq 1$ and $t_{1,k}$, $k = 1, \cdots, K_1$ are the switching time instants of $u_1(t)$, and

$$u_2(t) = \sum_{k=0}^{K_2} h_{2,k} \mathbf{1}(t - t_{2,k}),$$

where $K_2 \geq 1$ and $t_{2,k}$, $k = 1, \cdots, K_2$ are the switching time instants of $u_2(t)$. Such forms of u_i, $i = 1, 2$, cover many types of test signals such as steps, rectangular pulses, rectangular doublet pulses, PRBS signals and the relay feedback output.

To eliminate those derivatives in (9.31), we introduce a multiple integration operator,

$$P_j f(t) := \int_0^t \int_0^{\delta_{j-1}} \cdots \int_0^{\delta_1} f(\delta_0) \mathrm{d}\delta_0 \mathrm{d}\delta_1 \cdots \mathrm{d}\delta_{j-1}, \quad j \geq 1. \tag{9.32}$$

Integrating (9.31) with (9.32) n_i times yields

$$P_{n_i} y_i^{(n_i)}(t) + \sum_{k=0}^{n_i-1} a_{i,k} P_{n_i} y_i^{(k)}(t)$$

$$= \sum_{k=0}^{m_{i1}} b_{i1,k} P_{n_i} u_1^{(k)}(t-d_{i1}) + \sum_{k=0}^{m_{i2}} b_{i2,k} P_{n_i} u_2^{(k)}(t-d_{i2}) + P_{n_i} w_i(t). \tag{9.33}$$

Its left-hand side is

$$P_{n_i} y_i^{(n)}(t) + \sum_{k=0}^{n_i-1} a_{i,k} P_{n_i} y_i^{(k)}(t)$$

$$= y_i(t) + \sum_{k=0}^{n_i-1} a_{i,k} P_{n_i-k} y(t) + \sum_{k=0}^{n_i-1} \lambda_{i,k} t^k, \tag{9.34}$$

where the last term corresponds to the initial conditions of the output. In the right-hand side, it follows that

$$P_{n_i} u_1^{(p)}(t-d_{i1})$$

$$= \sum_{k=0}^{K_1} \frac{h_{1,k}(t-t_{1,k}-d_{i1})^{n_i-p}}{(n_i-p)!} \mathbf{1}(t-t_{1,k}-d_{i1}), \quad p = 0, 1, \cdots, m_{i1},$$

and

$$P_{n_i} u_2^{(p)}(t - d_{i2})$$
$$= \sum_{k=0}^{K_2} \frac{h_{2,k}(t - t_{2,k} - d_{i2})^{n_i - p}}{(n_i - p)!} \mathbf{1}(t - t_{2,k} - d_{i2}), \quad p = 0, 1, \cdots, m_{i2}.$$

Suppose that there holds

$$P_{n_i} w_i(t) = \sum_{k=0}^{q_i} v_{i,k} t^k, \tag{9.35}$$

where q_i is an integer. Equation (9.35) covers a wide range of disturbances [220] with its simplest as the static disturbance for which $w_i(t) = c\mathbf{1}(t)$, $P_{n_i} w_i(t) = ct^{n_i}/n_i!$ and $q_i = n_i$.

Equation (9.33) is then cast into the following regression linear in a new parameterization:

$$\phi_i^T(t)\theta_i = \gamma_i(t), \tag{9.36}$$

where $\gamma_i(t) = y_i(t)$,

$$\phi_i(t) = \begin{bmatrix} -P_1 y_i(t) \\ \vdots \\ -P_{n_i} y_i(t) \\ \sum_{k=0}^{K_1} h_{1,k} \mathbf{1}(t - t_{1,k} - d_{i1}) \\ \sum_{k=0}^{K_1} h_{1,k}(t - t_{1,k}) \mathbf{1}(t - t_{1,k} - d_{i1}) \\ \vdots \\ \sum_{k=0}^{K_1} h_{1,k}(t - t_{1,k})^{n_i} \mathbf{1}(t - t_{1,k} - d_{i1}) \\ \sum_{k=0}^{K_2} h_{2,k} \mathbf{1}(t - t_{2,k} - d_{i2}) \\ \sum_{k=0}^{K_2} h_{2,k}(t - t_{2,k}) \mathbf{1}(t - t_{2,k} - d_{i2}) \\ \vdots \\ \sum_{k=0}^{K_2} h_{2,k}(t - t_{2,k})^{n_i} \mathbf{1}(t - t_{2,k} - d_{i2}) \\ 1 \\ t \\ \vdots \\ t^{q_i} \end{bmatrix}, \quad \theta_i = \begin{bmatrix} \theta_{i,1} \\ \vdots \\ \theta_{i,n_i} \\ \theta_{i,n_i+1} \\ \theta_{i,n_i+2} \\ \vdots \\ \theta_{i,2n_i+1} \\ \theta_{i,2n_i+2} \\ \theta_{i,2n_i+3} \\ \vdots \\ \theta_{i,3n_i+2} \\ \theta_{i,3n_i+3} \\ \theta_{i,3n_i+4} \\ \vdots \\ \theta_{i,3n_i+3+q_i} \end{bmatrix}.$$

The first n_i elements in θ_i are the model parameter $a_{i,k}$:

$$\theta_{i,k} = a_{i,n_i-k}, \quad k = 1, \cdots, n_i. \tag{9.37}$$

$\theta_{i,k}, k = n_i + 1, \cdots, 2n_i + 1$ are functions of d_{i1} and $b_{i1,k}, k = 0, \cdots, m_{i1}$,

$$\theta_{i,k} = \sum_{p=\max(n_i-m_{i1},k-n_i-1)}^{n_i} \frac{(-d_{i1})^{p-k+n_i+1} \, b_{i1,n_i-p}}{(p-k+n_i+1)! \, (k-n_i-1)!}, \tag{9.38}$$

$$k = n_i+1, \cdots, 2n_i+1.$$

$\theta_{i,k}, k = 2n_i + 2, \cdots, 3n_i + 2$, are functions of d_{i2} and $b_{i2,k}, k = 0, \cdots, m_{i2}$,

$$\theta_{i,k} = \sum_{p=\max(n_i-m_{i2},k-2n_i-2)}^{n_i} \frac{(-d_{i2})^{p-k+2n_i+2} \, b_{i2,n_i-p}}{(p-k+2n_i+2)! \, (k-2n_i-2)!}, \tag{9.39}$$

$$k = 2n_i+2, \cdots, 3n_i+2.$$

$\theta_{i,k}, k = 3n_i + 3, \cdots, 3n_i + 3 + q_i$, account for the collective effects of the initial conditions and the disturbances. Note that all the elements in $\phi_i(t)$ should be mutually independent over the real number field to enable identifiability of the parameter vector, θ_i. This is not the case if $t_{1,k} = t_{2,k}$ for all k, for which $\sum_{k=0}^{K_1} h_{1,k}(t-t_{1,k})^p \mathbf{1}(t-t_{1,k}-d_{i1})$ and $\sum_{k=0}^{K_2} h_{2,k}(t-t_{2,k})^p \mathbf{1}(t-t_{2,k}-d_{i2})$, $p = 0, \cdots, n_i$, become dependent of each other. This should be avoided by the identification test design.

One invokes (9.36) for $t = t_0, \cdots, t_N$, to get

$$\Psi_i \theta_i = \Gamma_i, \tag{9.40}$$

where $\Psi_i = [\phi_i(t_0), \cdots, \phi(t_N)]^T$ and $\Gamma_i = [\gamma_i(t_0), \cdots, \gamma(t_N)]^T$. The ordinary least-squares method can be applied to find the solution

$$\hat{\theta}_i = \left(\Psi_i^T \Psi_i\right)^{-1} \Psi_i^T \Gamma_i.$$

In the presence of noise in the measurement of the process output, the instrumental variable (IV) method is adopted to guarantee the identification consistency. For our case, the instrumental variable $Z_i(t)$ is chosen as

$$Z_i(t) = \begin{bmatrix} \dfrac{1}{t^{n_i}} \\[1em] \vdots \\[1em] \dfrac{1}{t} \\[1em] \sum_{k=0}^{K_1} h_{1,k}\mathbf{1}(t - t_{1,k} - d_{i1}) \\[1em] \sum_{k=0}^{K_1} h_{1,k}(t - t_{1,k})\mathbf{1}(t - t_{1,k} - d_{i1}) \\[1em] \vdots \\[1em] \sum_{k=0}^{K_1} h_{1,k}(t - t_{1,k})^{n_i}\mathbf{1}(t - t_{1,k} - d_{i1}) \\[1em] \sum_{k=0}^{K_2} h_{2,k}\mathbf{1}(t - t_{2,k} - d_{i2}) \\[1em] \sum_{k=0}^{K_2} h_{2,k}(t - t_{2,k})\mathbf{1}(t - t_{2,k} - d_{i2}) \\[1em] \vdots \\[1em] \sum_{k=0}^{K_2} h_{2,k}(t - t_{2,k})^{n_i}\mathbf{1}(t - t_{2,k} - d_{i2}) \\[1em] 1 \\[1em] t \\[1em] \vdots \\[1em] t^{q_i} \end{bmatrix}.$$

It should be pointed out that for a selected t, the value of some elements of ϕ_i depend on d_{i1}, d_{i2}, which are to be identified and unknown. It is possible to estimate a range of d_{i1} and d_{i2} [220]. In many engineering applications, one can have simple reliable and probably conservative estimation of the range of time delay from knowledge of the process. For example, the range of transportation delay due to a long pipe can be easily estimated based on the pipe length and fluid speed range. Besides, one may start with a rough estimated delay range and use the proposed method to find $\hat{d}_{i1}$ and $\hat{d}_{i2}$, estimates of d_{i1} and d_{i2}. Then with $\hat{d}_{i1}$ and $\hat{d}_{i2}$, one retunes the ranges of time delays and apply the proposed method again to achieve a better estimation. Let d_{i1} and d_{i2} be in the ranges of $\left[\underline{d}_{i1}, \overline{d}_{i1}\right]$ and $\left[\underline{d}_{i2}, \overline{d}_{i2}\right]$, respectively. Define

$$\hat{T}_1 = \bigcup_{k=0}^{K_1-1} \left\{ t \,|\, t_{1,k} + \overline{d}_{i1} \leq t < t_{1,k+1} + \underline{d}_{i1} \right\} \bigcup \left\{ t \,|\, (t_{1,K_1} + \overline{d}_{i1} \leq t \leq T_{end} \right\},$$

and

$$\hat{T}_2 = \bigcup_{k=0}^{K_2-1} \{t \,|\, t_{2,k} + \overline{d}_{i2} \le t < t_{2,k+1} + \underline{d}_{i2}\} \bigcup \{t \,|\, t_{2,K_2} + \overline{d}_{i2} \le t \le T_{end}\},$$

where T_{end} is the ending time of the identification test. Then, t should be taken in the set of

$$T = \hat{T}_1 \bigcap \hat{T}_2,$$

to apply (9.40). There is no need to solve the estimation equation for each of the delay within the estimated range. Once the estimate ranges of time delays are given, time delays can be obtained by solving some polynomial equations without iteration. Then, all other parameters than delays are determined accordingly.

Once θ_i is estimated by applying the least-squares method or IV method, the model parameters can be recovered. From (9.38) for $k = 2n_i + 1 - m_{i1}, \cdots, 2n_i + 1$, $b_{i1,k}$, $k = 0, \cdots, m_{i1}$ can be expressed as the functions of d_{i1} and $\theta_{i,k}$,

$$b_{i1,k} = \sum_{p=0}^{k} \frac{(n_i - k + p)! \, \theta_{i,2n_i+1-k+p} \, d_{i1}^p}{p!}, \quad k = 0, 1, \cdots, m_{i1}. \tag{9.41}$$

Substitute $b_{i1,k}$, $k = 0, \cdots, m_{i1}$, into (9.38) for $k = 2n_i - m_{i1}$, and we have

$$\sum_{k=0}^{m_{i1}+1} \frac{(n_i - m_{i1} - 1 + k)! \, \theta_{i,2n_i-m_{i1}+k} \, d_{i1}^k}{k!} = 0. \tag{9.42}$$

Equation (9.42) is solved to get d_{i1} and $b_{i1,k}$, $k = 0, \cdots, m_{i1}$ are then obtained from (9.41). Similarly, we can find d_{i2} from the following algebraic equations:

$$\sum_{k=0}^{m_{i2}+1} \frac{(n_i - m_{i2} - 1 + k)! \, \theta_{i,3n_i+1-m_{i2}+k} \, d_{i2}^k}{k!} = 0.$$

$b_{i2,k}$, $k = 0, \cdots, m_{i2}$, are then calculated as

$$b_{i2,k} = \sum_{p=0}^{k} \frac{(n_i - k + p)! \, \theta_{i,3n_i+2-k+p} \, d_{i2}^p}{p!}, \quad k = 0, 1, \cdots, m_{i2}.$$

The proposed method will lead to $m_{ij} + 1$ estimates for d_{ij}, just like [157] and [220]. By inspecting the lag between the input and output signals, the selection can be made simply. The selection can be also made by virtue of the consistency between various sets of $b_{ij,k}$ and d_{ij} and those ignored relations [220].

9.3.2 Simulation Studies

Example 9.3. Consider the well-known Wood-Berry binary distillation column plant:

$$G(s) = \begin{bmatrix} \dfrac{12.8e^{-s}}{16.7s+1} & \dfrac{-18.9e^{-3s}}{21s+1} \\ \dfrac{6.6e^{-7s}}{10.9s+1} & \dfrac{-19.4e^{-3s}}{14.4s+1} \end{bmatrix}.$$

The equivalent differential equations are

$$350.7y_1^{(2)}(t) + 37.7y_1^{(1)}(t) + y_1(t)$$
$$= 268.8u_1^{(1)}(t-1) + 12.8u_1(t-1)$$
$$-315.63u_2^{(1)}(t-3) - 18.9u_2(t-3) + \hat{w}_1(t), \tag{9.43}$$

and

$$156.96y_2^{(2)}(t) + 25.3y_2^{(1)}(t) + y_2(t)$$
$$= 95.04u_1^{(1)}(t-7) + 6.6u_1(t-7)$$
$$-211.46u_2^{(1)}(t-3) - 19.4u_2(t-3) + \hat{w}_2(t). \tag{9.44}$$

Case A

Assume that $\hat{w}_1(t) = \mathbf{1}(t)$ and $\hat{w}_2(t) = 0.5\mathbf{1}(t)$ and the identification test starts from nonzero initial conditions: $y_1(0) = -1$, $y_1^{(1)}(0) = 1$, $y_2(0) = 0.5$ and $y_2^{(1)}(0) = 2$. The test signals, $u_1(t)$ and $u_2(t)$, are both pulse signals,

$$u_1(t) = \mathbf{1}(t) - \mathbf{1}(t - 60),$$

and

$$u_2(t) = \mathbf{1}(t) - \mathbf{1}(t - 30).$$

The process inputs and outputs are shown in Fig. 9.3 and the sampling interval is 0.02. Suppose that $0 \leq d_{11} \leq 2$, $0 \leq d_{12} \leq 6$. It leads to

$$\hat{T}_1 = \{t | 2 \leq t < 60, \text{ or } 62 \leq t < 100\},$$
$$\hat{T}_2 = \{t | 6 \leq t < 30, \text{ or } 36 \leq t < 100\},$$
$$T = \hat{T}_1 \bigcap \hat{T}_2 = \{t | 6 \leq t < 30, \text{ or } 36 \leq t < 60, \text{ or } 62 \leq t < 100\}.$$

$\hat{T}_1$ and $\hat{T}_2$ have some elements in common and these elements are included in T. In other word, the elements in T are members of both $\hat{T}_1$ and $\hat{T}_2$. This can be seen clearly in Fig. 9.4. Choose $t = t_0, \cdots, t_N$ in T, $n_1 = 2$, $m_{11} = m_{12} = 1$ and $q_1 = 2$. The proposed method leads to two estimates for d_{11}: one is -39.05 and the other is 1.02. The time delay must be positive so that we choose $\hat{d}_{11} = 1.02$. The proposed method also leads to two estimates for d_{12}: -29.11 and 3.02. For the same reason, we choose $\hat{d}_{12} = 3.02$. The first sub-process is then obtained as:

$$y_1^{(2)}(t) + 0.1079y_1^{(1)}(t) + 0.002867y_1(t)$$
$$= 0.7715u_1^{(1)}(t - 1.02) + 0.0367u_1(t - 1.02)$$
$$-0.9062u_2^{(1)}(t - 3.02) - 0.05418u_2(t - 3.02),$$

with

$$\hat{G}_{11} = \frac{0.7715s + 0.0367}{s^2 + 0.1079s + 0.002867}e^{-1.02s},$$
$$\hat{G}_{12} = \frac{-0.9062s - 0.05418}{s^2 + 0.1079s + 0.002867}e^{-3.02s}.$$

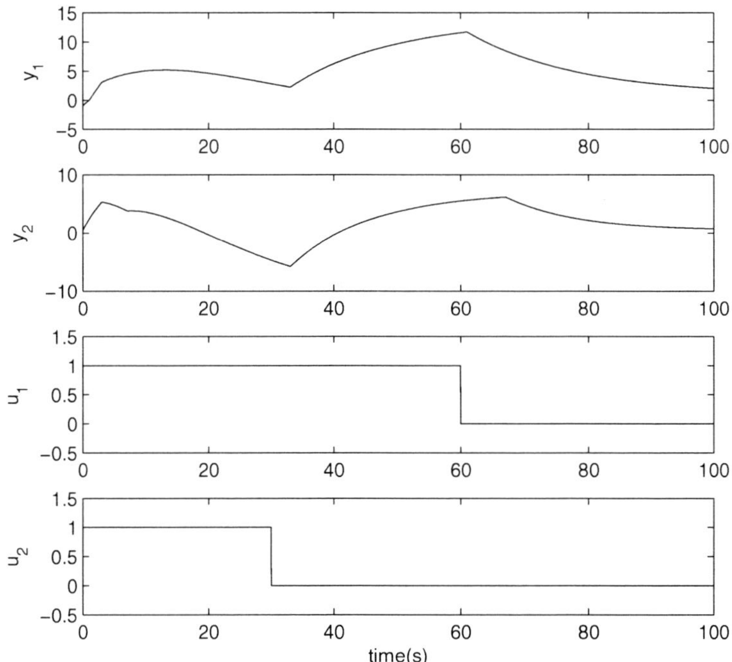

Fig. 9.3. Identification test of Example 9.3

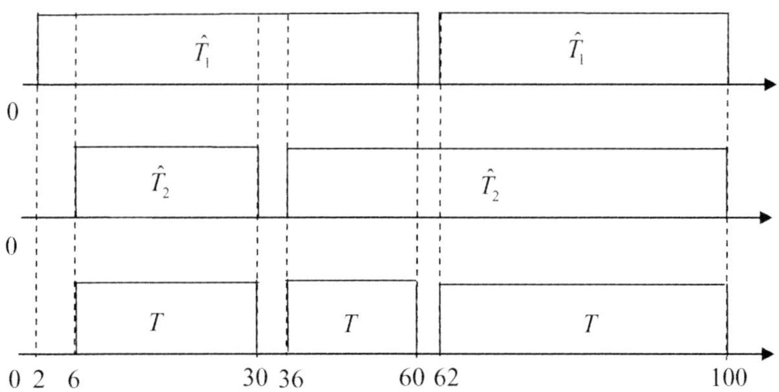

Fig. 9.4. Calculation of T

Suppose that $0 \le d_{21} \le 14$, $0 \le d_{22} \le 6$. The proposed method with $n_2 = 2$, $m_{21} = m_{22} = 1$ and $q_2 = 2$ leads to the second sub-process as:

$$y_2^{(2)}(t) + 0.162 y_2^{(1)}(t) + 0.006423 y_2(t)$$
$$= \; 0.6115 u_1^{(1)}(t - 7.02) + 0.04239 u_1(t - 7.02)$$
$$- 1.361 u_2^{(1)}(t - 3.02) - 0.1246 u_2(t - 3.02),$$

with

$$\hat{G}_{21} = \frac{0.6115s + 0.04239}{s^2 + 0.162s + 0.006423} e^{-7.02s},$$

$$\hat{G}_{22} = \frac{-1.361s - 0.1246}{s^2 + 0.162s + 0.006423} e^{-3.02s}.$$

The identification error, ERR $= \{\text{ERR}_{ij}\}$, is measured by the worst case error,

$$\text{ERR}_{ij} = \max \left| \frac{\hat{G}_{ij}(j\omega_k) - G_{ij}(j\omega_k)}{G_{ij}(j\omega_k)} \right|, \quad k = 1, \cdots, M, \tag{9.45}$$

where $\hat{G}_{ij}(j\omega_k)$ and $G_{ij}(j\omega_i)$ are the estimated frequency response and the actual ones. The Nyqusit curve for a phase ranging from 0 to $-\pi$ is considered, because this part is the most significant for control design. For this example, the identification error is

$$\text{ERR} = \begin{bmatrix} 4.04\% & 1.46\% \\ 0.95\% & 1.59\% \end{bmatrix}.$$

In real applications, numerical integration is employed to calculate the multiple integration of the output and this introduces errors. Better identification results can be obtain by sampling the process response with a small sampling interval. If the sampling interval is 0.2, the proposed method leads to the identification error as

$$\text{ERR} = \begin{bmatrix} 16.39\% & 5.90\% \\ 3.22\% & 6.42\% \end{bmatrix}.$$

In this case, the identification result is still acceptable. If the sampling interval is chosen as 1 and the identification error is obtained as

$$\text{ERR} = \begin{bmatrix} 83.31\% & 30.98\% \\ 19.53\% & 34.53\% \end{bmatrix}.$$

The identification error is very large. From these simulations, one can find that small sampling interval leads to good identification results. Generally, chemical processes have slow response. With the development of computer technologies, the sampling interval can be set very small and enough data can be obtained easily for use in process identification.

Case B

This is the same as Case A except that process outputs are subject to changing disturbances, where $\hat{w}_1(t)$ and $\hat{w}_2(t)$ are simulated by letting $\mathbf{1}(t)$ pass through the transfer functions of $1/(15s + 1)$ and $-3/(20s + 1)$, respectively. The proposed method, with $n_1 = n_2 = 2$, $m_{11} = m_{12} = m_{21} = m_{22} = 1$ and $q_1 = q_2 = 3$, leads to

$$y_1^{(2)}(t) + 0.1103y_1^{(1)}(t) + 0.003y_1(t)$$
$$= 0.7718u_1^{(1)}(t - 1.03) + 0.03851u_1(t - 1.03)$$
$$- 0.9058u_2^{(1)}(t - 3.03) - 0.05637u_2(t - 3.03),$$

with

$$\hat{G}_{11} = \frac{0.7718s + 0.03851}{s^2 + 0.1103s + 0.003}e^{-1.03s},$$

$$\hat{G}_{12} = \frac{-0.9058s - 0.05637}{s^2 + 0.1103s + 0.003}e^{-3.03s},$$

and

$$y_2^{(2)}(t) + 0.1528y_2^{(1)}(t) + 0.00582y_2(t)$$
$$= 0.6054u_1^{(1)}(t - 7) + 0.0376u_1(t - 7)$$
$$- 1.36u_2^{(1)}(t - 3.03) - 0.1119u_2(t - 3.03),$$

with

$$\hat{G}_{21} = \frac{0.6054s + 0.0376}{s^2 + 0.1528s + 0.00582}e^{-7s},$$

$$\hat{G}_{22} = \frac{-1.36s - 0.1119}{s^2 + 0.1528s + 0.00582}e^{-3.03s}.$$

The identification error is

$$\mathrm{ERR} = \begin{bmatrix} 4.15\% & 1.45\% \\ 1.91\% & 1.61\% \end{bmatrix}.$$

Case C

This is the same as Case B except that a white noise is added to corrupt the outputs. The noise-to-signal ratio defined by

$$\mathrm{NSR} = \frac{\mathrm{mean(abs(noise))}}{\mathrm{mean(abs(signal))}},$$

(denoted N_1) and

$$\mathrm{NSR} = \frac{\mathrm{variance(noise)}}{\mathrm{variance(signal)}},$$

(denoted N_2) are used to represent a noise level. Let the outputs be corrupted with noise of $N_1 = 15\%, 25\%$ and 40% or $N_2 = 3\%, 7\%$ and 18%, respectively. Suppose that the estimated ranges of time delays are $0.5 \leq d_{11} \leq 1.5$, $2 \leq d_{12} \leq 4$, $6 \leq d_{21} \leq 9$ and $2 \leq d_{22} \leq 4$. The identified parameters are expressed as the mean and standard deviation of each estimate from 20 noisy simulations and shown in Table 9.2.

Table 9.2. Estimated model parameters of Example 9.3

	$N_1 = 15\%$ ($N_2 = 3\%$)	$N_1 = 25\%$ ($N_2 = 7\%$)	$N_1 = 40\%$ ($N_2 = 18\%$)
$\hat{a}_{1,1}$	0.1101 ± 0.0076	0.1116 ± 0.0135	0.1126 ± 0.0225
$\hat{a}_{1,0}$	0.0029 ± 0.0007	0.0031 ± 0.0008	0.0032 ± 0.0013
$\hat{b}_{11,1}$	0.7746 ± 0.0235	0.7695 ± 0.0249	0.7284 ± 0.1765
$\hat{b}_{11,0}$	0.0389 ± 0.0064	0.0393 ± 0.0106	0.0395 ± 0.0187
$\hat{d}_{11}$	1.0232 ± 0.0801	1.0523 ± 0.1489	0.9909 ± 0.3228
$\hat{b}_{12,1}$	-0.9117 ± 0.0342	-0.9045 ± 0.0334	-0.9046 ± 0.0555
$\hat{b}_{12,0}$	-0.0549 ± 0.0076	-0.0573 ± 0.0086	-0.0579 ± 0.0143
$\hat{d}_{12}$	3.0499 ± 0.1254	3.0421 ± 0.1440	3.0561 ± 0.2381
$\hat{a}_{2,1}$	0.1554 ± 0.0097	0.1581 ± 0.0143	0.1607 ± 0.0242
$\hat{a}_{2,0}$	0.0060 ± 0.0005	0.0061 ± 0.0009	0.0063 ± 0.0015
$\hat{b}_{21,1}$	0.6066 ± 0.0345	0.6127 ± 0.0445	0.6126 ± 0.0753
$\hat{b}_{21,0}$	0.0403 ± 0.0056	0.0413 ± 0.0078	0.0429 ± 0.0133
$\hat{d}_{21}$	6.9337 ± 0.2134	6.9397 ± 0.2045	6.9233 ± 0.3372
$\hat{b}_{22,1}$	-1.3642 ± 0.0375	-1.3663 ± 0.0486	-1.3620 ± 0.0835
$\hat{b}_{22,0}$	-0.1130 ± 0.0096	-0.1156 ± 0.0122	-0.1180 ± 0.0206
$\hat{d}_{22}$	3.0548 ± 0.0841	3.0678 ± 0.1103	3.0796 ± 0.1857

In case of noise, we may also start with rough estimated delay ranges given in Case A and use the proposed method to find $\hat{d}_{ij}$, estimates of d_{ij}. Then with $\hat{d}_{ij}$, we retunes the ranges of time delays and apply the proposed method again to achieve a better estimation. For example, in case of $N_1 = 15\%$, one identification test is applied. The proposed method, with $0 \le d_{11} \le 2$, $0 \le d_{12} \le 6$, $0 \le d_{21} \le 14$ and $0 \le d_{22} \le 6$, leads to $\hat{d}_{11} = 1.08$, $\hat{d}_{12} = 2.88$, $\hat{d}_{21} = 7.23$ and $\hat{d}_{22} = 3.05$, with the identification error of

$$\mathrm{ERR} = \begin{bmatrix} 13.50\% & 9.66\% \\ 5.62\% & 6.42\% \end{bmatrix}.$$

We then retunes the ranges of the time delays as the above and the proposed method leads to a smaller identification error

$$\mathrm{ERR} = \begin{bmatrix} 6.85\% & 10.07\% \\ 5.63\% & 3.80\% \end{bmatrix}.$$

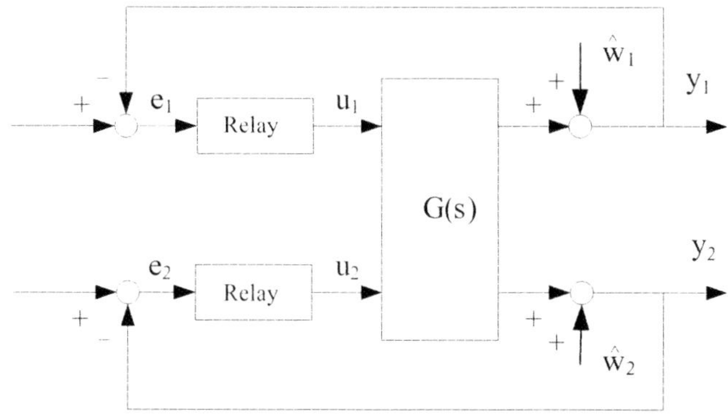

Fig. 9.5. Relay feedback experiment

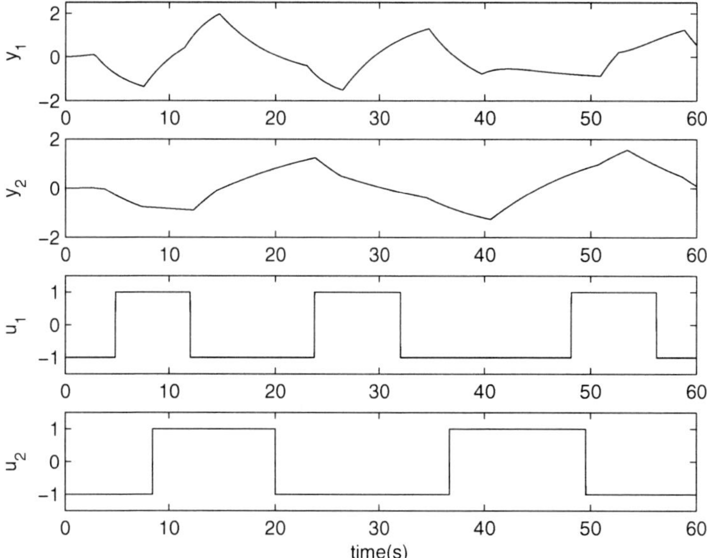

Fig. 9.6. Identification test of Example 9.4

Example 9.4. Consider a TITO system,

$$G(s) = \begin{bmatrix} \dfrac{2e^{-2.7s}}{5s+1} & \dfrac{0.5e^{-3s}}{2s+1} \\ \dfrac{0.4e^{-2.5s}}{6s+1} & \dfrac{2.2e^{-3.8s}}{10s+1} \end{bmatrix}.$$

A closed-loop relay feedback is applied on this example. The relay feedback system is shown in Fig. 9.5. The relay unit is described as

$$u(t) = \begin{cases} u_+, & \text{if } e(t) > \varepsilon_+, \text{ or } e(t) \geq \varepsilon_- \text{ and } u(t_-) = u_+, \\ u_-, & \text{if } e(t) < \varepsilon_-, \text{ or } e(t) \leq \varepsilon_+ \text{ and } u(t_-) = u_-, \end{cases} \tag{9.46}$$

where $e(t)$ and $u(t)$ are the relay input and output, respectively. The relay experiment is applied at $t = 0$ with $u_+ = 1$, $u_- = -1$, $\varepsilon_+ = 0.8$ and $\varepsilon_- = -0.8$ under zero initial conditions and nonzero static disturbances of $\hat{w}_1 = \hat{w}_2 = 0.51(t)$. The process inputs and outputs are shown in Fig. 9.6 and the sampling interval is 0.02. Suppose that $2 \leq d_{11} \leq 3$, $2 \leq d_{12} \leq 3$, $2 \leq d_{21} \leq 3$ and $3 \leq d_{22} \leq 4$. The proposed method, with $n_1 = n_2 = 2$, $m_{11} = m_{12} = m_{21} = m_{22} = 1$ and $q_1 = q_2 = 2$, leads to

$$\hat{G}(s) = \begin{bmatrix} \dfrac{0.4067s + 0.1947}{s^2 + 0.6886s + 0.09707}e^{-2.7s} & \dfrac{0.2489s + 0.04814}{s^2 + 0.6886s + 0.09707}e^{-2.99s} \\ \dfrac{0.06743s + 0.007456}{s^2 + 0.2814s + 0.01911}e^{-2.52s} & \dfrac{0.2177s + 0.04014}{s^2 + 0.2814s + 0.01911}e^{-3.8s} \end{bmatrix},$$

with the identification error as follows

$$\text{ERR} = \begin{bmatrix} 1.21\% & 0.82\% \\ 2.45\% & 4.52\% \end{bmatrix}.$$

9.3.3 General MIMO Processes

The TITO identification method is now extended to a general MIMO process. Consider a process with l inputs and m outputs,

$$Y(s) = G(s)U(s),$$

where $Y(s) = [Y_1(s), \cdots, Y_i(s), \cdots, Y_l(s)]^T$ is the output vector, $U(s) = [U_1(s), \cdots, U_j(s), \cdots, U_m(s)]^T$ is the input vector, and $G(s) = \{G_{ij}(s)\} = \{\alpha_{ij}(s)e^{-d_{ij}s}/\beta_{ij}(s)\}$, with $i = 1, \cdots, l$ and $j = 1, \cdots, m$, is the process transfer function matrix. The given MIMO process may be decomposed into l sub-processes, which can be described as

$$\begin{aligned} Y_i(s) &= \begin{bmatrix} G_{i1}(s) & \cdots & G_{ij}(s) & \cdots G_{im}(s) \end{bmatrix} U(s) \\ &= \begin{bmatrix} \dfrac{\alpha_{i1}(s)}{\beta_{i1}(s)}e^{-d_{i1}s} & \cdots & \dfrac{\alpha_{ij}(s)}{\beta_{ij}(s)}e^{-d_{ij}s} & \cdots & \dfrac{\alpha_{im}(s)}{\beta_{im}(s)}e^{-d_{im}s} \end{bmatrix} U(s), \\ & \qquad i = 1, \cdots, l. \end{aligned}$$

Let the common denominator of all G_{ij}, $j = 1, \cdots, m$ be $\beta_i^*(s)$. We have

$$\beta_i^*(s)Y_i(s) = \begin{bmatrix} \alpha_{i1}^*(s)e^{-d_{i1}s} & \cdots & \alpha_{ij}^*(s)e^{-d_{ij}s} & \cdots & \alpha_{im}^*(s)e^{-d_{im}s} \end{bmatrix} U(s),$$
$$i = 1, \cdots, l.$$

The equivalent differential equations are

$$y_i^{(n_i)}(t) + \sum_{k=0}^{n_i-1} a_{i,k} y_i^{(k)}(t) = \sum_{j=1}^{m} \sum_{k=0}^{m_{ij}} b_{ij,k} u_j^{(k)}(t - d_{ij}) + w_i(t), \quad i = 1, \cdots, l. \qquad (9.47)$$

The inputs under considerations are

$$u_j(t) = \sum_{k=0}^{K_j} h_{j,k} \mathbf{1}(t - t_{j,k}), \quad j = 1, \ldots, m,$$

where $t_{j,k}$ is the kth switch instant of $u_j(t)$.

Integrating (9.47) with (9.32) n_i times yields

$$P_{n_i} y_i^{(n_i)}(t) + \sum_{k=0}^{n_i-1} a_{i,k} P_{n_i} y_i^{(k)}(t) = \sum_{j=0}^{m} \sum_{k=0}^{m_{ij}} b_{ij,k} P_{n_i} u_j^{(k)}(t - d_{ij}) + P_{n_i} w_i(t). \qquad (9.48)$$

The left-hand side is (9.34) again. For the right-hand side, it follows that

$$P_{n_i} u_j^{(p)}(t - d_{ij}) = \sum_{k=0}^{K_j} \frac{h_{j,k}(t - t_{j,k} - d_{ij})^{n_i-p}}{(n_i - p)!} \mathbf{1}(t - t_{j,k} - d_{ij}), \ \ p = 0, 1, \ldots, m_{ij}.$$

Equation (9.48) can be rearranged as

$$\phi_i^T(t)\theta_i = \gamma_i(t), \qquad (9.49)$$

where $\gamma_i(t) = y_i(t)$,

$$\phi_i(t) = \begin{bmatrix} -P_1 y_i(t) \\ \vdots \\ -P_{n_i} y_i(t) \\ \sum_{k=0}^{K_1} h_{1,k} \mathbf{1}(t - t_{1,k} - d_{i1}) \\ \sum_{k=0}^{K_1} h_{1,k}(t - t_{1,k}) \mathbf{1}(t - t_{1,k} - d_{i1}) \\ \vdots \\ \sum_{k=0}^{K_1} h_{1,k}(t - t_{1,k})^{n_i} \mathbf{1}(t - t_{1,k} - d_{i1}) \\ \vdots \\ \sum_{k=0}^{K_m} h_{m,k} \mathbf{1}(t - t_{m,k} - d_{im}) \\ \sum_{k=0}^{K_m} h_{m,k}(t - t_{m,k}) \mathbf{1}(t - t_{m,k} - d_{im}) \\ \vdots \\ \sum_{k=0}^{K_m} h_{m,k}(t - t_{m,k})^{n_i} \mathbf{1}(t - t_{m,k} - d_{im}) \\ 1 \\ t \\ \vdots \\ t^{q_i} \end{bmatrix}, \quad \theta_i = \begin{bmatrix} \theta_{i,1} \\ \vdots \\ \theta_{i,n_i} \\ \theta_{i,n_i+1} \\ \theta_{i,n_i+2} \\ \vdots \\ \theta_{i,2n_i+1} \\ \vdots \\ \theta_{i,m(n_i+1)} \\ \theta_{i,m(n_i+1)+1} \\ \vdots \\ \theta_{i,m(n_i+1)+n_i} \\ \theta_{i,(m+1)(n_i+1)} \\ \theta_{i,(m+1)(n_i+1)+1} \\ \vdots \\ \theta_{i,(m+1)(n_i+1)+q_i} \end{bmatrix}.$$

Note that the first n_i elements of θ_i are the same as (9.37). $\theta_{i,k}$, $k = j(n_i + 1) + 1, \ldots,$ $j(n_i + 1) + n_i$, and $j = 1, \ldots, m$, are combinations of the model parameters $b_{ij,k}$, $k = 0, \cdots, m_{ij}$ and d_{ij}, and are given by

$$\theta_{i,k} = \sum_{p=\max(n_i - m_{ij}, k - j(n_i + 1))}^{n_i} \frac{(-d_{ij})^{p-k+j(n_i+1)} \, b_{ij,n_i-p}}{(p - k + j(n_i + 1))! \, (k - j(n_i + 1))!},$$

$$k = j(n_i + 1), \cdots, j(n_i + 1) + n_i. \tag{9.50}$$

$\theta_{i,k}$, $k = (m + 1)(n_i + 1), \cdots, (m + 1)(n_i + 1) + q_i$ account for the effects of the aforementioned nonzero conditions and the disturbances.

Suppose that d_{ij}, $j = 1, \cdots, m$ are in the ranges of $[\underline{d}_{ij}, \overline{d}_{ij}]$. Define

$$\hat{T}_j = \bigcup_{k=0}^{K_j-1} \{t \, | \, t_{j,k} + \overline{d}_{ij} \le t < t_{j,k+1} + \underline{d}_{ij}\} \bigcup \{t \, | \, (t_{j,K_j} + \overline{d}_{ij} \le t \le T_{end}\},$$

$$j = 1, \ldots, m.$$

Then, t should be taken in the set of

$$T = \bigcap_{j=1}^{m} \hat{T}_j.$$

One invokes (9.49) for t in T with $t = t_0, t_1, \cdots, t_N$, and they give

$$\Psi_i \theta_i = \Gamma_i, \tag{9.51}$$

where $\Psi_i = [\phi_i(t_0), \cdots, \phi_i(t_N)]^T$ and $\Gamma_i = [\gamma_i(t_0), \cdots, \gamma_i(t_N)]^T$. The ordinary least-squares method can be applied to find the solution; in the presence of noise in the measurement of the process output, the instrumental variable (IV) method is adopted to guarantee the identification consistency. Once θ_i is estimated by applying the least-squares method or IV method, the model parameters can be recovered. We can recover d_{ij} from $\theta_{i,k}$, $k = j(n_i + 1), \cdots, j(n_i + 1) + n_i$, using the following algebraic equations:

$$\sum_{k=0}^{m_{ij}+1} \frac{(n_i - m_{ij} - 1 + k)! \, \theta_{i,j(n_i+1)+n_i-1-m_{ij}+k} \, d_{ij}^k}{k!} = 0,$$

$$i = 1, \cdots, l, \; j = 1, \cdots, m.$$

Once d_{ij} are obtained, the parameter $b_{ij,k}$ are then calculated as

$$b_{ij,k} = \sum_{p=0}^{k} \frac{(n_i - k + p)! \, \theta_{i,j(n_i+1)+n_i-k+p} \, d_{ij}^p}{p!},$$

$$k = 0, 1, \cdots, m_{ij}, \; i = 1, \cdots, l, \; j = 1, \cdots, m.$$

Example 9.5. Consider a system in [229]

$$G(s) = \begin{bmatrix} \dfrac{119e^{-5s}}{21.7s + 1} & \dfrac{40e^{-5s}}{337s + 1} & \dfrac{-2.1e^{-5s}}{10s + 1} \\[2ex] \dfrac{77e^{-5s}}{50s + 1} & \dfrac{76.7e^{-3s}}{28s + 1} & \dfrac{-5e^{-5s}}{10s + 1} \\[2ex] \dfrac{93e^{-5s}}{50s + 1} & \dfrac{-36.7e^{-5s}}{166s + 1} & \dfrac{-103.3e^{-4s}}{23s + 1} \end{bmatrix}.$$

The equivalent differential equations are

$$73129y_1^{(3)}(t) + 10900y_1^{(2)}(t) + 368.7y_1^{(1)}(t) + y_1(t)$$
$$= 401030u_1^{(2)}(t-5) + 41293u_1^{(2)}(t-5) + 119u_1(t-5)$$
$$+ 8680u_2^{(2)}(t-5) + 1268u_2^{(2)}(t-5) + 40u_2(t-5)$$
$$- 15357u_3^{(2)}(t-5) - 753.27u_3^{(2)}(t-5) - 2.1u_3(t-5) + \hat{w}_1(t),$$

$$14000y_2^{(3)}(t) + 2180y_2^{(2)}(t) + 88y_2^{(1)}(t) + y_2(t)$$
$$= 21560u_1^{(2)}(t-5) + 2926u_1^{(2)}(t-5) + 77u_1(t-5)$$
$$+ 38350u_2^{(2)}(t-3) + 4602u_1^{(2)}(t-3) + 76.7u_2(t-3)$$
$$- 7000u_3^{(2)}(t-5) - 390u_1^{(2)}(t-5) - 5u_2(t-5) + \hat{w}_2(t),$$

and

$$190900y_3^{(3)}(t) + 13268y_3^{(2)}(t) + 239y_3^{(1)}(t) + y_3(t)$$
$$= 355074u_1^{(2)}(t-5) + 17577u_1^{(2)}(t-5) + 93u_1(t-5)$$
$$- 42205u_2^{(2)}(t-5) - 2679.1u_1^{(2)}(t-5) - 36.7u_2(t-5)$$
$$- 857390u_3^{(2)}(t-4) - 22313u_1^{(2)}(t-4) - 103.3u_2(t-4) + \hat{w}_3(t).$$

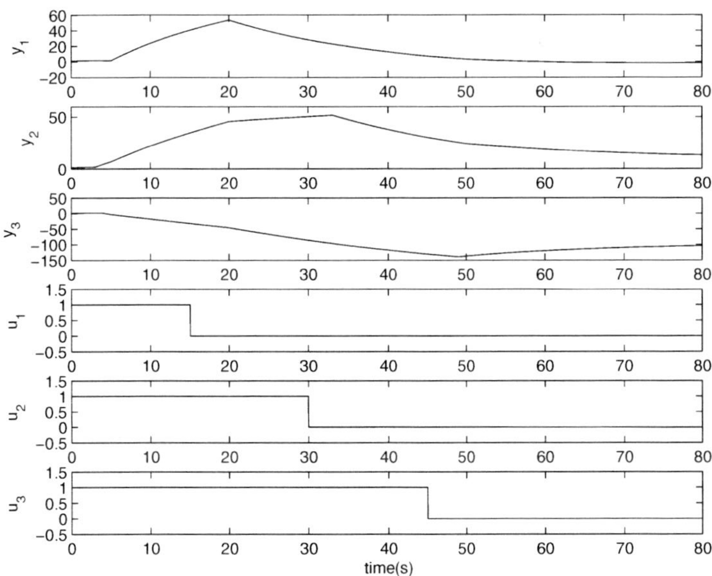

Fig. 9.7. Identification test of Example 9.5

Suppose that $\hat{w}_1(t) = 100\,\mathbf{1}(t)$, $\hat{w}_2(t) = 20\,\mathbf{1}(t)$ and $\hat{w}_3(t) = 100\,\mathbf{1}(t)$ and the identification test starts from nonzero initial conditions: $y_1(0) = y_2(0) = y_3(0) = 1$, $y_1^{(1)}(0) = y_2^{(1)}(0) = y_3^{(1)}(0) = 0.5$ and $y_1^{(2)}(0) = y_2^{(2)}(0) = y_3^{(2)}(0) = -0.2$. The process inputs and outputs are shown in Fig. 9.7 and the sampling interval is 0.02. Let $0 < d_{11} < 7$, $0 < d_{12} < 7$, $1 < d_{13} < 6$, $0 < d_{21} < 7$, $0 < d_{22} < 5$, $0 < d_{23} < 6$, $2 < d_{31} < 7$, $1 < d_{32} < 7$ and $0 < d_{33} < 7$. The proposed method with $n_i = 3$ $m_{ij} = 2$ and $q_i = 3$, where $i = 1,2,3$ and $j = 1,2,3$, leads to the following MIMO transfer function matrix

$$\hat{G}(s) = \begin{bmatrix} \hat{g}_{11}(s) & \hat{g}_{12}(s) & \hat{g}_{13}(s) \\ \hat{g}_{21}(s) & \hat{g}_{22}(s) & \hat{g}_{23}(s) \\ \hat{g}_{31}(s) & \hat{g}_{32}(s) & \hat{g}_{33}(s) \end{bmatrix},$$

where

$$\hat{g}_{11}(s) = \frac{5.506s^2 + 0.5666s + 0.001628}{s^3 + 0.1494s^2 + 0.005058s + 1.375 \times 10^{-5}} e^{-5.02s},$$

$$\hat{g}_{12}(s) = \frac{0.1192s^2 + 0.01741s + 0.0005498}{s^3 + 0.1494s^2 + 0.005058s + 1.375 \times 10^{-5}} e^{-5.02s},$$

$$\hat{g}_{13}(s) = \frac{-0.2107s^2 - 0.01034s - 2.8 \times 10^{-5}}{s^3 + 0.1494s^2 + 0.005058s + 1.375 \times 10^{-5}} e^{-5.02s},$$

$$\hat{g}_{21}(s) = \frac{1.547s^2 + 0.2097s + 0.005514}{s^3 + 0.156s^2 + 0.006295s + 7.132 \times 10^{-5}} e^{-5.02s},$$

$$\hat{g}_{22}(s) = \frac{2.751s^2 + 0.3297s + 0.005481}{s^3 + 0.156s^2 + 0.006295s + 7.132 \times 10^{-5}} e^{-3.02s},$$

$$\hat{g}_{23}(s) = \frac{-0.5019s^2 - 0.02793s - 0.0003587}{s^3 + 0.156s^2 + 0.006295s + 7.132 \times 10^{-5}} e^{-5.02s},$$

$$\hat{g}_{31}(s) = \frac{1.864s^2 + 0.09219s + 0.0004873}{s^3 + 0.06955s^2 + 0.001253s + 5.244 \times 10^{-6}} e^{-5.02s},$$

$$\hat{g}_{32}(s) = \frac{-0.2215s^2 - 0.01405s - 0.0001917}{s^3 + 0.06955s^2 + 0.001253s + 5.244 \times 10^{-6}} e^{-5.02s},$$

$$\hat{g}_{33}(s) = \frac{-4.499s^2 - 0.117s - 0.0005396}{s^3 + 0.06955s^2 + 0.001253s + 5.244 \times 10^{-6}} e^{-4.02s}.$$

The identification error is

$$E = \begin{bmatrix} 0.68\% & 0.63\% & 3.02\% \\ 0.67\% & 1.09\% & 1.92\% \\ 0.64\% & 1.89\% & 2.36\% \end{bmatrix}.$$

9.3.4 Real Time Testing

The proposed method is also applied to a laboratory scale temperature chamber system. The experiment setup consists of two parts: a thermal chamber set (which is made by

Fig. 9.8. Temperature chamber set

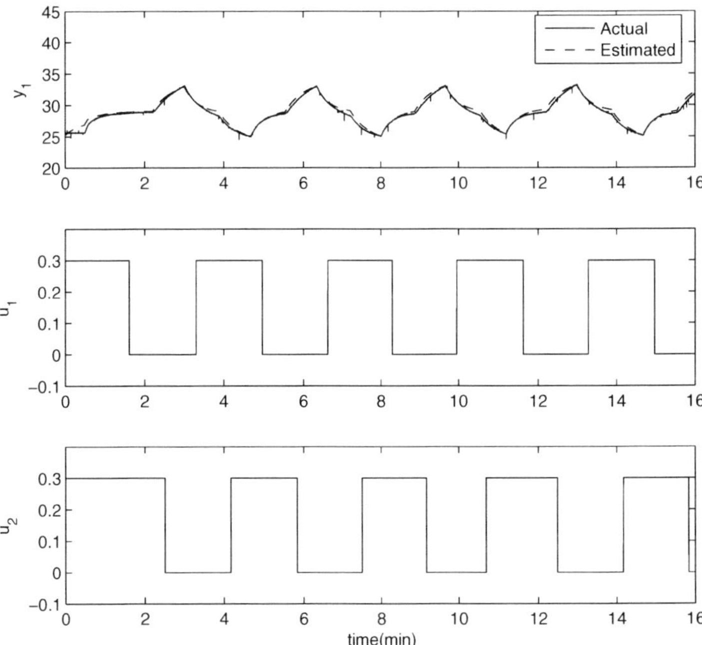

Fig. 9.9. Identification test of the thermal control system

National Instruments Corp. and shown in Fig. 9.8) and a personal computer with data acquisition cards and LabView software. The system has two inputs: one is to control 12 V Light with 20 W Halogen Bulb, the other is to control 12 V Fan. The system output is the temperature of the temperature chamber. Extra transport delays are simulated by using Labview software. An identification test is applied at $t = 0$. The process inputs

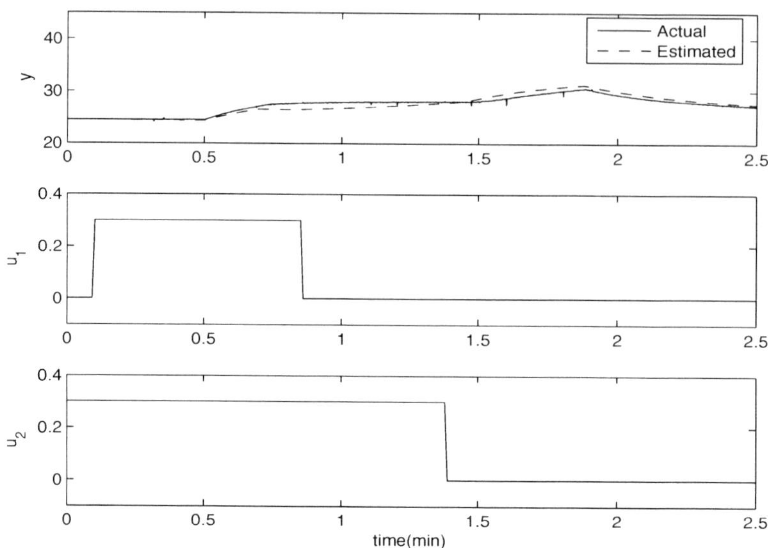

Fig. 9.10. Measured output response of the thermal control system and the model prediction

and the output are given in Fig. 9.9 and the sampling interval is 0.1 second. $u_1(t)$ in Fig. 9.9 is used to control the fan speed, and $u_2(t)$ is used to control the light intensity.

First, we estimate the range of time delays roughly: $0 \leq d_{11} \leq 0.8$ and $0 \leq d_{12} \leq 0.8$. Applying the proposed method with $n_1 = 2$, $m_{11} = m_{12} = 1$ and $q_1 = 2$, the estimated time delays are obtained as $\hat{d}_{11} = 0.555$ and $\hat{d}_{12} = 0.354$. Based on these estimated time delays, we can de-tune the ranges of time delay more accurately: $0.3 \leq d_{11} \leq 0.7$ and $0.2 \leq d_{12} \leq 0.6$. Applying the proposed identification method again and one obtains the model as follows,

$$y^{(2)}(t) + 3.333y^{(1)}(t) + 1.089y(t)$$
$$= -32.39u_1^{(1)}(t - 0.58) - 29.76u_1(t - 0.58)$$
$$+49.52u_2^{(1)}(t - 0.495) + 32.12u_2(t - 0.495).$$

If the disturbance is static, the initial conditions can be then estimated [220]. The estimated response of the model and the real one are shown in Fig. 9.9 for comparison. After the identification test, another cross-evaluation test is applied and the actual response is recorded. The model obtained in the above is used to predict the process response. The actual response and the predicted one are then compared in Fig. 9.10. The effectiveness of the proposed method is obvious.

9.4 Conclusion

Most industrial processes are multivariable in nature and have time delays. Implementation of modern advanced controllers such as model predictive control explicitly makes use of process models. Thus, identification of multivariable processes with time delay is in great demand. In this chapter, an integral identification method has been presented for both single-variable continuous-time delay systems and multivariable processes with multiple time delays. It adopts the integral technique and can work under non-zero initial conditions and dynamic disturbances. The unknown multiple time delays can be obtained without iteration. The permissible identification tests include all popular tests used in applications. Only a reasonable amount of computations is required. It is shown through simulation and real time implementation that the proposed method can yield accurate and robust identification results.

References

1. Koivo, H.N., Tanttu, J.T.: Tuning of PID controllers: survey of SISO and MIMO techniques. In: Intelligent Tuning and Adaptive Control, Selected Papers from the IFAC Symposium, pp. 75–80 (1991)
2. Åström, K.J., Hägglund, T.H.: PID Controllers: Theory, Design and Tuning, 2nd edn., Research Triangle Park, NC. Instrument Society of America (1995)
3. Tan, K.-K., Wang, Q.-G., Hang, C.C.: Advances in PID Control. Springer, London (1999)
4. Hang, C.C., Tan, K.K., Ong, S.L.: A comparative study of controller tuning formulae. IEE Proceedings, Part D 138(2), 111–118 (1979)
5. Miller, J.A., Lopez, A.M., Smith, C.L., Murrill, P.W.: A comparison, of controller tuning techniques. Control Engineering (1967)
6. Cohen, G.H., Coon, G.A.: Theoretical consideration of retarded control. Trans. ASME 75 (1953)
7. Gawthrop, P.J.: Self-tuning PID controllers: Algorithm and implementation. Automatic Control, IEEE Transactions, 31 (1986)
8. Wang, Q.-G., Fung, H.W., Lee, T.H.: PID tuning for improved performance. IEEE Trans. Control System Tech. 7(4), 457–465 (1999)
9. Chang, C.C., Sin, K.K.: An online auto-tuning method based on cross-correlation. IEEE Transactions on Industrial Electronics 38(6) (1991)
10. Ho, W.K., Hang, C.C., Cao, L.S.: Tuning of PID controller based on gain and phase margin specifications. Automatica 31(3), 497–502 (1995)
11. Fung, H.W., Wang, Q.-G., Lee, T.H.: PI tuning in terms of gain and phase margins. Automatica 34(9), 1145–1149 (1998)
12. Wang, Q.-G., Fung, H.W., Zhang, Y.: PID tuning with exact gain and phase margins. ISA Transactions 38, 243–249 (1999)
13. Shinskey, F.G.: Process Control Systems. McGraw-Hill, New York (1988)
14. Åström, K.J., Panagopoulos, H., Hägglund, T.: Design of PI controllers based on non-convex optimization. Automatica 34(5), 585–601 (1998)
15. MacFarlane, A.G.J., Belletrutti, J.J.: The characteristic locus design method. Automatica 9, 575–588 (1973)
16. Mayne, D.Q.: The design of linear multivariable systems. Automatica 9(1) (1973)
17. Bryant, G.F., Yeung, L.F.: Multivariable Control System Design Techniques: Dominance and Direct Methods. John Wiley & Sons, Chichester, England (1996)
18. Falb, P.L., Wolovich, W.A.: Decoupling in the design and synthesis of multivariable control systems. IEEE Trans. Automat. Control 12(6) (1967)
19. Rosenbrock, H.H.: State Space and Multivariable Theory. John Wiley, New York (1970)
20. Luyben, W.L.: Simple method for tuning siso controllers in multivariable systems. Ind. Eng. Chem. Proc. Des. Dev. 25, 654–660 (1986)
21. Palmor, Z.J., Halevi, Y., Kradney, N.: Automatic tuning of decentralized PID controller for tito process. Automatica 31, 1001–1010 (1993)
22. Wang, Q.-G., Huang, B., Guo, X.: Auto-tuning of tito decoupling controllers from step tests. ISA Transactions 39(4), 407–418 (2000)
23. Åström, K.J., Johansson, K.H., Wang, Q.-G.: Design of decoupled PI controllers for two-by-two systems. IEE Proceedings, Part D: Control Theory and Applications 149(1), 74–81 (2002)

24. Loh, A.P., Hang, C.C., Quek, C.X., Vasnani, V.U.: Autotuning of multiloop proportional-integral controllers using relay feedback. Ind. Eng. Chem. Res. 32, 1102–1107 (1993)
25. Zhang, Y., Wang, Q.-G., Astrom, K.J.: Dominant pole placement for multi-loop control systems. Automatica 38(7), 1213–1220 (2002)
26. Wang, Q.-G., Zhou, B., Lee, T.-H., Bi, Q.: Auto-tuning of multivariable PID controllers from decentralized relay feedback. Automatica 33(3), 319–330 (1997)
27. Huang, C.-J., Yu, C.-C., Shen, S.-H.: Selection of measurement locations for the control of rapid thermal processor. Automatica 36(5), 705–715 (2000)
28. Wang, Q.-G., Liu, M., Hang, C.C., Tang, W.: Robust process identification from relay tests in the presence of nonzero initial conditions and disturbance. Industrial and Engineering Chemistry Research 45(12), 4063–4070 (2006)
29. Grosdidier, P., Morari, M.: A computer aided methodology for the design of decentralized controllers. Comput. Chem. Eng. 11(423) (1987)
30. Chiu, M.S., Arkun, Y.: Decentralized control structure selection based on integrity considerations. Ind. Eng. Chem. Res. 29, 369 (1990)
31. Seborg, D.E., Edgar, T.F., Mellichamp, D.A.: Process Dynamics and Control. John Wiley and Sons, New York (1989)
32. Bristol, E.H.: On a new measure of interactions for multivariable process control. IEEE Trans. Autom. Control 11(133) (1966)
33. Wolff, E.A., Skogestad, S.: Operation of integrated three-product (petlyuk) distillation columns. Ind. Eng. Chem. Res. 34(2094) (1995)
34. Hansen, J.E., Jørgensen, S.B., Heath, J., Perkins, J.D.: Control structure selection for energy integrated distillation column. J. Process Control 8(3), 185–195 (1998)
35. Schaper, C.D., Kailath, T., Lee, Y.J.: Decentralized control of water temperature for multizone rapid thermal processing systems. IEEE Trans. Semicond. Manuf. 12(2), 193–199 (1999)
36. Grosdidier, P., Morari, M.: Closed-loop properties from steady-state gain information. Ind. Eng. Chem. Fundam. 24, 221 (1985)
37. Niederlinski, A.: A heuristic approach to the design of linear multivariable interacting subsystems. Automatica 7, 691 (1971)
38. McAvoy, T.J.: Interaction Analysis, ISA, Research Triangle Park, NC (1983)
39. Skogestad, S., Lundstrom, P., Jacobsen, E.W.: Selecting the best distillation control configuration. AIChE J. 36, 753 (1990)
40. Zhu, Z.-X.: Stability and integrity enforcement by integrating variable pairing and controller design. Chem. Eng. Sci. 53, 1009 (1998)
41. Hovd, M., Skogestad, S.: Simple frequency dependent tools for control system analysis, structure selection and design. Automatica 28, 989 (1992)
42. Witcher, M., McAvoy, T.J.: Interacting control systems: Steady state and dynamic measurement of interaction. ISA Trans. 16, 35 (1977)
43. Bristol, E.H.: Recent results on interactions in multivariable process control, Chicago, IL. AIChE Annual Meeting (1978)
44. Tung, L., Edgar, T.: Analysis of control-output interactions in dynamic systems. AIChE J. 27, 690 (1981)
45. Grosdidier, P., Morari, M.: Interaction measures for systems under decentralized control. Automatica 22, 309 (1986)
46. Grosdidier, P., Morari, M.: The μ interaction measures. Ind. Eng. Chem. Res. 26, 1193 (1987)
47. Asano, K., Morari, M.: Interaction measure of tensionthickness control in tandem cold rolling. Control Eng. Pract. 6, 1021 (1998)

48. Skogestad, S., Postlethwaite, I.: Multivariable Feedback Control: Analysis and Design, 2nd edn. John Wiley and Sons, New York (2005)
49. Morari, M.: Robust stability of systems with integral control. IEEE Trans. Autom. Control 30(6), 574–577 (1985)
50. Manousiouthakis, V., Savage, R., Arkun, Y.: Synthesis of decentralized process control structures using the concept of block relative gain. AIChE J. 32, 991 (1986)
51. Yu, C.C., Luyben, W.L.: Design of multiloop siso controllers in multivariable processes. Ind. Eng. Chem. Res 25, 498 (1986)
52. Skogestad, S., Morari, M.: Variable selection for decentralized control. AIChE Annual Meeting, Washington, DC (November 1988)
53. Yu, C.C., Fan, M.K.H.: Decentralized integral controllability and d-stability. Chem. Eng. Sci 45, 3299 (1990)
54. Campo, P.J., Morari, M.: Achievable closed-loop properties of systems under decentralized control: Conditions involving the steady-state gain. IEEE Trans. Autom. Control 39, 932 (1994)
55. Morari, M., Zafiriou, E.: Robust Process Control. Prentice-Hall, Englewood Cliffs (1989)
56. Yamanaka, K., Kawasaki, N.: Criterion for conditional stability of mimo integral control. Int. J. Control 50, 1071 (1989)
57. Lee, J., Edgar, T.F.: Conditions for decentralized integral controllability. Journal of Process Control 12, 797 (2002)
58. Lee, J., Edgar, T.F.: Computational method for decentralized integral controllability of low dimensional processes. Computers and Chemical Engineering 24, 847 (2000)
59. Ogunnaike, B.A., Ray, W.H.: Process Dynamics, Modeling, and Control. Oxford University Press, New York (1994)
60. Bristol, E.H.: On a philosophy of interaction in a multiloop world. In: ISA Chemical & Petroleum Division Chempid Symposium, May 1967, St. Louis, Missouri (1967)
61. Zhu, Z.-X.: Structural analysis and stability conditions of decentralized control systems. Ind. Eng. Chem. Res. 35, 736 (1996)
62. Zhu, Z.-X.: Variable pairing selection based on individual and overall interaction measures. Ind. Eng. Chem. Res. 35, 4091 (1996)
63. Zhu, Z.-X.: Steady state structural analysis and interaction characterization for multivariable control systems. Ind. Eng. Chem. Res. 36, 3718 (1997)
64. Arkun, Y., Downs, Y.J.: A general method to calculate input-output gains and the RGA for integrating processes. Comput. Chem. Eng. 14, 1101 (1990)
65. Nering, E.D.: Linear Algebra and Matrix Theory. Willey, Chichester (1970)
66. He, M.-J., Cai, W.-J.: New criterion for control loop configuration of multivariable processes. Ind. Eng. Chem. Res. 43, 7057 (2004)
67. Chiang, T.P., Luyben, W.L.: Comparison of the dynamic performance of three heat-integrated distillation configurations. Ind. Eng. Chem. Res. 27, 99 (1988)
68. Nwokah, O.D.I.: The stability of linear multivariable systems. International Journal of Control 37, 623–629 (1983)
69. Nwokah, O.D.I., Perez, R.A.: On multivariable stability in the gain space. In: Proceedings of the 29th conference on Decision and Control, Honolulu, Hawaii, December 1990, pp. 328–333 (1990)
70. Datta, A., Ho, M.-T., Bhattacharyya, S.P.: Structure and Synthesis of PID Controllers. Springer, London (2000)

71. Silva, G.J., Datta, A., Bhattacharyya, S.P.: PID Controllers for Time-delay Systems, Birkhauser, Boston (2005)
72. Wang, Q.-G.: Book review of Structure and Synthesis of PID Controllers by A. Datta, M.-T. Ho and S. P. Bhattacharyya. Automatica 39(4), 758 (2003)
73. Wang, Q.-G.: Book review: PID Controllers for Time-Delay Systems by G.J. Silva, A. Datta and S.P. Bhattacharyya. The SIAM Review (to be appear, 2005)
74. Safonov, M.G., Athans, M.: A multiloop generalization of the circle criterion for stability margin analysis. IEEE Transaction on Automatic Control 26(2), 415–422 (1981)
75. Yaniv, O.: Synthesis of uncertain MIMO feedback systems for gain and phase margins at different channel breaking points. Automatica 28(5), 1017–1020 (1992)
76. Li, Y., Lee, E.B.: Stability robustness characterization and related issues for control systems design. Automatica 29(2), 479–484 (1993)
77. Ho, W.K., Lee, T.H., Gan, O.P.: Tuning of multiloop PID controllers based on gain and phase margin specifications. Ind. Eng. Chem. Res. 36(6), 2231–2238 (1997)
78. Doyle, J.C.: Analysis of feedback systems with structured uncertainties. IEE Proceedings, Part D 129(6), 242–250 (1982)
79. Zheng, F., Wang, Q.-G., Lee, T.H.: On the design of multivariable PID controllers via LMI approach. Automatica 38, 517–526 (2002)
80. Lin, C., Wang, Q.-G., Lee, T.H.: An improvement on multivariable PID controller design via iterative LMI approach. Automatica 40, 519–525 (2004)
81. Guo, L., Wang, H.: Applying constrained nonlinear generalized PI strategy to PDF tracking control through square root b-spline models. International Journal of Control 77(17), 1481–1492 (2004)
82. Guo, L., Wang, H.: PID controller design for output PDFs of stochastic systems using linear matrix inequalities. IEEE Trans. Systems, Man and Cybernetics-Part B: Cybernetics 35(1), 65–71 (2005)
83. Guo, L., Wang, H.: Generalized discrete-time PI control of output PDFs using square root b-spline expansion. Automatica 41, 159–162 (2005)
84. Tsinias, J., Kalouptsidos, N.: Output feedback stabilization. Automatic Control, IEEE Transactions on 35(8), 951–954 (1990)
85. Trofino-Neto, A., Kucera, V.: Stabilization via static output feedback. Automatic Control, IEEE Transactions on 38(5), 764–765 (1993)
86. Kucera, V., de Souza, C.E.: A necessary and sufficient condition for output feedback stabilizability. Automatica 31(9), 1357–1359 (1995)
87. Wang, J., Wang, J., Li, H.-Y.: Nonlinear PI control of chaotic systems using singular perturbation theory. Chaos, Solitons & Fractals 25(5), 1057–1068 (2005)
88. Bernussou, J., Peres, P.L.D., Geromel, J.C.: A linear programming oriented procedure for quadratic stabilization of uncertain systems. Systems & Control Letters 13(1), 65–72 (1989)
89. Boyd, S.P., El Ghaoui, L., Feron, E., Balakrishnan, V.: Linear Matrix Inequalities in System and Control Theory. Siam Studies in Applied Mathematics, Philadelphia, PA (1994)
90. Cao, Y.-Y., Lam, J., Sun, Y.-X.: Static output feedback stabilization: An ILMI approach. Automatica 34(12), 1641–1645 (1998)
91. Crusius, C.A.R., Trofino, A.: Sufficient LMI conditions for output feedback control problems. Automatic Control, IEEE Transactions 44(5), 1053–1057 (1999)
92. Geromel, J.C., de Oliveira, M.C., Hsu, L.: LMI characterization of structural and robust stability. Linear Alg. Appl. 285, 69–80 (1998)

93. He, Y., Wu, M., She, J.-H.: Improved bounded-rela-lemma representation and H_∞ control of systems with polytopic uncertainties. IEEE Trans. Circuits and Systems-II: Express Briefs 52(7), 380–383 (2005)

94. Peaucelle, D., Arzelier, D., Bachelier, O., Bernussou, J.: A new robust $\mathcal{D}$-stability condition for real convex polytopic uncertainty. Syst. Control Lett. 40, 21–30 (2000)

95. Ramos, D.C.W., Peres, P.L.D.: An LMI condition for the robust stability of uncertain continuous-time linear systems. IEEE Trans. Autonat. Contr. 47(4), 675–678 (2002)

96. Gahinet, P., Nemirovski, A., Laub, A.J., Chilali, M.: LMI Control Toolbox for Use with Matlab, Natick, MA. Math Works (1995)

97. Wang, Q.-G.: Decoupling Control. Springer, New York (2003)

98. Maciejowski, J.M.: Multivariable Feedback Design. Addison-Wesley, Reading (1989)

99. Lin, C., Wang, Q.-G., Lee, T.H.: Robust normalization and stabilization of uncertain descriptor systems with norm-bounded perturbations. IEEE Trans. Automat. Contr. 50(4), 515–520 (2005)

100. Dai, L.: Singular Control Systems. Springer, Berlin (1989)

101. Masubuchi, I., Kamitane, Y., Ohara, A., Suda, N.: H_∞ control for descriptor system: A matrix inequalities approach. Automatica 33, 669–673 (1997)

102. Lin, C., Lam, J., Wang, J.-L., Yang, G.H.: Analysis on robust stability for interval descriptor systems. Systems and Control Letters 42, 267–278 (2001)

103. Lin, C., Wang, J.-L., Soh, C.B.: Maximum bounds for robust stability of linear uncertain descriptor systems with structured perturbations. International Journal of Systems Science 34(7), 463–467 (2003)

104. Barmish, B.R.: New Tools for Robustness of Linear Systems. Macmillan, New York (1994)

105. Lee, L., Fang, C.-H., Hsieh, J.-G.: Exact unidirectional perturbation bounds for robustness of uncertain generalized state-space systems: Continuous-time case. Automatica 33(10), 1923 (1997)

106. Horowitz, I.M.: Synthesis of Feedback Systems. Academic, New York (1963)

107. Bar-on, J.R., Jonckheere, E.A.: Phase margins for multivariable control systems. Int. J. Control 52(2), 485–498 (1998)

108. Wang, Q.-G., Lin, C., Ye, Z., Wen, G., He, Y., Hang, C.C.: A quasi-LMI approach to computing stabilizing parameter ranges of multi-loop PID controllers. Journal of Process Control 17, 59–72 (2007)

109. He, Y., Wu, M., She, J.H., Liu, G.P.: Delay-dependent robust stability criteria for uncertain neutral systems with mixed delays. Syst. Contr. Lett. 51(1), 57–65 (2004)

110. He, Y., Wu, M., She, J.H., Liu, G.P.: Parameter-dependent lyapunov functional for stability of time-delay systems with polytopic-type uncertainties. IEEE Trans. Automat. Contr. 49(5), 828–832 (2004)

111. Wu, M., He, Y., She, J.H., Liu, G.P.: Delay-dependent criteria for robust stability of time-varying delay systems. Automatica 40(8), 1435–1439 (2004)

112. He, Y., Wang, Q.-G., Lin, C., Wu, M.: Augmented Lyapunov functional and delay-dependent stability criteria for neutral systems. Int. J. Robust Non. Control 15(18), 923–933 (2005)

113. He, Y., Wu, M., She, J.H.: Delay-dependent stability criteria for linear systems with multiple time delays. IEE Proc.-Control Theory Appl. 153(4), 447–452 (2006)

114. Park, P.: A delay-dependent stability criterion for systems with uncertain time-invariant delays. IEEE Trans. Automat. Contr. 44(4), 876–877 (1999)

115. Moon, Y.S., Park, P., Kwon, W.H., Lee, Y.S.: Delay-dependent robust stabilization of uncertain state-delayed systems. Int. J. Control 74(14), 1447–1455 (2001)

116. Fridman, E., Shaked, U.: An improved stabilization method for linear time-delay systems. IEEE Trans. Automat. Contr. 47(11), 1931–1937 (2002)

117. Fridman, E., Shaked, U.: Delay-dependent stability and H_∞ control: constant and time-varying delays. Int. J. Control 76(1), 48–60 (2003)

118. Kharitonov, V.L., Niculescu, S.-I.: On the stability of linear systems with uncertain delay. IEEE Trans. Automat. Contr. 48(1), 127–132 (2003)

119. Han, Q.L.: On robust stability of neutral systems with time-varying discrete delay and norm-bounded uncertainty. Automatica 40(6), 1087–1092 (2004)

120. Han, Q.L.: A descriptor system approach to robust stability of uncertain neutral systems with discrete and distributed delays. Automatica 40(10), 1791–1796 (2004)

121. Xu, S., Lam, J.: Improved delay-dependent stability criteria for time-delay systems. IEEE Trans. Automat. Contr. 50(2), 384–387 (2005)

122. Xu, S., Lam, J., Zou, Y.: Simplified descriptor system approach to delay-dependent stability and performance analyses for time-delay systems. IEE Proc.-Control Theory Appl. 152(2), 147–151 (2005)

123. Bertsekas, D.P.: Constrained Optimization and Lagrange Multiplier Methods. Academic Press, New York (1982)

124. Lancaster, P., Tismenetsky, M.: The Theory of Matrices. Academic Press, Orlando, Florida (1985)

125. Grasse, K.A., Bar-on, J.R.: Regularity properties of the phase for multivariable systems. SIAM J. Control Optim. 35(4), 1366–1386 (1997)

126. Wang, Q.-G., He, Y., Ye, Z., Lin, C., Hang, C.C.: On loop phase margins of multivarible control systems. Journal of Process Control (Submitted 2006)

127. Ziegler, J.G., Nichols, N.B.: Optimum settings for automatic controllers. Trans. ASME 64, 759 (1942)

128. Rivera, D.E., Morari, M., Skogestad, S.: Internal model control. 4. PID controller design. Ind. Eng. Chem. Process Des. Dev. 25, 252 (1986)

129. Skogestad, S.: Simple analytic rules for model reduction and PID controller tuning. Journal of Process Control 13, 291 (2003)

130. Monica, T.J., Yu, C.C., Luyben, W.L.: Improved multiloop single-input/single-output (siso) controllers for multivariable processes. Ind. Eng. Chem. Res. 27, 969 (1988)

131. Chien, I.L., Huang, H.P., Yang, J.C.: A simple multiloop tuning method for PID controllers with no proportional kick. Ind. Eng. Chem. Res. 38, 1456 (1999)

132. Hovd, M., Skogestad, S.: Sequencial design of decentralized controllers. Automatica 30, 1601 (1994)

133. Shiu, S.J., Hwang, S.H.: Sequencial design method for multivariable decoupling and multiloop PID controllers. Ind. Eng. Chem. Res. 37, 107 (1998)

134. Lee, D.Y., Lee, M., Lee, Y., Park, S.: Mp criterion based multiloop PID controllers tuning for desired closed loop responses. Korean J. Chem. Eng. 20, 8 (2003)

135. Lee, M., Lee, K., Kim, C., Lee, J.: Analytical design of multiloop PID controllers for desired closed-loop responses. AIChE J. 50, 1631 (2004)

136. Bao, J., Forbes, J.F., McLellan, P.J.: Robust multiloop PID controller design: a successive semidefinite programming approach. Ind. Eng. Chem. Res. 38, 3407 (1999)

137. Vlachos, C., Williams, D., Gomm, J.B.: Genetic approach to decentralised PI controller tuning for multivariable processes. IEE Proc. Control Theory Appl. 146, 58 (1999)

138. Economou, M., Morari, M.: Internal model control: 6. multiloop design. Ind. Eng. Chem. Proc. Des. Dev. 25, 411–419 (1986)

139. Hovd, M., Skogestad, S.: Improved independent design of robust decentralized controllers. Journal of Process Control 3, 43 (1993)

140. Chen, D., Seborg, D.E.: Design of decentralized PI conttrol systems based on nyquist stability analysis. Journal of Process Control 13, 27 (2003)

141. Lee, J., Cho, W., Edgar, T.F.: Multiloop PI controller tuning for interaction multivariable processes. Computers and Chemical Engineering 22, 1711 (1998)

142. Wang, Q.G., Lee, T.H., Zhang, Y.: Multi-loop version of the modified ziegler-nichols method for two input two output process. Ind. Eng. Chem. Res. 37, 4725 (1998)

143. Huang, H.P., Jeng, J.C., Chiang, C.H., Pan, W.: A direct method for multi-loop PI/PID contoller design. Journal of Process Control 13, 769 (2003)

144. He, M.-J., Cai, W.-J.: Evaluation of decentralized closed-loop integrity for multi-variable control system. Ind. Eng. Chem. Res. 44, 3567 (2005)

145. Witcher, M., McAvoy, T.J.: Interacting control systems: Steady state and dynamic measurement of interactions. ISA Trans. 16, 35 (1977)

146. Lee, J., Edgar, T.F.: Dynamic interaction measures for decentralized control of multivariable processes. Ind. Eng. Chem. Res. 43, 283 (2004)

147. Huang, H.P., Jeng, J.C.: Monitoring and assessment of control performance for single loop systems. Ind. Eng. Chem. Res. 41, 1297 (2002)

148. Ingimundarson, A., Hägglund, T.: Performance comparison between PID and dead-time compensating controllers. Journal of Process Control 12, 887–895 (2002)

149. Åström, K.J., Wittenmark, B.: Computer-Controller Systems: Theory and Design. Pricentice Hall, New Jersey (1997)

150. Michiels, W., Engelborghs, K., Vansevenant, P., Roose, D.: Continuous pole placement for delay equations. Automatica 38, 747–761 (2002)

151. Persson, P., Åström, K.J.: Dominant pole design – a unified view of PID controller tuning. In: Adaptive Systems in Control and Signal Processing 1992 IFAC Symposia Series, Tarrytown, NY, USA, pp. 377–382 (1993)

152. Coelho, C.A.D.: Dominant pole placement with maximum zero/poie ratio phase-lead controllers. In: Proceedings of the American Control Conference, Philadelphia, Pennsylvania, USA, pp. 1159–1164 (1998)

153. Franklin, G.F., Powell, J.D., Workman, M.: Digital Control of Dynamic Systems, 2nd edn. Addison Wesley, USA (1990)

154. Wang, Q.G., Lee, T.H., Lin, C.: Relay Feedback: Analysis, Identification and Control. Springer, London (2003)

155. Ang, K.H., Chong, G., Li, Y.: PID control system analysis, design, and technology. IEEE Trans. Control Systems Technology 13(4) (2005)

156. Liu, M., Wang, Q.G., Huang, B., Hang, C.C.: Improved identification of continuous-time delay processes from piecewise step tests. Journal of Process Control 17, 51–57 (2007)

157. Wang, Q.-G., Zhang, Y.: Robust identification of continuous systems with dead-time from step responses. Automatica 37, 377–390 (2001)

158. Wang, Q.-G., Hang, C.C., Yang, X.-P.: Single-loop controller design via IMC principles. Automatica 37, 2041–2048 (2001)

159. Bernstein, D.S.: Some open problems in matrix theory arising in linear systems and control. Linear Algebra and its Applications, 162–164, 409–432 (1992)
160. Syrmos, V.L., Abdallah, C.T., Dorato, P., Grigoriadis, K.: Static output feedback — a survey. Automatica 33, 125–137 (1997)
161. Fu, M., Luo, Z.Q.: Computational complexity of a problem arising in fixed order output feedback design. Systems & Control Letters 30, 209–215 (1997)
162. Gao, H., Lam, J., Wang, C., Wang, Y.: Delay-dependent output-feedback stabilisation of discrete-time systems with time-varying state delay. IEE Proc-Control Theory Appl. 151, 691–698 (2004)
163. Yu, J.T.: A convergent algorithm for computing stabilizing static output feedback gains. IEEE Trans. Automat. Contr. 49, 2271–2275 (2004)
164. Cao, Y.Y., Sun, Y.X., Mao, W.J.: A new necessary and sufficient condition forout-put feedback stabilizability and comments on stabilization via static output feedback. IEEE Trans. Automat. Contr. 43, 1110–1111 (1998)
165. Geromel, J.C., de Souza, C.C., Skelton, R.E.: LMI numerical solution foroutput feedback stabilization. In: Amer. Control Conf., pp. 40–44 (1994)
166. Iwasaki, T., Skelton, R.E.: The xy-centring algorithm for the dual LMI problem: A new approach to fixed-order control design. Int. J. Control 62 (1995)
167. Benton, R.E., Smith, D.: Static output feedback stabilization with prescribed degree of stability. IEEE Trans. Automat. Contr. 43 (1998)
168. Cao, Y.Y., Sun, Y.X., Lam, J.: Simultaneous stabilization via static output feedback and state feedback. IEEE Trans. Automat. Contr. 44, 1277–1282 (1999)
169. Ghaoui, L.E., Oustry, F., AitRami, M.: A cone complementarity linearization algorithm for static output-feedback and related problems. IEEE Trans. Automat. Contr. 42(8), 1171–1176 (1997)
170. Leibfritz, F.: An LMI-based algorithm for designing suboptimal static h_2/h_∞ output feedback controllers. SIAM J. Control Optimal 39, 1711–1735 (2001)
171. Rotunno, M., de Callafon, R.A.: A bundle method for solving the fixed order control problem. In: 41st Conf. Decision and Control, Las Vegas, Nevada USA, December 2002, pp. 3156–3161. IEEE, Los Alamitos (2002)
172. Fujimori, A.: Optimization of static output feedback using substitutive LMI formulation. IEEE Trans. Automat. Contr. 49(16), 995–999 (2004)
173. Zhou, K., Doyle, J.C., Glover, K.: Robust and Optimal Control. Prentice-Hall, New Jersey (1996)
174. He, Y., Wang, Q.G.: An improved ILMI method for static output feedback control with application to multivariable PID control. IEEE Trans. on Automatic Control 51(10), 1678–1683 (2006)
175. Liu, W.Q., Yan, W.Y., Teo, K.L.: On initial instantaneous jumps of singular systems. IEEE Trans. Automat. Contr. 40, 1650–1655 (1995)
176. Takaba, K.: Robust H^2 control of descriptor system with time-varying uncertainty. Int. J. Control 71, 559–579 (1998)
177. Pecora, L.M., Carroll, T.L.: Synchronization in chaotic systems. Physical Review Letters 64(8), 821–824 (1990)
178. Chen, G., Dong, X.: From chaos to order-perspectives, methodologies, and applications. World Scientific, Singapore (1998)
179. Yang, X.S.: Concepts of synchronization in dynamical systems. Physics Letters A 260(5), 340–344 (1999)
180. Femat, R., Solis-Perales, G.: On the chaos synchronization phenomena. Physics Letters A 262(1), 50–60 (1999)

181. Wen, G.L., Wang, Q.G., Lin, C., Han, X., Li, G.Y.: Synthesis for robust synchronization of chaotic systems under output feedback control with multiple random delays. Chaos, Solitons and Fractals 29(5), 1142–1146 (2006)

182. Suykens, J.A.K., Curran, P.F., Chua, L.O.: Robust synthesis for master-slave synchronization of lur'e systems. IEEE Trans. Circuits Syst. I 46(7), 841–850 (1999)

183. Curran, P.F., Suykens, J.A.K., Chua, L.O.: Absolute stability theory and master-slave synchronization. Int. J. Bifurcation Chaos 7, 2891 (1997)

184. Kawata, J., Nishio, Y., Dedieu, H., Ushida, A.: Performance comparison of communication systems using chaos synchronization. IEICE Trans. Fundamentals E82A, 1322 (1999)

185. Bowong, S.: Stability analysis for the synchronization of chaotic systems with different order: application to secure communications. Physics Letters A 326(1–2), 102–113 (2004)

186. Alvarez, G., Heránez, L., Muñoz, J., Montoya, F., Li, S.: Security analysis of communication system based on the synchronization of different order chaotic systems. Physics Letters A 345(4–6), 245–250 (2005)

187. Cuomo, K.M., Oppenheim, A.V.: Circuit implementation of synchronized chaos with applications to communications. Phys. Rev. Lett. 71(1), 65–68 (1993)

188. Itoh, M., Murakami, H.: New communication systems via chaotic synchronizations and modulations. IEICE Trans. Fundamentals E78-(A3), 285–290 (1995)

189. Masoller, C., Zanette, D.H.: Different regimes of synchronization in nonidentical time-delayed maps. Physica A 325(3–4), 361–370 (2003)

190. Puebla, H., Alvarez-Ramirez, J.: Proportionalcintegral feedback demodulation for secure communications. Physics Letters A 276(5–6), 245–256 (2000)

191. Yalcin, M.E., Suykens, J.A.K., Vandewalle, J.: Master-slave synchronization of lur'e systems with time-delay. Int. J. Bifurcation Chaos 11(6), 1707–1722 (2001)

192. Jiang, G.P., Zheng, W.X.: An LMI criterion for linear-state-feedback based chaos synchronization of a class of chaotic systems. Chaos, Solitons and Fractals 26(2), 437–443 (2005)

193. Parmananda, P.: Synchronization using linear and nonlinear feedbacks: a comparison. Physics Letters A 240(1–2), 55–59 (1998)

194. Liu, F., Ren, Y., Shan, X.M., Qiu, Z.L.: A linear feedback synchronization theorem for a class of chaotic systems. Chaos, Solitons and Fractals 13(4), 723–730 (2002)

195. Yassen, M.T.: Controlling chaos and synchronization for new chaotic system using linear feedback control. Chaos, Solitons and Fractals 26(3), 913–920 (2005)

196. Yassen, M.T.: Adaptive synchronization of rossler and lü systems with fully uncertain parameters. Chaos, Solitons and Fractals 23(5), 1527–1536 (2005)

197. Morgul, O., Solak, E.: Observer based synchronization of chaotic systems. Phys. Rev. E 54(5), 4803–4811 (1996)

198. Busawon, K.K., Kabore, P.: Disturbance attenuation using proportional integral observers. Int. J. Control 74(6), 618–627 (2001)

199. Hua, C.C., Guan, X.P.: Synchronization of chaotic systems based on PI observer design. Physics Letters A 334(5–6), 382–389 (2005)

200. Grassi, G., Mascolo, S.: Nonlinear observer design to synchronize hyperchaotic systems via a scalar signal. IEEE Trans. Circuits Syst. I 44(10), 1011–1014 (1997)

201. Wen, G.L., Xu, D.: Observer-based control for full-state projective synchronization of a general class of chaotic maps in any dimension. Physics Letters A 333(5–6), 420–425 (2004)

202. Wen, G.L., Xu, D.: Nonlinear observer control for full-state projective synchronization in chaotic continuous-time systems. Chaos, Solitons and Fractals 26(1), 71–77 (2005)
203. Femat, R., Alvarez-Ramirez, J., Zarazua, M.A.: Laplace domain controllers for chaos control. Physics Letters A 252(1–2), 27–36 (1999)
204. Femat, R., Campos-Delgado, D.U., Martinez-lopez, F.J.: A family of driving forces to suppress chaos in jerk equations: Laplace domain design. chaos 15(4), 043102 (2005)
205. Khalil, H.K.: Nonlinear Systems. Macmillan Publishing Company, New York (1993)
206. Luyben, W.L.: Process Modeling, Simulation and Control for Chemical Engineers, 2nd edn. McGraw-Hill Publishing Company, USA (1990)
207. Wu, M., He, Y., She, J.H., Liu, G.P.: New delay-dependent stability criteria and stabilizing method for neutral systems. IEEE Trans. Automat. Contr. 49(12), 2266–2271 (2004)
208. Ljung, L.: System Identification: Theory for the User, 2nd edn. Prentice Hall, Upper Saddle River (1999)
209. Gawthrop, P.J.: Parametric identification of transient signals. IMA journal of Mathematical Control and Information 1 (1984)
210. Unbehauen, H., Rao, G.P.: Identification of Continuous Systems. North-Holland, Amsterdam (1987)
211. Wang, L., Gawthrop, P.J.: On the estimation of continuous time transfer functions. International Journal of Control 74, 889–904 (2001)
212. Garnier, H., Gilson, M., Richard, A.: Continuous-time model identification from sampled data, implementation issues and performance evaluation. International Journal of Control 76(13), 1337–1357 (2003)
213. Wang, Q.-G., Hang, C.-C., Bi, Q.: Process frequency response estimation from relay feedback. Control Engineering Practice 5(9), 1293–1302 (1997)
214. Wang, Q.-G., Bi, Q., Zhou, B.: Use of fft in relay feedback systems. Electronics Letters 33(12), 1099–1100 (1997)
215. Bi, Q., Wang, Q.-G., Hang, C.-C.: Relay-based estimation of multiple points on process frequency response. Automatica 33(9), 1753–1757 (1997)
216. Wang, Q.-G., Hang, C.-C., Bi, Q.: A technique for frequency response identification from relay feedback. IEEE Trans. Control Systems Technology 7(1), 122–128 (1999)
217. Diamessis, J.E.: A new method for determining the parameters of physical systems. Proceedings of IEEE 53, 205–260 (1965)
218. Whitfield, A.H., Messali, N.: Integral-equation approach to system identification. Int. J. Control 45, 1431–1445 (1987)
219. Golubev, B., Wang, L.W.: Plant rational transfer function approximation from input ouput data. Int. J. Control 36, 711–723 (1982)
220. Hwang, S.H., Lai, S.T.: Use of two-stage least-square algorithms for identification of continuous systems with time delay based on pulse responses. Automatica 40, 1561–1568 (2004)
221. Wang, Q.-G., Liu, M., Hang, C.C.: Simplified identification of time delay systems with non-zero initial conditions from pulse tests. Industrial & Engineering Chemistry Research 44(19), 7591–7595 (2005)
222. Ahamed, S., Huang, B., Shah, S.L.: Parameter and delay estimation of continuous-time models using a linear filter. Journal of Process Control 16, 323–331 (2006)
223. Ikonen, E., Najim, K.: Advanced Process Identification and Control. Marcel Dekker, New York (2002)

224. Cott, B.J.: Introduction to the process identification workshop at the 1992 Canadian chemical engineering conference. International Journal of Control 5(2), 67–69 (1995)
225. Zhu, Y.C.: Multivariable process identification for MPC: the asymptotic method and its applications. Journal of Process Control 8(2), 101–115 (1998)
226. Wang, Q.-G., Zhang, Y.: A novel FFT-based robust multivariable process identification. Industrial & Engineering Chemistry Research 40, 2485–2494 (2001)
227. Garnier, H., Gilson, M., Young, P.C., Huselstein, E.: An optimal iv technique for identifying continuous-time transfer function model of multiple input systems. Control Engineering Practice 46(15), 471–486 (2007)
228. Sinha, N.K., Rao, G.P.: Identification of Continuous-Time Systems. Kluwer academic pulisher, Netherlands (1991)
229. Vasnani, U.V.: Towards Relay Feedback Auto-tuning of Multi-loop Systems. PhD thesis, National University of Singapore, Singapore (1995)

Index

Lecture Notes in Control and Information Sciences

Edited by M. Thoma, M. Morari

Further volumes of this series can be found on our homepage:
springer.com

Vol. 351: Lin, C.; Wang, Q.-G.; Lee, T.H., He, Y.
LMI Approach to Analysis and Control of
Takagi-Sugeno Fuzzy Systems with Time Delay
204 p. 2007 [978-3-540-49552-9]

Vol. 350: Bandyopadhyay, B.; Manjunath, T.C.;
Umapathy, M.
Modeling, Control and Implementation of Smart
Structures 250 p. 2007 [978-3-540-48393-9]

Vol. 349: Rogers, E.T.A.; Galkowski, K.;
Owens, D.H.
Control Systems Theory
and Applications for Linear
Repetitive Processes
482 p. 2007 [978-3-540-42663-9]

Vol. 347: Assawinchaichote, W.; Nguang, K.S.;
Shi P.
Fuzzy Control and Filter Design
for Uncertain Fuzzy Systems
188 p. 2006 [978-3-540-37011-6]

Vol. 346: Tarbouriech, S.; Garcia, G.; Glattfelder,
A.H. (Eds.)
Advanced Strategies in Control Systems
with Input and Output Constraints
480 p. 2006 [978-3-540-37009-3]

Vol. 345: Huang, D.-S.; Li, K.; Irwin, G.W. (Eds.)
Intelligent Computing in Signal Processing
and Pattern Recognition
1179 p. 2006 [978-3-540-37257-8]

Vol. 344: Huang, D.-S.; Li, K.; Irwin, G.W. (Eds.)
Intelligent Control and Automation
1121 p. 2006 [978-3-540-37255-4]

Vol. 341: Commault, C.; Marchand, N. (Eds.)
Positive Systems
448 p. 2006 [978-3-540-34771-2]

Vol. 340: Diehl, M.; Mombaur, K. (Eds.)
Fast Motions in Biomechanics and Robotics
500 p. 2006 [978-3-540-36118-3]

Vol. 339: Alamir, M.
Stabilization of Nonlinear Systems Using
Receding-horizon Control Schemes
325 p. 2006 [978-1-84628-470-0]

Vol. 338: Tokarzewski, J.
Finite Zeros in Discrete Time Control Systems
325 p. 2006 [978-3-540-33464-4]

Vol. 337: Blom, H.; Lygeros, J. (Eds.)
Stochastic Hybrid Systems
395 p. 2006 [978-3-540-33466-8]

Vol. 336: Pettersen, K.Y.; Gravdahl, J.T.;
Nijmeijer, H. (Eds.)
Group Coordination and Cooperative Control
310 p. 2006 [978-3-540-33468-2]

Vol. 335: Kozłowski, K. (Ed.)
Robot Motion and Control
424 p. 2006 [978-1-84628-404-5]

Vol. 334: Edwards, C.; Fossas Colet, E.;
Fridman, L. (Eds.)
Advances in Variable Structure and Sliding Mode
Control
504 p. 2006 [978-3-540-32800-1]

Vol. 333: Banavar, R.N.; Sankaranarayanan, V.
Switched Finite Time Control of a Class of
Underactuated Systems
99 p. 2006 [978-3-540-32799-8]

Vol. 332: Xu, S.; Lam, J.
Robust Control and Filtering of Singular Systems
234 p. 2006 [978-3-540-32797-4]

Vol. 331: Antsaklis, P.J.; Tabuada, P. (Eds.)
Networked Embedded Sensing and Control
367 p. 2006 [978-3-540-32794-3]

Vol. 330: Koumoutsakos, P.; Mezic, I. (Eds.)
Control of Fluid Flow
200 p. 2006 [978-3-540-25140-8]

Vol. 329: Francis, B.A.; Smith, M.C.; Willems,
J.C. (Eds.)
Control of Uncertain Systems: Modelling,
Approximation, and Design
429 p. 2006 [978-3-540-31754-8]

Vol. 328: Loría, A.; Lamnabhi-Lagarrigue, F.;
Panteley, E. (Eds.)
Advanced Topics in Control Systems Theory
305 p. 2006 [978-1-84628-313-0]

Vol. 327: Fournier, J.-D.; Grimm, J.; Leblond, J.;
Partington, J.R. (Eds.)
Harmonic Analysis and Rational Approximation
301 p. 2006 [978-3-540-30922-2]

Vol. 326: Wang, H.-S.; Yung, C.-F.; Chang, F.-R.
H_∞ Control for Nonlinear Descriptor Systems
164 p. 2006 [978-1-84628-289-8]

Vol. 325: Amato, F.
Robust Control of Linear Systems Subject to
Uncertain
Time-Varying Parameters
180 p. 2006 [978-3-540-23950-5]

Vol. 324: Christofides, P.; El-Farra, N.
Control of Nonlinear and Hybrid Process Systems
446 p. 2005 [978-3-540-28456-7]

Vol. 323: Bandyopadhyay, B.; Janardhanan, S.
Discrete-time Sliding Mode Control
147 p. 2005 [978-3-540-28140-5]

Vol. 322: Meurer, T.; Graichen, K.; Gilles, E.D.
(Eds.)
Control and Observer Design for Nonlinear Finite
and Infinite Dimensional Systems
422 p. 2005 [978-3-540-27938-9]

Vol. 321: Dayawansa, W.P.; Lindquist, A.;
Zhou, Y. (Eds.)
New Directions and Applications in Control
Theory
400 p. 2005 [978-3-540-23953-6]